D0712458

THE
FUTURE OF
TECHNOLOGICAL
CIVILIZATION

THE
FUTURE OF
TECHNOLOGICAL
CIVILIZATION

Victor Ferkiss

GEORGE BRAZILLER

NEW YORK

Copyright © 1974 by Victor Ferkiss

Published simultaneously in Canada by Doubleday Canada, Limited
All rights reserved.

For information, address the publisher:
George Braziller, Inc.
One Park Avenue
New York, N.Y. 10016

Standard Book Number: 0-8076-0738-x
Library of Congress Catalog Card Number: 73-90926

FIRST PRINTING
Printed in the United States of America
DESIGNED BY VINCENT TORRE

For my parents,
who never liked the way
"they" were running things either

Acknowledgments

So many people—colleagues and students, friends and acquaintances, readers and audiences—have influenced my thinking directly and indirectly that it would be impossible to give due credit to all of them for their assistance both in ideas and in moral support. Some of the preliminary research—which goes back many years—was supported by a grant from the Rockefeller Foundation program in Political and Legal Philosophy and by a summer faculty research grant from Georgetown University. The year the final draft of this work was being prepared was spent as a Visiting Scholar at the Institute of International Studies of the University of California at Berkeley, and I owe a special debt of gratitude for the hospitality of its director, Professor Ernest B. Haas. Penny Scharf, Moira Gort, Judy Uptigrove, and Chris Stamelias of the support staff of the Political Science, Sociology and Anthropology department of Simon Fraser University generously assisted in typing the final draft; and our friends Thomas and Helen Donovan of Vancouver, Washington, provided much appreciated refuge and hospitality during the last frantic days of editing. Edwin Seaver, my editor at George Braziller, Inc., has been, as always, patient and encouraging. My wife Barbara has been not only a source of vital substantive and literary assistance but a valiant companion on the sometimes troubled personal and intellectual odyssey which this book represents.

V.F.

Contents

I

Liberalism and Beyond

1. The Invasion of the Human World   3
2. The End of the Modern Era   8
3. Liberal Ideology and the Assault on Nature   22
4. American Liberalism in Triumph and Decay   31
5. Technology and the End of Liberalism   52
6. Roads to Nowhere: Conservatism, Socialism, Anarchism, and the Antipolitics of Despair   61

II

Ecological Humanism

7. The Real and the Ideal: Philosophy and Politics   87
8. Man in Nature and the Nature of Man   102
9. Nature and Human Values   120
10. The Goodness of Man and the Primacy of Politics   137
11. Beyond Liberalism to Freedom   156
12. Human Needs in Social Perspective   165
13. The Common Good: Present and Future   186
14. Ecological Humanism: A Political Philosophy for Technological Man   206

III

A Planet Fit for Man

15. Necessary Utopia I: The Politics of
    Ecological Humanism   213
16. Necessary Utopia II: The Economics of
    Ecological Humanism   230
17. Politics and Culture in a Humane Society   248

IV

The Emergent Future

18. Getting There from Here: The Immanent Revolution   259
19. Ecological Humanism and Planetary Society:
    The Restoration of Earth and Beyond   276

Epilogue   291
Notes   295
Bibliography   325
Index   361

THE
FUTURE OF
TECHNOLOGICAL
CIVILIZATION

I

Liberalism and Beyond

1

_____

The Invasion of the Human World

"THE Martians are coming, the Martians are coming" was the cry heard across the American continent one night a generation ago when the famed actor-producer Orson Welles inadvertently terrorized much of the nation with his radio dramatization of H. G. Wells's *War of the Worlds*. Today we can be virtually certain that whatever life it may hold, the planet Mars supports no creatures capable of an armed invasion of Earth. But there are other sources of alien invasion and, as the continuing popularity of science fiction attests, such a theme has a basic appeal to the human mind. It is a Jungian archetype speaking to our deepest fears.

One familiar science-fiction plot adds a special twist of terror. Somehow the hero discovers that Earth has been secretly invaded by mysterious powers who have taken over the bodies of men and women and are using these slaves to weaken the human race's defenses, so that the inhuman masters can more easily seize control. The alien thralls are indistinguishable from normal men and women except for one unobtrusive feature, such as a small red mark behind the left ear. The hero, a noted scientist, tries to alert his fellow human beings to their peril, but is scorned as mad. Just as he is about to lose all hope, he is granted a special meeting with the President and his advisers, who are concerned about the alarm the scientist's charges have created among the population and are anxious to restore public calm by going through the motions of holding a hearing. The meeting goes well. The hero's hopes that it is not too late to save humanity begin to rise. It is then that he notices what he should have looked for when he first came

3

into the room. Wherever he can see the back of a listener's head, clearly visible is a small red spot behind the left ear.

This vision of reality—of being the only one to be aware of desperate peril when virtually everyone else sees nothing unusual and no one will listen, of sounding the alarm when the established rulers of society are secret agents of the enemy—is a staple of mystery stories as well as of science fiction. There is a special horror in danger concealing itself behind normality, in the alien masquerading as someone familiar.

In the final decades of the twentieth century it has become increasingly difficult for those concerned with the future of humanity not to feel that human society is the scene of a breakthrough of sinister forces on a planetary scale, of agents of destruction hiding behind or abetted by the normal guardians of our social life. The war against the earth—against nature and humanity—is being carried on both by those in power who have been transformed into alien agents and by their innocent dupes and victims among the world's peoples. Even events which are in themselves normal and lifegiving—the breaking of lands to the plow, the building of cities, the work of scientists and technicians, even the birth of children— assume a sinister cast when viewed in the perspective of a new realism about the limits of the earth and the needs of humanity.

The world of humanity has been invaded—invaded not by aliens from without but by aliens within—by those who, although indistinguishable in most cases from ordinary men and women, are doing the work of forces which, if unchecked, will destroy human civilization and create in its place a crowded, ugly, mechanized, regimented, and totally dehumanized world.

Human history as we have known it is coming to an end. The story of humankind has been the saga of a species which has sought to bring to fulfillment its inherent potentialities—to enable its members to overcome hunger and disease so as to live a normal lifespan, to increase their ability to move about and communicate so as to develop their intellectual and social gifts, and to create cultures in which men and women could strive to come to terms with their uniqueness as individuals and as human beings. In order to do all these things the human animal has utilized its unparalleled intellectual powers to influence or dominate its fellow living things and its total physical environment by means of the techniques and tools which collectively we call technology. The result is that today humanity is on the verge of being able to reshape its own social and biological heritage.

But the powers which made these triumphs possible now threaten the physical and cultural survival of human civilization. The technology created by humanity as an adaptive mechanism for coping with an often hostile world has become increasingly maladaptive to the point where it is today fully capable of physically destroying the human race and its habitat, the planet Earth. Though the complacent may continue to believe that there is nothing new under the sun, that the more things change the more they stay the same, and that the future will be like the past only more so, humankind is today at a crossroads in its development for which there is no historical precedent.

A star is a planet whose inhabitants have learned to create atomic energy, the philosopher Alan Watts has said.[1] This is poetry, nonsense as literal scientific truth, but full of meaning. The Genghis Khans and Adolf Hitlers of the past have killed human beings by the millions, and conquerors have sown salt in their enemies' fields, but never before have human beings had the power to physically destroy human and plant life on a worldwide scale, a power now well within their grasp. Human societies have fouled their habitats and used resources faster than they could be recreated or new ones found, but never has the rising tide of pollution and the growing shortage of key raw materials threatened society on a planetary scale in the way it does in our contemporary global economy.[2]

Nor is the threat to the future of humanity purely physical. Even if social means are found to avert global war and technological means are found to deal with the complex, interrelated problems of population growth, environmental degradation, and resource depletion, the technology which man has created may still destroy him spiritually and culturally. Jacques Ellul, among others, has argued that increasingly we live in a society in which the inherent dynamics of technology will overcome the inherent dynamics of humanism, and the machines we have created to enlarge our human existence will enslave us and recreate us in their image.[3] While it is untrue that technology determines the future independently of human volition, there is no question but that human individuals and human society are increasingly under pressure to conform to the demands of technological efficiency, and there is a real possibility that the essence of humanity will be lost in the process, that human history will come to an end and be converted into a mere prelude to the history of a posthuman society in which machines rather than men rule. A related possibility is that our growing ability to manipulate our genetic nature will lead to the creation of a society neither of machines nor of men but of new biological forms of our own creation, a society of beings which previous generations would aptly have called monsters.[4] If either of these developments comes to pass, human history will be at an end as surely as if someone were to set off the ultimate weapon and reduce the planet to a charred cinder.

Many voices have been raised against the forces threatening mankind. The alarm against the invaders is being sounded with ever more urgency. Not only have individual women and men, scientists and humanists, religious and political leaders, called attention to the drift of events, but millions of average citizens (housewives, students, voters, even children) have enlisted in movements against war and mechanization and environmental degradation. But in their struggles they suffer from a fatal handicap.

Stacked against them is not only the physical and economic power of those who wittingly or not are collaborating in the perversion of the future of humanity, but the power of entrenched ideas as well. Suggestions and demands for changes which will protect humankind are brushed aside not only on the grounds that they are unnecessary, unfeasible, or too costly in terms of present vested interests, but on the ground that they run counter to our basic philosophy of politics and life. Needed reforms are alleged to threaten our most cherished liberties and ideals, to imperil morally mandated rights, privileges, and procedures.

The intellectual bulwarks behind which the forces of destruction hide are especially strong in the democratic nations of the West—in the capitalist, liberal, industrial societies. The philosophy of liberalism which dominates these nations, and above all the United States, has ruled as out of order the goals which must be achieved and in some cases the means which must be used in order to turn back the invasion of the human world by posthuman forces. The invaders can operate safely from behind the power of the established institutions and ideas, as in a city where all the police are criminals in disguise. Yet the only hope for a human future lies in the overthrow of the power and pretensions of these alien forces.

What has to be done to prevent the destruction of humanity by war, universal poverty, or cultural and biological perversion? The only hope lies in a revolution in political philosophy.

Such a statement may sound as absurd as advocating the use of band-aids to cure an epidemic of bubonic plague, but it is true nevertheless. The essence of humanity's current crisis is that we have allowed our collective destiny to be determined by the political philosophy usually called liberalism, which holds that the prime purpose of human society is to encourage individual self-aggrandizement. It engages human society in a fatal attempt to conquer rather than cooperate with nature, licensing technology as the unfettered instrument of that conquest. Unfortunately, liberalism's major ideological rivals, conservatism and socialism, lead to essentially the same results. The only way in which we can return to a more human course of evolutionary development and avert the destruction of human culture and identity toward which our liberal attitudes and institutions are leading us is to seize collective control of our destiny. But collective control necessarily involves political action and the creation of new political institutions. Political action and institutions are in turn necessarily an expression of values. If liberalism and its warring cousins do not provide the values needed to promote—or even permit—the defense of mankind, then a new political philosophy conducive to the continuance of human history must be created.

Experts may argue about the precise nature of our problems and what technical means are required to cope with them; citizens may mobilize to support measures of change and reform. But before their efforts can bear fruit on the scale necessary to make the continued existence of a human world possible, old ways of looking at nature and society must be discarded and replaced by new perspectives on the basis of which needed changes can be justified. Any revolution in policy must necessarily be preceded by a revolution in ideas.

The root failure of liberalism, and of conservatism and socialism as well, is that they restrict the concern of political philosophy to power relationships among people: They are inherently incapable of taking note of the fact that we have entered a new stage of development in which humanity's collective relationship to nature and to technology will determine our future destiny. To talk about power relationships among people while ignoring the power which nature and technology have over the fate of human beings, and the way in which relationships among human beings influence nature and technology, is to ignore two of the three basic actors in the drama of human history.

Liberalism, the political philosophy which undergirds the modern world, came into being at a time when the political sphere was necessarily limited, when the continuance of the life processes of the planet and the continuity of the species could be taken for granted. Now that technology has given us the power to destroy these life processes and to alter the nature of the human species, *every decision is intrinsically political*: The future is completely open to human decisionmaking, whether for good or ill. Nothing at all can be taken for granted. Even the continued existence of human society as such is a matter for debate and decision. No philosophy of society which fails to recognize this fact makes sense or has anything useful to say to us. The various symptoms of political, social, and physical malaise throughout world society—pollution, crime and disorder, environmental degradation, overpopulation, poverty and racism, mechanization and alienation, and the specter of nuclear war—are all signs that the era of liberalism has come to an end. Humanity must find a philosophical substitute for liberalism if it is to survive.

Any contemporary political philosophy has a threefold task. It must help us to understand why liberalism and its proffered substitutes are inadequate. It must set forth the premises of a value-system which will enable us to deal effectively with the humanity-technology-nature relationship. It must offer some insight into the kinds of institutions and dynamics that might follow from such a revolution in political values. This obviously cannot be the work of one book or of one person. But if we are going to ensure a human future for our descendants in the necessarily technological civilization of the future, we must begin somewhere; and we must begin now. With every passing day the grip of the invaders tightens and the struggle to dislodge them becomes more difficult.

2

The End of the Modern Era

THAT liberalism has been the dominant political philosophy of modern industrial civilization is no historical accident. Liberalism came into being with the advent of modernity; it is obsolescent and dangerous today because the era which gave it birth is coming to an end.

The modern age has been distinguished by two essential characteristics. In the intellectual sphere modernity has involved the growth of rationalism and the attempt to understand and conquer nature through mathematical reason; this rationalism has been accompanied by an emphasis on the dualism between mind and body and between humanity and nature and by a downgrading of feeling and affect at the individual level and of tradition and community at the social level. In the physical realm, modernization has been characterized by the unlimited growth of human population and the unlimited impact of human artifice—of technology— upon the earth and by the creation of a worldwide technological society.

The modern era is drawing to a close because the process of creating a planetary society is virtually complete, at least in its physical, economic, and technological aspects, and the combined pressures of population growth and of the technologically conditioned use of the earth's resources are approaching the point where further expansion will be impossible. Not growth but recycling, miniaturization, and ephemeralization are called for if humanity is to survive. As we approach the limits of expansion, modernity in culture will naturally and necessarily come to an end, resulting in the healing of the rupture between mind and body, individual and society, and man and nature, and in attempts to reorder technological development toward these ends. Just as the modern era first flourished in the most technologically advanced areas of the world, so the first stirrings of the

8

postmodern era will be found in the world's most technologically and socially modernized nations, especially in the United States.

In order to understand the contrast between premodern and postmodern civilization and why it is that liberal political philosophy is unsuitable as a guide for the future, it is useful to take an overall look at the world of liberalism's founding and at the radically changed world of its decline.

The Lost World of John Locke

John Locke was, as we all are, a child of his age. But like few men before or since, he was the father of ages to come. Born in 1632 and dead by 1704, he was a man of the seventeenth century, which had one foot in the Renaissance and the Reformation and was but a step away from the Middle Ages. But the century's other foot was firmly planted on the road to modernity, and John Locke helped put it there.

Though an Englishman whose work in philosophy and politics has had significance for all western and westernized men throughout the world, he is a figure of special importance to Americans. Although his direct influence on the American Revolution and the Constitution was perhaps somewhat less than is generally claimed by scholars, in a broad sense he was the father of the liberal tradition which has dominated the whole history of American political thought.

What was the seventeenth-century world of John Locke like, and how did it differ from the world we are creating for our children?[1]

First of all, it was a much smaller world. In ancient times there had been contacts between Rome and China, and between China and Africa and, for all we know, perhaps even between the civilizations of the Old and the New Worlds. But in Locke's day the expansion of Europe, which was in modern times to make the world one society, was still only in midcourse. Spain and Britain had established colonies in the New World, and the Dutch especially were active in the Far East. But the Ottoman Empire dominated the Near East; China and Japan lived in effective isolation; Africa was still the Dark Continent; the British conquest of India was only in its early stages; and even the physical contours of the globe were only partially known.

More important, the nations of the world lived apart from one another. International trade, powered by the joint dynamic of war and exploration, was conducted not to secure the necessities of life but only special luxuries or conveniences. In the West only Norway and parts of the Low Countries were dependent on imported food, and under the dominant doctrine of mercantilism even the major powers such as France and England sought economic self-sufficiency. In 1678 England prohibited all imports from France, a law not formally repealed until after Locke's death. Means of international travel and communication were primitive. The world was not one. Even in the economically advanced nations people lived their lives within small geographical areas and small social settings, much as do villagers in most of the less developed part of the world today.

The world of John Locke was smaller not only in terms of geography —of effective international intercourse—but in population.[2] The population of England and Ireland together was only 5.5 million in 1600; that of the whole United Kingdom was only 7.5 million in 1700—little more than the population of Los Angeles County today. Birth rates were high, but so were death rates. It took three hundred years—from 1350 to 1650—for the population of England to double. France, with 19 million inhabitants in 1700, was the largest nation in Europe, with the possible exception of distant, isolated Russia. Europe as a whole had a population of only about 100 million, having barely tripled in size since the time of Christ. The vast Indian subcontinent into which British imperialism was making inroads had an estimated population of 125 million, apparently no larger than in 500 B.C. America north of the Rio Grande was home to one million human beings in 1650. The planet as a whole sustained some 500 million people. All of these figures are of course rough estimates at best, since little systematic attention was paid to population size or dynamics. Insofar as men thought about population at all, it was to worry about its small size or its supposed decline in particular areas. It was simply assumed that the carrying capacity of the earth as a whole was, for all practical purposes, unlimited and that there was room for endless expansion.

The seventeenth century was not an era of rapid technological progress, though slow accretions continued to be made in mankind's ability to manipulate nature. Women and men lived lives not radically different from those of their ancestors of previous centuries in terms of available tools and techniques. They would have been incapable of grasping the idea of "future shock."[3] Though the factory system and modern industrial organization were beginning to emerge, they were in their infancy. In an economy still based on tradition, there was little room for innovation. Sawmills were introduced into England in 1663 but were abandoned because of opposition; they did not become established until decades later. Science, though the age was one of scientific breakthroughs, played virtually no role in technological innovation, and the modern "invention of invention" was centuries in the future.[4]

The technological development of public life was rudimentary. In Locke's England there was no national agency for roadbuilding. Modern postal service only began with the London Penny Post in 1681. The paving and lighting of streets had barely started. Public health measures (save for the almost instinctive notion of quarantine, practiced by human groups throughout history independently of the development of the modern germ theory of disease), were virtually nonexistent. Such problems as pollution, conservation, and environmental degradation were hardly thought of, and were dealt with on a purely ad hoc basis if at all. There was some concern about the deforestation of England caused by shipbuilding (especially for the Royal Navy) and the increasing industrial use of timber but, although mercantilist theory recoiled, timber could be imported. From Elizabethan times on there had been attempts to do something about the noxious airs of London (including at one point imposing the death penalty for the burning of coal under certain circumstances), but the nation as a whole had plenty of fresh air and water. The cities of Britain and the continent were as crowded, dirty and unsanitary as parts of Asia and Africa are today but, despite the distaste of the more fastidious, people learned to live with filth just as contemporary New Yorkers are learning to live with the rising tide of canine excrement. No one thought of pollution or noise as more than inconveniences. That economic growth and developing technology could be a source of human ills as well as of human well-being, was almost unthinkable.

The economy of Locke's day was not free but controlled by the state and its agents, with the regulatory powers formerly exercised by guilds, towns, and local lords now centralized in the monarchy. Stability, not competition, was the norm for business, although regulation and enforcement were crude and inefficient. Just as free trade in goods was suppressed, so was free trade in ideas. Attempts to enforce political and religious conformity were universal, though the censorship and repression were often as clumsy and ineffective as mercantilist controls were in the economic sphere. Despite the autocratic proclivities of rulers, the power of the state to control the lives of its subjects was drastically limited by the crudity of the available technology and by the economic and physical decentralization of society, and neither territorial sovereignty nor national frontiers had the absolute character they were to assume in later centuries.

The seventeenth century remained an era of ideological and social cohesion. Belief in religion, however tenuous among the educated (including Locke himself), was the strong foundation-stone of social control. The superiority of Christianity was taken for granted (the generalized belief in European cultural superiority long antedates the Victorian concept of the "white man's burden"). Slavery was vital to European commercial and industrial expansion, and only in 1671 do we find Quakers beginning to express the first religiously grounded doubts about its justice within the English-speaking world. If the subordination of the darker peoples was generally accepted, so also was that of the working class. There were sporadic outbursts of protest but no real workers' movements in Locke's century.[5] Proletarian dissent was repressed in England as on the Continent, albeit perhaps a bit less ruthlessly. Effective political and economic power was of course exclusively in male hands. Locke casually refers, *inter alia*, to the husband being able to rule his wife because he is the "abler and stronger."[6] Bourgeois western values were everywhere ascendant.

At the heart of the bourgeois value-system was optimism about the progress of man, an optimism increasingly based on the progress of science.[7] This was the age of mathematical reasoning. Tradition and church law had impeded the study of medicine, but Harvey's discovery of the nature of the circulation of the blood, published in 1628, was a harbinger of knowledge—and of powers—to come. Robert Boyle's *The Skeptical Chymist* in 1662 effectively destroyed the respectability of alchemy. Halley's calculations concerning the comet of 1682 helped put astrological superstition to rout among the educated. Theories about the nature of science itself were also being elaborated. Descartes published his *Discourse on Method* in 1637, laying the groundwork for the mind-body dualism which has dominated modern culture. Newton's *Principia Mathematica* appeared in 1687. Though modern atomic physics was centuries away, insofar as theories and techniques are concerned, it was already prefigured in the materialist philosophies of science which were coming into vogue.

Drawing upon the philosophical atomism of Democritus, Lucretius, and other ancients, the thinkers of the era conceived of the universe as composed of hard objects, existing independently of their relationships with one another both in terms of their origin and their characteristics. They were wholly isolated from each other except insofar as they directly exerted various kinds of force upon one another. Monistic materialism and mathematical rationalism went hand in hand. Matter in motion, atoms in a void were the starting point for Newtonian physics and liberal political philosophy alike. These developments in science and what was still called natural philosophy primarily affected the thinking of the informed upper classes. The last English execution for sorcery took place as late as 1712, years after Locke's death, but a hopeful new era was coming into being. Man was grasping new powers; he was about to be liberated.

Yet the twentieth-century fruition of these powers would have been almost inconceivable to Locke and his contemporaries. The science of one era is bound to appear to be magic to men of earlier days. Just as seventeenth-century scientists would have had difficulty imagining the contemporary problems of pollution and population, they could not have predicted the revolution in physics which, in creating nuclear weaponry, ironically vindicated the alchemists by making the transmutation of elements possible; or the revolution in psychology and pharmacology which has enabled us to control the human mind and has thus restored respectability to the witch doctor; or the revolution in biology and genetics which is making it possible to control the shape of the human body and so to play the sorcerer.

To be a liberal in an age of political monarchy, economic mercantilism, religious and cultural orthodoxy, and nascent science made sense. It was an affirmation of belief in human potential. It still might make sense in parts of the developing or socialist world which have not yet become open, secular, industrial societies even today. But for the mainstream of mankind the world of Locke is lost forever.

The World of the Twenty-First Century

CAN WE KNOW THE FUTURE?

The study of the future is hardly an exact science. Yet we can dismiss as empty hyperbole the contention of some (usually reactionary) social critics that the future is largely unknowable. Even before primitive man first developed agriculture, humanity operated on the premise that many important aspects of life are subject to regular patterns discernible by man and that, barring catastrophe, we can make useful predictions about future conditions.[8]

Some aspects of the future are predictable because—like the growth patterns of human beings which dictate that most males develop beards at a certain age—they are preprogrammed by nature itself. Other characteristics of the future are knowable because they are based on conditions which change slowly if at all—the amount of energy sent to the earth by the sun or the geological relationship of the continents to each other.[9] Still others are knowable because they are based on cumulative trends in which, other things being constant, the new is added to the old: The more we smoke cigarettes the more our lungs deteriorate; the more metal we dump in the junkyard the higher the pile gets.

Sometimes cumulative trends come to an abrupt halt because of changes in the context in which they exist. People in the late nineteenth century, who worried about how the nation was going to feed its increasing horse population and rid the streets of horse manure, were right to be concerned. Yet the invention of the automobile, an event external to the specific problem of feeding or cleaning up after horses, made their particular fears moot. Some trends reach a peak because of their own intrinsic limitations. New products sell with increasing rapidity for a time, then sales level off when a finite market is saturated. Some cumulative trends run counter to other trends. At the rate certain scientific occupations were growing during the past several decades, a simple extrapolation of growth curves would have indicated that virtually all Americans would be physicists by the early years of the twenty-first century, a manifest absurdity since chemists were also increasing at roughly the same rate. When trends intersect so as to contravene one another, one or the other must yield or, what is more likely in the real world, both yield to other—perhaps previously unnoted—trends. Many trends come to an end because they de-

pend on external factors which are finite, and thus are necessarily asympto-
tic. Animal populations increase until food supplies are exhausted and then
often almost die off. Or, alternatively, their increase spurs an increase of
predators which limit their numbers. The speed of interhuman communication across distances increases until it reaches the speed of light and then
can apparently increase no further. A basic reason why liberalism is increasingly unviable as a philosophy of society is that the growth curves
which were in their early stage in Locke's time have reached the point at
which the laws of nature have brought them to a halt.

Predicting the future, therefore, does not mean simply assuming that
developmental patterns of the past will persist but being alert to radically
new developments which may alter current trends and even change their
direction.[10] It means being able to calculate which trends are reaching
their natural climaxes and which trends will necessarily conflict with
others. It means being able to spot trends which show a probability of long-
term growth though their present importance seems negligible. Some areas
of human life and behavior are easier to predict than others. The more a
trend is based on factors involving large statistical masses (birth and accident rates, monetary inflation), the more it is linked with natural patterns
of growth and decay, the easier it is to predict. The more aspects of the
future depend on individual volition or discrete historical contingencies
(the quality of presidential leadership, the death of a ruler in an automobile
accident), the less predictable these are.

While it is always wise to assume that the one thing we can be certain
of is some kind of surprise, as Herman Kahn asserts,[11] we are far from
helpless in the face of the unknown. Some surprises are more surprising
than others. Barring all-out nuclear war or some unthinkable catastrophe
such as a collision with a body from outer space, the population of the
world will be greater in the next century than it is today. Barring destruction of civilization as we know it, we will have more rather than less
scientific information at our disposal, be able to communicate no less rapidly than we do today, and so on. In the relatively short run, few surprises
about the basic aspects of human life are probable. The twenty-first century is less than thirty years (a generation in conventional usage) away
from today. In many nations, if not in the world as a whole, most of the
people who will be adults in the year 2000 have already been born. Most of
the buildings that will be standing then have probably already been built.
Even most of the technology of the year 2000 already exists or is on the
drawing boards.

More significantly, it can be argued that we have already succeeded in
"inventing the future,"[12] just as the previous century invented invention.
The human species has reached such a level of imagination and logical
sophistication that virtually every possibility has already been conceived of,
even if we have no way of knowing which possibilities will be actualized,

or to what degree or, equally important, in what combination. Is there any technical device that might exist in the twenty-first century or any form of marriage or political rule which has not already been speculated about by professional futurologists in government or industry or at least in the science-fiction magazines? Probably not. There will be few particular features of the twenty-first century that today's children—the adults of that era—will not already have discussed in the classroom or at least have seen in television fantasies. Not the components but the overall shape of the future is what is in doubt and hazard.

THE ON-COMING SOCIETY

What will the world of the twenty-first century be like? Given what we know about current scientific developments, about current social trends, and about the regularities and possibilities of human social and physical nature, in what ways will the future be most likely to differ from the lost world of John Locke?

In a physical sense it will be a larger world than that of the seventeenth century; in a psychological and social sense it will be a smaller one. There will be no unexplored continents and no uncharted seas. Already earth satellites are being used to take the census in underdeveloped nations (extrapolating village populations from the number of huts) and to inventory the world's agricultural and mineral resources through the use of infrared photography and other techniques. The equivalent of the Domesday Book of William the Conqueror (a detailed mapping and tallying of eleventh-century England) will be available on a worldwide scale by the twenty-first century.

The economic interdependence of the world has long been a commonplace of elementary school social studies: the bananas on the breakfast table from Latin America, the rubber for galoshes from Asia or Africa, and so on. But what was true of marginal aspects of the consumption patterns of the favored classes in favored nations will tomorrow be true of the whole world, even of the poorer countries. Today the forests of America are being cut down to provide packing crates for the Japanese to send us TV sets in, and at one point in the recent past one quarter of American wheat was being used for emergency rations for India. The continued operation of technological society in Western Europe, Japan, and even the United States, is largely dependent on oil from the Arab world. Political détente between the Soviet Union and China and the West makes it likely that the socialist economies will soon be integrated with that of the planet as a whole. Barring a return to a Communist autarchy (which was never complete anyway), by the twenty-first century the world economy will be one—despite whatever distinctions continue to exist between rich and poor within and among nations—whether in harmony or conflict, through international cooperation or international exploitation.

The world is already becoming one in terms of transportation and communication. Henry Kissinger has replaced Captain Cook. The Concorde supersedes the sailing ship. While the world of 2000-plus may not be the "global village" heralded by Marshall McLuhan and his followers[13] (many linguistic, cultural and political barriers to full planetary cultural integration may still exist), the communications satellite already offers the necessary technology, and its use will doubtless increase. Americans of the twenty-first century will be no more (or no less) surprised by a peasant revolt in Latin America than by a student rebellion in Berkeley. Not lack of information but information overload, not dearth of stimuli, but inability to respond to too many stimuli will be the communications problem of the future. Not producing and expanding communications but managing and making sense of them will be the task of a future planetary society.

The world of the twenty-first century will be larger than that of Locke in terms of population as well as in geographic size, but therefore smaller in the sense of being more crowded and confined. Great Britain with its 50 million people probably will not grow very much larger,[14] and in the United States, the population, if present trends continue, could level off at less than 300 million. But the world as a whole—which is now inhabited by almost 4 billion people, about eight times as many as in Locke's day—will almost certainly double in population by the first decades of the twenty-first century.[15] This is one growth curve that is approaching its asymptote.

Current worldwide concern about overpopulation reflects a recognition that growth must come to an end at some point—and sooner rather than later—if not only the bare necessities of physical existence but the requirements of human cultural and spiritual life are to be made available to all these billions. There is bitter disagreement over the population-re-sources-pollution equation and over the level of population the planet can support with the technology which will be available by the twenty-first century, as well as about the extent to which extraterrestrial opportunities are relevant to the problem, but there is no question that population size and its relationship to economic capabilities and cultural forms will be a major—if not *the* major—political concern of the foreseeable future. For all practical purposes the great economic and social problem of the twenty-first century will be the proper management of man's only home, the fragile spaceship Earth.

The twenty-first century will be a technological society—not necessarily as the phrase is used in the nightmarish projections of Ellul and his followers, but in the simple sense that the major conditioning factors of human individual and social life will be the characteristics, distribution, and use of machines and techniques developed on the basis of scientific knowledge. How the people of the future eat, sleep, breathe, shelter themselves, interact socially and culturally, and order their societies will depend on their technologies.

This will be true even though the nature of technology is itself undergoing fundamental change. Even today many important technological devices are no longer machines in the traditional sense, that is, arrangements of moving parts designed to perform work.[16] The TV set and the computer have no moving parts of any significance and do not perform work in the sense of nineteenth-century engineering. Increasingly our technology does not alter states of matter except incidentally. Instead it processes information, alters human consciousness, or directs human activity, as do computers, television apparatus, credit cards, and mood-controlling drugs. In the twenty-first century not production but communication will be the prime function of technology, and the social impact of technology will be even greater than it is today.

Women and men of the future will not be able to escape from the all-pervading influence of technology—whether of the work-performing or communicating variety—for the simple reason that by the twenty-first century the world will be incapable of supporting the vast population which is already inevitable (barring worldwide catastrophe) without the use of highly advanced technology. The coming 7 to 8 billion human beings— almost half of whom are already here—cannot all be fed by organically grown foods marketed locally by small farmers, cannot be clothed in homespun animal or vegetable fibers, and cannot be sheltered in local wood or stone. Not only will most of humankind be primarily dependent for the production of the necessities of life on a highly advanced technology, but in order to get products to people and to manage an increasingly interdependent world economy plagued by scarcities of key resources, an increasingly complex system of management and communications will be required.

The alternative to technological growth and development is even more widespread poverty than now exists, while to enable the masses of the world of 2000-plus to raise their living standards and to participate in the cultural life of the global village (as basic a requirement for human dignity as physical survival) will also require an advanced technology. Even to provide the nontechnological amenities of life—clean air and water, wilderness and access to nature, maximization of the option to choose simpler and more natural lifestyles—will paradoxically require more technologically based knowledge, controls, and management. The problem of the twenty-first century will not be whether to accept or reject technological civilization as such but how to order it to truly human ends.

The women and men of the twenty-first century, and all the men and women involved in preparing for that era, must therefore deal with a dual problem unknown to Locke and the other founders of liberalism: how society can be ordered so as to take into account the increasingly apparent limits to humanity's economic and demographic growth, and how society can be ordered so as to enable humanity to control the technology which is necessary to sustain human life on this planet. Put another way, the dual issues are how man should relate to nature and how he should relate to the machine.

Unlike the situation in the seventeenth century, when the intent of rulers to control the economic and personal lives of their subjects was frustrated by the weakness of the technologies of surveillance and control available to them, the world of the twenty-first century will be one in which human freedom must be planned for and, indeed, created, through social regulation of the powerful technologies of surveillance and control that will then exist. It will be a world in which an almost unlimited amount of information about every human being will be a matter of public record: medical and psychological history, school and employment records, titles of books borrowed from libraries, purchases made by credit cards, telephone calls made and received—the possible list is endless. To preserve privacy and autonomy in such a society will require detailed rules about how information is to be made accessible, and to whom, and how it is to be used and the invention of technologies that will make such rules practicable and enforceable.

It will be a world in which human beings will be involved in a seamless and ubiquitous web of communication with each other, one in which social rules and technological means will be required to keep such communications pluralistic and to enable every individual to speak as well as listen. It will be a world in which people will live their lives in a complex matrix of organizational and institutional relationships and in which regimentation and alienation will have to be fought through social and technological invention.

The people of Locke's time could identify restriction of freedom with particular political institutions such as the absolute monarchy. Today it is obvious that restrictions on freedom are inherent in the structure of technological society itself. Contemporary technology has implicit dynamics which transcend existing political ideologies and institutions and, as current ideological hangups are surmounted, the twenty-first century will come to realize and accept this fact. The liberal ideal sought freedom through separation and identified individual autonomy with isolation: The ideal was the Whig squire of Locke or the yeoman farmer of Jefferson, able to retire to his acres, live off the soil, and tell rulers and rabble alike to leave him alone. In the postliberal society of the twenty-first century, human autonomy and integrity is going to have to be found within rather than apart from the larger community. It is going to have to be found not by limiting society's impact on the individual but, instead, by consciously influencing the nature of that impact.

Centralization, bureaucracy, and regimentation of the individual are spurred by war and the passions war excites. The importance of war as a social institution has increased rather than decreased since Locke's time, and not only will humanity have to create a peaceful world if it is to survive physically into the twenty-first century, but it will have to break the power of the military-technological complex and dismantle the garrison state if it is to eliminate poverty, preserve ecological balance, and escape slavery to the machine. War is indeed, in Randolph Bourne's famous phrase, the "health of the state"; it is also the dynamic behind many of the most antihumanistic distortions of scientific and technological endeavor. No political philosophy which makes room for major international conflicts can have meaning for the next century.

The planetary society of that century will have to be organized and guided in a cultural setting which lacks the common ideology and mores of Locke's time. Traditional religious and social codes have lost most of their control over the belief and behavior of upper classes and masses alike, and there is no reason to believe this trend can be reversed. The whole world is becoming a culturally pluralistic society in which different ideas about human life and conduct exist side by side. Just as western rationalism has penetrated the cultures of the eastern world and conquered, at least superficially, the indigenous primitive cultures of developing nations, the philosophy of the East is having more influence in the West today than the Sinophiles of the seventeenth century could have imagined, as ever more westerners seek to find in other cultures correctives to the psychological desiccation bequeathed to the West by Cartesian dualism.

As historian Lynn White points out, not only is the Greco-Roman hold on the West being broken, but the hierarchy of values within western culture has been shaken and the value of the nonrationalistic aspects of human communication and action are being reasserted.[17] Bodies are in revolt against mind, the nonverbal against the verbal, at the same time the poor within and among nations contest the domination of the rich, the darker-skinned that of the lighter, and—in some industrial societies— women rebel against the domination of men and "masculine" values. The world of the twenty-first century will not be able to take moral and intellectual consensus for granted, but will have to syncretistically create order out of cultural diversity and harmony out of cultural and group conflict. It must transcend dualism, healing the breach between mind and body if it is to restructure the relationship between man and nature and man and the machine. The capitalist industrial society, of which Lockean liberalism was a willing midwife, conquered the world, at least temporarily, for bourgeois values, but the next century will march to different drummers.

The greatest difference between the world of the twenty-first century and the lost world of John Locke will be the valuation placed on science and technology. Science must continue to add to man's knowledge and technology must continue to develop if mankind is to keep the earth habitable for growing numbers of human beings. But the naive belief of the seventeenth and succeeding centuries, the "technophiliac" faith in science and technology as the prime and certain source of human happiness and the good life for all—the Baconian dream of felicity—is already under open attack in the heartland of industrial civilization, however much it may still flourish in less developed nations.

This disenchantment has been spurred by developments within science itself. The more the physicists try to identify the fundamental constituents of matter, the more elusive and vain the search seems to become as additional particles continue to be found. The discovery of antimatter has led Buckminster Fuller to argue, not unseriously, that it proves the universe is really spiritual since everything balances out to nothing.[18] Certainly the simple materialism of an earlier age is compelled to become ever more subtle as matter and energy are determined to be the same thing under different forms. The Newtonian world of matter in motion (hard atoms in a void) is replaced by an almost unverbalizable picture of the universe in which reality is not "things" but a *pattern*, the constituents of which we can hardly define, in which relationship is everything, determining even the nature of the things related, and quality, which modern science sought to exorcise by reducing it to quantity, in effect sneaks in again through the back door under the guise of "structure" or "levels of complexity" or even as quantity itself. The victory of reductionism in science becomes more pyrrhic every decade.

As science increasingly abandons the quest to discover the essential meaning of the universe, dismissing the question as unscientific, individuals turn to other sources in their search for ultimate reality—not only to traditional western religious and political values but also to ideas from the East and even to systems such as astrology, which the men of Locke's time thought they were destroying. The study of the I Ching and the use of tarot found in the vicinity of such bastions of scientific scholarship as Berkeley and M.I.T. suggest that the formerly close relationship between science and magic may perhaps be undergoing some kind of renewal. Loss of faith in science as the key to reality is accompanied by growing skepticism about science and technology as the keys to felicity. This skepticism takes two forms. At the personal level, the "new naturalism"—especially prevalent among the young—is exemplified by increased acceptance of sensuality and by interest in primitive nature, by the "encounter" movement, concern with mind-altering drugs and body awareness, a growing interest in spiritualism and eastern religions—all that is involved in the "counterculture" and the prematurely heralded "greening of America." Increasingly, the educated young are rejecting the belief that the use of technological means to dominate nature is the high road to individual human happiness. At the social level, skepticism about the benefits of science and technology takes the form of fear—fear based on concern about humanity's ability to survive in a world of nuclear weapons, concern about how human freedom can exist in an automated, computerized society, and concern about the problem of maintaining human identity and individuality given an ever more crowded world and the capacity for manipulation of the human person inherent in the genetic revolution and in mind-controlling drugs and electronic devices.

But our technological development has not only rewarded us with a sense of psychological unease, it is also beginning to fail to deliver even on its own terms. Even our material well-being is threatened by our runaway technology. If pollution endangers health as well as making life unpleasant, if crowding leads to wasted time as well as psychological pressures, if food is poisoned as well as tasteless, if automobiles are unsafe at any speed, if the energy to run our appliances is in scarce supply, then the machine age is not delivering the goods, even if one assumes that the goods it is designed to deliver are the right ones.

That one can be unhappy though rich has been a moral commonplace throughout history, but only in the last decades of the twentieth century are we coming to realize that one can be poorer though apparently richer. The ability of technology to create even material bliss has been rendered permanently problematic.

Liberalism made economic and technological growth the ultimate ends of human society; a viable postliberal society would regard them as merely instrumental values, as only means to ends. But if the primacy of the goals of economic growth and technological change are to be effectively challenged, they must be replaced by something else.

The difference between the lost world of John Locke and the coming world of the twenty-first century will hinge most of all on what that something else is, on the image of human happiness that is substituted for liberalism's fading dreams.

3

Liberal Ideology and the Assault on Nature

LIBERALISM, conceived in an era vastly different from the one we are creating for our children, is an outmoded political philosophy, unable to meet the challenges posed by man's radically new relationship to nature and technology. But what is liberalism?

That the answer to this question should be a problem may seem strange. Don't we all know who the liberals are? They are people like Hubert Humphrey and George McGovern who, unlike such conservatives as Barry Goldwater and Ronald Reagan, favor massive government intervention in economic life and greater racial and social equality. Such a use of the term *liberal*, though pressed by some conservative intellectuals and tacitly accepted by many "liberal" political publicists, is false to standard usage throughout most of modern history. It is completely misleading outside the United States and has the result of obscuring the extent to which virtually all Americans, including most of those who style themselves conservatives, are in fact liberals.

Historically liberalism is linked to individualism and opposition to state management of the economy. The Liberal parties of England and the Continent still remain faithful by and large to this usage. Many Americans who call themselves liberals would be called social democrats or even socialists abroad.

In the United States in the early twentieth century, many of those who called themselves liberals became more tolerant of governmental action in behalf of economic stability or social equality, and set in motion a complex historical process which twisted the meaning of the word almost unrecognizably.[1] So thorough has been the triumph of the newer usage in America, especially among the general public, that those most tenaciously faithful to the beliefs of the earlier liberalism have been forced to invent new names for themselves, such as "individualists" or "libertarians," in order to distinguish their position from those whose views, especially on economic issues, they most deplore.[2]

But the problem is not simply a semantic one of concern only to nitpicking scholars or debaters seeking to score points. The contemporary American confusion over the meaning of liberalism masks the extent to which virtually all Americans—whether self-styled liberals, conservatives, or socialists—accept the essential tenets of the liberal faith, and the extent to which our thinking about political and social options is crushed by the heavy hand of our basically liberal orientation. When a bright young journalist wrote a bestselling book describing Richard Nixon as the "authentic voice of the surviving American liberalism," he was not engaging in paradox for its own sake but illustrating a basic and significant fact about American political culture which the extent of Nixon's reelection victory only underlined.[3]

The Philosophical Foundations of Liberalism

In order to understand what liberalism is and what it implies, it is necessary to begin at the beginning. We need to go back to the ideas of John Locke, whose teachings are so faithfully enshrined in American thinking and institutions. Locke built his philosophy upon foundations dug by Machiavelli and above all by Hobbes. Hobbes "constructed the individualist basis on which the architects of the British and American political systems built structures designed by John Locke and other . . . liberal political theorists."[4]

The classic political philosophers such as Plato, Aristotle, and Cicero, and the medieval scholastics took it for granted that man was a social animal and did not seek to justify the existence of political society but to prescribe how rights, duties, and rewards ought to be apportioned within society—the problem of justice. These philosophers assumed that political society was natural, a social analogue of the order of nature, which they regarded as the measure and source of value. Machiavelli abandoned the quest for justice, substituting for it a search for the regularities of human behavior, knowledge of which could give greater power to princes. Hobbes, more systematic and philosophically inclined, was in agreement that the search for justice was meaningless, but sought once again to base political authority on nature. But it was a nature radically different from that of his predecessors: It was a nature devoid of moral significance.

Strictly speaking, while Hobbes embraced the new physics according to which everything is reducible to "bodies moving in space"—a world in which everything is "explicable in terms of motions of physical particles"[5] —his political ideas were not based on the empirical science of his own day (such as it was), though they are clearly compatible with it. Hobbes was not an empiricist but a rationalist. For him the model of the sciences was geometry. What then were Hobbes's axioms? For Hobbes, the primary fact about human existence was man's unbridled egoism, his constant search for more and more power—power as an instrument to gain pleasure and above all safety, to secure freedom from pain and the postponement of death.

On the basis of this belief in the essential and inevitable egoism of men and its corollary, their implacable hostility to one another, Hobbes erected most of the fundamental tenets of modern liberalism: human legal equality, natural rights, the social contract, and the abandonment of the quest for justice. Locke was to add a fifth: hostility to the state, at least as an agent of control over property.

Hobbes rested his system on his concept of the state of nature. Much ink has been spilled over whether Hobbes and Locke believed the "state of nature" was a period of human history or simply used it as a logical construct. In either case, to adopt such a model as a starting point for a social theory meant taking man at his most socially impoverished, under what even Hobbes and Locke would have to admit were extreme circumstances, and making this the norm from which man's nature was to be deduced. It meant forcing men to think of themselves as less social beings than they in fact manifest themselves to be. And it created a kind of self-fulfilling prophecy for the future of liberal society.

In Hobbes's thought the state of nature overlaps another concept, the state of war, and he often treated these two as synonymous. The life of man in the state of nature was, in Hobbes's classic phrase, "nasty, solitary, poor, brutish and short."[6] Because human beings are equal (even the weakest can kill the strongest when the latter is asleep), no security is possible in the state of nature. Therefore men seek to achieve personal safety through the social contract, which creates a political society ruled by a single sovereign. Since rights exist only when they can be enforced, there are no rights in the state of nature. Because return to the state of nature would destroy all rights, the worst sovereign is preferable to none at all as long as he does not destroy his subjects' lives. Whatever the sovereign decrees is therefore legitimate.

Hobbes's teachings were obviously a perfect formula for royal absolutism. But because of his reputation for atheism and his attacks on established philosophical and religious schools of thought, he was not taken up by the rulers whose pretensions his principles justified, while his acceptance of absolutism made him anathema to the middle-class Whigs who were disputing royal claims to power, especially over their property. For

Hobbes's philosophy to become an ideological vehicle for the rising bourgeoisie it had to be modified, and this was the task of John Locke.[7]

Scholars disagree about what Locke really said and meant but, since he has influenced modern, and especially American, political thought and practice mostly at second hand and through popularizations, it would seem legitimate to summarize his teachings in a conventional and somewhat simplified manner.[8] While the relationship of Locke's political teachings to his other writings and to the currents of thought of his day is somewhat clouded, there is no question that both his political ideas and his theory of knowledge are perfectly consistent with a dualistic, mechanistic view of man and of the physical universe. F. S. C. Northrop therefore would seem to be justified in making Locke the archetype and in large measure the originator of the modern western worldview which attempts to reduce all knowledge to the laws of physics; separates mind and body, man and society, humanity and nature;[9] and stands in sharp contrast to the more holistic eastern view of the world as, in Northrop's phrase, an "undifferentiated aesthetic continuum" which can be understood in several ways simultaneously.[10]

Locke accepted the Hobbesian concept of the state of nature, but he could not accept Hobbes's political corollary, the justification of the absolute power of the sovereign, since this was contrary to the position of the Whigs, whose views he shared, explicated, and defended. Consequently he argued that human rights existed in the state of nature, prior to the creation of civil society. The creation of a political community with a common superior was necessary not to create these rights but to make possible the peaceful adjudication of disputes arising from them and to secure individual rights against aggression by other individuals. Thus Locke's social contract differed from that of Hobbes in that it created not rights but a mechanism for protecting the rights which already existed in the state of nature.

Locke had an importance for American political thought that can hardly be exaggerated, and "there is probably no better short summary of the ideas of Locke than the American Declaration of Independence."[11]

What is it that Jefferson is so eloquently stating in the Declaration of Independence?

"We hold these truths to be self-evident" (not subject to empirical argument)

". . . that all men are created equal and that they are endowed by their Creator" (Jefferson like many influential men of his time was a deist, a believer in a Newtonian kind of supreme being who had created a world analogous to a Swiss watch, wound it up, and had sense enough to leave it alone)

". . . with certain inalienable rights" (they could never be surrendered, even voluntarily)

"... among which rights are life, liberty, and the pursuit of happiness" (property, the third element in the Lockean triad, had too blatant a sound for Jefferson and his colleagues; besides, Locke himself sometimes used the term *property* in a broad sense to include reputation, status, and most of what made life worth living)

"... that to secure these rights" (not of course to create them)

"... governments are instituted among men" (not to secure justice or any other similar lofty or future-oriented aim)

"... deriving their just powers from the consent of the governed" (when government fails to secure these rights it is the right of the people to overthrow or abolish it, following Locke's theory of the right of revolution if the hypothetical social contract with the sovereign is broken).

Certain difficulties are immediately apparent. Who has the right to institute a government? Must the decision be unanimous? Why and how are the politically powerless—women, children, slaves, the poor, and future generations—automatically subject to the effects of governmental actions? Who defines the specific content of the rights to life, liberty, and property—surely they are not absolute? Who has the right to declare when revolution is necessary? The people shall judge, has been the liberal formula, but who are the people? Most of the literature on Locke and liberalism—discussion of majority rule, government by law, executive-legislative relations, and so on—is devoted to an explication of these issues, which have been the stuff of day-to-day political conflict in liberal societies. The purpose of this book is not and cannot be to solve these problems. What is especially important to note about liberalism as formulated by Locke and Jefferson is the way in which it diverges from earlier political thought. Its assumption is that there is a social order among men and that there are recognized rights to life, liberty, and property which antedate the political community and are not created by or in it.

What Locke and his followers have never been able to make clear is the difference between the existence of a social order capable of defining human rights and their content and the existence of a political order. All human beings have, because of their nature, certain claims upon whatever is necessary to enable them to live as human beings; we all need access to food, air, water, shelter, and so on. But these needs can normally be met only in a particular social context, to a certain degree and in a certain fashion, with the active cooperation or passive tolerance of other human beings. In order to ensure that our needs are fulfilled in a particular way—in order to make sure that our rights are protected—we have to be able to depend on certain regularities in the behavior of others. The specific content of human rights in a particular society and the specific way in which they are exercised may be defined by custom rather than by consciously created law, but the exercise of human rights is always threatened by deviant behavior, and therefore society must both define and control deviance if rights are to be respected and needs fulfilled. The specification of the content of rights, and the enforcement of them against behavior which threatens their exercise, is inherently political; it involves the "authoritative allocation of values," in the language of contemporary political science.[12]

Every social order which recognizes and protects rights is inherently a political order. Natural rights leave the realm of the abstract and general and become concrete and particular only within a political context. To talk of specific rights such as private property as existing in a prepolitical context, outside of "civil society" or the "social contract" in Lockean terms, is to talk nonsense. The right to the goods necessary to sustain life is natural, but the right to grow crops on any particular plot of land is a creation of the state. Particular rights cannot preexist civil society, and therefore civil society can neither come into existence in order to protect them nor can its legitimacy be based on its protection of them.

Locke on Property, Nature, and Technology

Locke's intention was to justify a state which would ensure not rights in the abstract but the perpetuation of the particular inequalities which supposedly existed in the state of nature, to ratify existing power relationships and patterns of domination, to which, he held, men had somehow freely given their consent. His concern was with the danger of a tyrant disturbing the social status quo, especially as it related to property. According to Locke, the coming into existence of property was the major reason men had to leave the state of nature.

But the function of civil society is not merely to preserve property which already exists. Even more important, civil society is necessary so that property may grow and man's power to dominate nature may be fully unleashed. "The increase of lands and the right employing of them is the greatest art of government," Locke tells us.[13] It is, in Leo Strauss's words, not "static" but "dynamic" property that Locke's state serves.[14] The loose umbrella of feudal society was able to protect the former, but the liberal state was necessary to create the latter. The liberal state is not a neutral umpire among players in an already existing game; it is a booster club for the sport itself.

Locke's teachings on property are central to liberalism, because they contain an implicit philosophy of the proper relationship of man to nature and to technology. The autodetermination and virtual omnipotence of technology and the unlimited growth of the GNP characteristic of modern societies are not incidental consequences of Locke's principles; "the central theme of Locke's whole political teaching . . . is . . . increase."[15] The purpose of the liberal state is to promote the subjugation of nature.[16]

Prior to modern times various kinds of property rights existed. Under

Roman law there was almost unlimited freedom of use of property not unlike the situation in capitalist societies today. But the medieval world hedged property rights with an elaborate structure of corollary duties and restrictions; and greed was condemned as a vice by Christian theology. God had given the earth to mankind as a whole for its nurture and enjoyment. Private property could be justified as a social convenience in allocating resources and responsibilities, but all ownership was a stewardship of what really belonged to God. Thomas Aquinas held that a starving poor man who was forced to steal from the rich in order to survive committed no sin, since the superfluities of the rich really belonged to the poor. In the late Middle Ages the revival of Roman law (a powerful instrument in the growth of the centralized nation-state) meant a revival of freedom of property from social control. But with Locke the free use of property became the very foundation of society.

Locke's teaching, though he hedged it about with many bows to tradition, differed radically from that of his philosophical predecessors. Nature provided for mankind as a whole, but it provided in a niggardly fashion. Natural things had no intrinsic worth. What made the gifts of nature valuable was human industry. Though Locke was fully aware of the wretchedness of the poor in his own society, he still compared them favorably with primitive peoples, and wrote that among the Indians of America, a handful of people in a lavishly endowed continent, there was penury: "A king of a large and fruitful territory there feeds, lodges, and is clad worse than a day-laborer in England."[17] Why were the English so much better off? They had conquered nature. It was a physical technology—agriculture —and a social technology—money—that made men prosperous.

For technology to flourish private property was necessary. The belief that technology led to well-being was directly related to Locke's justification for unfettered private ownership. Not nature but man, capitalist man, was the source of wealth:

> He who appropriates land to himself by his labor does not lessen but increases the common stock of mankind. For the provisions serving to the support of human life produced by one acre of enclosed and cultivated land are . . . ten times more than those which are yielded by an acre of land of an equal richness laying waste in common. And therefore he that encloses land, and has a greater plenty of the conveniences of life from ten acres than he could have from a hundred left to nature, may truly be said to give ninety acres to mankind.[18]

Private property means increased wealth for all.

But are there no limits to the amount of private property one can legitimately seek to amass, as the medieval thinkers had argued? None at all. In the state of nature one could not rightly take more than one could use; that would be spoilage and waste. If I have harvested all the bananas by my labor they are mine, not those of idlers. But what if I cannot eat them and they go rotten? That would be a waste of a natural good. But this problem is easily solved by the invention of money. Others give me money for my bananas and they eat them, preventing spoilage. I become wealthy. Therefore the more I labor and the more wealthy I become, the better off everyone is. That bananas might spoil because some men do not have the money to buy them, or that my greed might cause me not to be content with harvesting already growing bananas but to plant still more trees and impoverish the soil, are not possibilities which disturb Locke. Civil society is so rich, in contrast with the state of nature, that even the natural-law prohibition against waste is no longer binding.[19] Nor could Locke conceive that it might be desirable to leave land uncultivated. Nature should never be left alone, for "land that is left wholly to nature . . . is called, as indeed it is, waste."[20] In Locke's ideal world there would be billboards on the sides of the Grand Canyon. Mankind's mission is to subdue niggardly nature through technology, to increase wealth as much as possible. Mother Nature is a bitch, but she can be raped. For Locke, man is "effectively emancipated from the bonds of nature,"[21] and "the negation of nature is the way toward happiness."[22]

Thus it follows that the primary purpose of the political community is to protect the institution of property, to promote technological progress and the use of money, and to safeguard the power relations which result from this process. Locke knew full well that some men would succeed in amassing property while lesser men would fail, and the world would be divided into the haves and the have-nots. Government had as its purpose the protection of the former from the latter.[23] Locke was not, as many believe, for weak government as such. He was in favor of a government incapable of threatening the position of the rising bourgeoisie but strong enough to protect that position against others, and willing and able to assist the bourgeoisie in the process of capital accumulation and technological advance. That prince, as Robert Goldwin notes, is called "godlike" who "by established laws of liberty secures protection and encouragement to the honest industry of mankind."[24] Society is above all a structure based on economics.

Society for Hobbes existed to protect life, to control relations between egoistic, hostile, physically aggressive men. By contrast, for Locke "society consists of relations among proprietors."[25] But for neither was the end of government justice or any common good beyond protection against death or the promotion of increased wealth. For Hobbes society was a refuge against murderers, for Locke a machine for encouraging acquisitiveness. For neither was citizen a proud title.

Locke cannot leave nature alone, not only because left alone it is a waste to mankind in general, but also because left alone it provides no field for the growth of a class of wealthy and powerful men. Locke's state, far from being a neutral umpire among contending forces, was designed as an instrument for enabling the bourgeoisie to control and exploit nature and in the process become rich and powerful. The "domination of nature" has always meant the domination of some men over others,[26] and technology has been an instrument for the creation of an elite of power and privilege. But Locke's theory of "possessive individualism" has a dynamic of its own.[27] In addition to its function as an instrument of class domination, it has provided the basis for an unrelenting and remorseless assault on nature. Liberalism could well be styled the political theory of technological messianism, or as one American cleric has aptly labeled its American variant, "the vandal ideology."[28]

4

American Liberalism in
Triumph and Decay

John Locke has been called America's philosopher, our king in the only way a philosopher can ever be the king of a great nation. We, therefore, more than any other people in the world, have the right and the duty and the experience to judge the rightness of his teaching.[1]

OUR verdict must be that Locke's teachings have bestowed on us a ravaged continent and a disordered society, and threaten us with the loss of our humanity in enslavement to technology. But it would be unfair to put all the blame on Locke. We freely conferred the royal crown and scepter.

"America was promises," wrote poet Archibald MacLeish in a nationalistic ode published during World War II, when intellectuals suddenly began once again to discover that they too were Americans and started to search for their roots. But promises of what? From its beginnings—even, in a sense from the time before it was settled by Europeans—the New World of which America became the most powerful nation held a dual promise of felicity and terror; it was seen as both a promised land to be embraced and an alien stronghold to be reduced and looted. The Renaissance image of the New World, Howard Mumford Jones writes in O Strange New World, "was compounded of both positive and negative elements that attract and repel. We inherit the image, and its elements haunt us still."[2]

From our earliest days we were the creation of would-be saints and all-too-successful pirates, and the United States as a nation has confounded the world by its special blend of moral idealism and ruthless cynicism. Our dismal adventure in Southeast Asia is only the latest illustration of the terrible ambivalence of our national character as we burned children with napalm, then tenderly healed their wounds. This ambivalence has not only divided groups of Americans and schools of American thought from our earliest history—the satisfied and complacent from the angry and despairing—but it divides us all internally.

A few years ago Joan Didion, an insightful essayist and novelist, wrote in an article about financier Howard Hughes, "There has always been that divergence between our official and our unofficial heroes . . . the apparently bottomless gulf between . . . what we officially admire and secretly desire, between . . . the people we marry and the people we love." Hughes, she notes, is "brilliantly anti-social . . . the dream we no longer admit."[3] But we do admit it, and always have. We are taught in school that the Pilgrims came for religious freedom or the nineteenth-century Germans immigrated to escape conscription, but we know in our hearts that most of our ancestors came because they believed the streets were paved with gold—gold for the taking, by whoever could grab it first, the symbol of a society not even of proprietors but of prospectors, and a lawless society at that.

A noted advocate of the quintessential, historically given liberalism of American society writes of the "master assumption of American political thought . . . the reality of atomistic social freedom" as being "instinctive to the American mind."[4] Even if we know nothing of political philosophy we know what he means. We have, as another political scientist puts it, endowed ourselves with "the inalienable right to clutter the landscape with as many automobiles, children, and tin cans as [we] see fit."[5] "It's a free country, isn't it," we learn to cry from childhood. We glory in our freedom and our wealth, linking them together in our psyches. Lyndon Johnson once told American troops in Korea that there are only a few millions of us and billions of Asians and other poor people, and that American soldiers were fighting to make sure they didn't take our riches away from us. Gold in the streets, beer cans on the highway, guns in the closet, together constitute one powerful version of the American dream.

A "people of plenty," as one historian has described us,[6] we care little for authority. The story is told of some American war prisoners in Korea drinking polluted water by the roadside and being admonished by a passing American senior officer, himself a prisoner, to stop lest they contract dysentery. "Screw you, Jack," they replied, "you're no better than us now." The still largely unrecognized root problem of American liberal society today is the question of how long we can go on doing as we individually please without suffering the inevitable natural consequences.

But our attitudes are not new. We have been self-centered and aggrandizing from the beginning of our history. The search for El Dorado on the Spanish Main brought pillage and torture and death to people by hands other than ours,[7] but our North American New World was also bloody and corrupt and we have—more democratically, more systematically, perhaps more hypocritically—always searched for El Dorados of our own, both as individuals and as a nation.

Much of the political turmoil of the sixties and the self-doubts that affected some Americans as a result stemmed from the rise to cultural consciousness and political activity of those groups among us which had been assaulted, or simply left behind, in the headlong search for power and wealth, the Lockean drive for increase. Blacks, Chicanos, American Indians have asserted their claims against those who run American society or are content with its goals and procedures. There is no question that many of these claims are justified and that American society has been and is racist and oppressive. But if the Indian tribes had been wiped out by smallpox after their first contact with Europeans, or had never existed, if the first slave had never been landed, if the Mexican War had never been fought, the United States would still be a troubled land. Liberalism as a political and economic creed has contributed to creating our minority problems and giving them their present shape, but it did so incidentally (however necessarily) as part of its broader impact on American society. Internal and external imperialism and their effects on subject populations are elements of a more general liberal assault on the order of humanity, society, and nature.[8]

The Poisoned Womb

Locke's view of nature—that it was a practically limitless bounty but sheer waste unless converted into private property and exploited by means of technology—has shaped American life from the beginning. This attitude was especially strong in New England, where Puritanism with its opposition to sensuality and its belief that the wilderness was a repository of Satanic peoples and powers lacked even the gentlemanly gardener's respect for cultivated nature found in Penn's Philadelphia or Cavalier Virginia.[9] Nature was there to be used—privately. To "the English immigrants all of North America was a vast preserve of resources to be processed."[10] Postrevolutionary state constitutions sought to forbid all restrictions on private hunting and fishing as undemocratic tyranny. After all, hadn't Robin Hood been a poacher? The goal of the farmers was to cut down trees and drain land, a natural desire in the circumstances but carried on from the beginning with an unnatural frenzy. The Indians had to be driven away, and ever more land made subject to the frontiersmen. In such a society it was hardly to be expected that either private or public moral standards would be high.

The American revolution was waged (after a brief unsatisfactory excursion into Burkean rhetoric about the prescriptive rights of Englishmen) with Locke's words, and its aims were largely economic. Smugglers or would-be industrialists wanted out from under the restrictions which mercantilism placed on free enterprise, debt-ridden farmers understandably chafed under the mercantilist fetish about the value of gold, and frontiersmen were impatient with the British government's attempt to stabilize the boundaries of western expansion. One does not need to question the individual legitimacy of many of these desires in order to note that the revolution was not merely a war against distant British control but was in large measure waged on behalf of the absolute autonomy of individual desires and against any form of social discipline.

The Founding Godfathers

But once the revolution was won, a new order had to be created. Federalist polemicists and later historians have exaggerated the difficulties of life under the Articles of Confederation, and there were many who, like the Jeffersonians, would have been reasonably content to live under a kind of agrarian anarchism in a republic of yeoman farmers going about their business under minimal government. Controversy among historians still clouds the question of the extent to which the framing and ratification of the American Constitution was a coup on behalf of a minority of the rich and against the interests and wishes of the largely rural majority.[11] But regardless of the political dynamics involved, the essential truth is expressed in Louis Hartz's argument that one reason the Constitution worked was that, contrary to the fears of many of its framers, there was no mob to guard property against.[12] Virtually all Americans were Lockean Whigs at heart, who believed in the rights of property whether or not they yet had much of their own. James Madison and the other framers were simply stating what were for virtually all Americans truisms—that men are unequal by talent, that inequality of talent gives rise to differences in wealth, and that government exists to secure, indeed to enhance, these differences.

Put in theoretical terms, the major struggle over the Constitution and its interpretation has been the issue of how far the political order should go in adhering to Locke's logic and creating a government which positively seeks to promote increase, a government able to bring into being a national market and give direct encouragement to industry. Jeffersonians were philosophically wary of industry because of their agrarian bias derived in part from the teachings of the French physiocrats, and at the political level the poorer classes were unhappy about the idea of creating a national debt, especially since speculators had bought up Continental Congress and state indebtedness at sacrifice prices and stood to reap fortunes from federal assumption of these obligations.

But Alexander Hamilton, the major Federalist ideologue, was a truer

Lockean even than Jefferson. He openly and actively pushed for the creation of a national debt in order to bring into existence a political community based on common economic entanglements rather than on any common concept of virtue or justice.[13] The Constitution reflects his success. He wanted a national bank (a vehicle for the creation and management of the national debt) in order to provide a firm foundation for American capitalism. He succeeded through the collaboration of Federalist Chief Justice John Marshall, whose opinion in *McCullough* vs. *Maryland* in 1819 is a virtual transcript of Hamilton's arguments of 1791 on the scope of national powers. The expansion of the constitutional limits of federal powers was necessary for the creation of a capitalistic society dedicated to the Lockean imperatives of competition and increase.

Hamilton also set the stage for America's love affair with the machine, the worship of technology as an end in itself. He did not want men's powers to go to waste any more than nature's bounty. His "Report on Manufactures" successfully advocated the establishment of industry based largely on tariff protection (a tax on consumers), not simply in order to increase the material well-being of the average American but in order to give full scope to individual human powers and differences in ability. His panegyric on the benefits of child labor is a beautiful expression of the Lockean lust to make everything useful: Time spent by children in play is just as evil as land that is not put to the plough.[14]

Obviously Hamilton succeeded only because he was far from alone in his liberal beliefs about the proper relationship of man, nature, and the machine. Jefferson and John Taylor of Caroline could celebrate the old pastoral dream of America, its destiny as the garden of humanity which, in counterpoint to the vandal ideology, had done much to inspire discoverers and settlers. But when the chips were down, technology was irresistible. John Adams paid sincere tribute to frugality and even praised sumptuary legislation, but only for ulterior motives—as a source of revenue for war, and a cure for vanities and fopperies which warred against all "great, manly, and warlike virtues."[15] For Adams the test of government was how much of a contribution it made to material prosperity, which (pity the shade of Aristotle) he held to be identical with happiness. "The form of government which communicates ease, comfort, security, or, in one word, happiness to the greatest number of persons in the greatest degree is the best," he wrote, long before Bentham's utilitarianism became the vogue.[16]

Though Jefferson was the political opponent of Hamilton and Adams, he too, on becoming President, adopted their principles both as a Lockean and as a nationalist, though not without regret—a prototype of American ambivalence about power and decency. Industrialization was necesary so that the United States could hold its own in time of war and become a great power, though, as Jefferson admitted during the war of 1812, "Our enemy has indeed the consolation of Satan on removing our first parents from Paradise; from a peaceful and agricultural nation, he makes us into a military and manufacturing one."[17] How much American expansionism was responsible for the war, he neglects to ask. Jefferson—in an attempt to have his cake and eat it as well—even convinced himself that, because of our agricultural background, industrialization would not produce in America the despicable urban mobs of Europe and, unlike the more far-sighted Madison, held that Parson Malthus, though right about Europe, was wrong about America.[18] Since America would never suffer the population pressures that produced poverty and class war in the Old World, we could have democracy, as all Americans were believers in the sacred rights of property. "Everyone, by his prosperity, or by his satisfactory situation, is interested in the support of law and order. And such men may safely and advantageously reserve to themselves a wholesome control over their public affairs, and a degree of freedom which, in the hands of the canaille of the cities of Europe, would be instantly perverted to the demolition and destruction of everything public and private."[19]

Moreover, Jefferson as President accepted the Hamiltonian principle of the loose interpretation of the Constitution. He justified the purchase of Louisiana as necessary to ensure that the other bank of the Mississippi would be settled by "our own brethren and children" rather than by strangers. It was Jefferson who, in thus stretching presidential authority in the field of foreign affairs, laid the foundation for the military-industrial complex and the Vietnam war.

Democratic Empire

During the period between the adoption of the American constitutional system and the Civil War a partnership among nationalism, capitalism, and science made possible the rapid extension of human material comfort over a wide area with no thought to possible negative consequences.[20] The eagle screamed, the wagons rolled, and the machines began to roar in earnest.

The only serious political argument was over how egalitarian the political and economic system would be. The dominant financial interests backed the National Whigs, and Henry Clay's "American System" provided for a cooperative exploitation of the continent by government, business, and technology through government subsidies for such "internal improvements" as roads and canals.

The great political achievement of Andrew Jackson was to bring to power a petty capitalist class, poorer and more rustic than the Whigs, but just as eager to get rich quickly, and willing to curb the power of the would-be Whig aristocrats in the process. Jackson's battle with Nicholas Biddle and the Bank of the United States was not just a fight on behalf of the "little man," however; it was at the same time a fight for even more frenetic economic growth. This is typical of the historic pattern of American liberalism: short-circuiting problems of economic inequality not by decreasing the economic power of those with the largest shares but by increasing the size of the pie to be cut. Jacksonian democracy wanted cheap credit as a means to expansion. What was resented was not the wealth of the rich but their exclusiveness. The common enemy was whoever stood in the way of physical expansion, like the Indians who had to be ruthlessly cleared from the path. The Cherokee nation's "trail of tears" was the inevitable consequence of the triumph of the common man. "Philanthropy could not wish to see the continent restored to the condition in which it was found by our forefathers," Jackson wrote.[21]

The America so aptly described by the French visitor Alexis de Tocqueville in *Democracy in America* was not only a democratic nation in which social status meant little. It was, par excellence, the society of proprietors postulated by the possessive individualism of Hobbes and Locke, in which the arts and sciences and even religion were acceptable only insofar as they were useful—useful in subduing the earth for the sake of human comfort. And to its inhabitants it was a wonderful place. Its poet Walt Whitman heard many voices singing but the loudest was the voice of technology. "Give," he asks, "but a passing glance at the fat volume of Patent Offices Reports and bless your stars that fate has cast your lot in the year of our Lord 1857."[22]

Slavery was important, not merely as a source of supposedly cheap labor for capitalist expansion but, paradoxically, as a bulwark of liberal democracy. Jefferson had had forebodings about slavery and saw in the repatriation of freed slaves to Africa a way of avoiding otherwise dire conflict.[23] De Tocqueville brooded about the anomaly of slavery in a democratic nation. But it was possible for a long time to expect slavery to fall of its own expensive weight. This hope was frustrated by technology. Eli Whitney's invention of the cotton gin made slavery seem once again economically profitable. The ravages inflicted upon nature by the cotton monoculture of the Old South made its extension to the trans-Mississippi states inevitable and raised the specter of a slavery-inspired American imperialism in Mexico, the Caribbean islands, and even the Spanish Main. Slaves were property, not human persons, but when Chief Justice Taney reiterated this in the Dred Scott case he not only acerbated American sectional rivalries but underlined the concept of America as a society of owners and echoed the Lockean belief in the inevitable differences among men and their stations.

Slavery was a necessary byproduct of liberalism. The Swedish economist Gunnar Myrdal was wrong when in his classic and influential *An American Dilemma* he spoke of a basic conflict of values between democracy and racism. For, as one anthropologist has perceptively pointed out, in terms of social psychology a lower class was—and is—a functional and almost necessary aspect of a competitive liberal society.[24] If life is a bitter competitive struggle in which some win and some lose, if liberal society is one in which money replaces or creates status, as de Tocqueville observed, and some become rich and powerful while others are poor and obscure, terrible tensions result. There must therefore be a floor somewhere below which the contestants cannot fall, or else the struggle is as unbearable as Hobbes's state of nature. No matter how badly a white man fared he could not become a slave—nor could a black, after emancipation. So he could bear to continue to participate in the war of all against all. An underclass—not only poorer but of lesser claims to rights and dignity—is necessary to make the status of the losers bearable. Today in America there are constant attempts to make the recipients of welfare serve the function of such an underclass.

Just as it had been necessary to clear the Indians out of the way of progress, so liberal society sought to drive other powers—the Mexicans, the native Hawaiians, even ultimately the Spanish in the Pacific—from control of lands which it could more efficiently utilize. The immediate cause of the Mexican War of 1845 was the need of the slave economy for more land; later expansion into the Pacific had specific economic motives and owed much to historical accident, but the westward movement of the "course of empire" was rooted in a desire even deeper than that of mere imperial domination. It stemmed from the obsession with dominating nature itself.

The words and actions of the leaders and proponents of American imperialism during the nineteenth century can be interpreted in racist terms: Lesser breeds stood in the path of America, the great Teutonic power. But what was the basis of the American claim to superiority? Not merely race, religion, or republican institutions, but the ability to better exploit the earth was what made us superior. We had the technology, the contempt for nature, and the economic system best designed to promote human progress. Locke had seen unused nature as a waste; Hamilton had deplored nonworking children. How then could their spiritual descendants stand by while other peoples who wasted their time in sleep, or fiestas, or traditional agriculture and crafts, occupied vast, still undeveloped regions? There was always in American imperialism a strong strain of moral indignation at opportunities for wealth and power going to waste.

Prior to the Civil War there were, of course, countercurrents in American life and thought. Emerson opposed the materialism of Jackson's America and deplored the fact that "the power of love, as the basis of a state, has never been tried."[25] Thoreau opposed the Mexican War on moral grounds. James Fenimore Cooper in his novels of the frontier often pictured the pioneers as squalid squatters and ignorant brutes and as generally inferior to the Indians whose lands they were invading. Some utopians and intellectuals took exception to the general worship of the machine, wealth, and power. But outside the South there was no meaningful political opposition to the technostructure of northern industrialization which was being raised on the foundation of our liberal political tradition.

The Southern Myth

The conservatism of the South was always more myth than reality. Despite their aristocratic pretensions and the later legends built around the world of the plantation, the southern ruling class was just another gang of petty-bourgeois Whigs, with no feeling for or love of the land, no culture of any consequence beyond a fondness for dogs and horses, and a lust for the fast buck: "The succession of exhausted soils that marks the passage of plantation society from the Tidewater to Texas was simply the clinching proof that the planters were business people and not gentlemen."[26] The cotton kingdom was as ruthless a form of soil-mining in defiance of ecological sanity as any corporate "factory in the field" today.

A motley group of intellectuals and publicists tried to pretend otherwise, creating a gracious southern aristocracy in books where none existed in reality, a legend immortalized in the motion-picture version of *Gone with the Wind*. Southern apologist George Fitzhugh bitterly condemned northern industrialism and liberal economics in terms worthy of Marx's classic descriptions of the lot of the English working class. He argued that as long as class-divided societies were, as he assumed, inevitable, it made sense to treat the lower class decently—which the South allegedly did, unlike the North—and at the same time forbid them any civil or political rights. Ironically his description of northern horrors was more accurate than his legends of southern bliss, but only southerners took him seriously.

At a more systematic and pretentious level John C. Calhoun, statesman as well as propagandist, attempted without success to create a viable political theory to serve southern interests. In his much-touted *Disquisition on Government* he trenchantly exposed the social contract as the foolish antisocial myth that it was and is, and argued that society is natural to man:

> Man is so constituted as to be a social being. His inclinations and wants, physical and moral, irresistibly impel him to associate with his kind; and he has, accordingly, never been found, in any age or country, in any other state than the social. In no other, indeed, could he exist, and in no other—were it possible for him to exist—could he attain to a full development of his moral and intellectual faculties or raise himself in the scale of being, much above the level of the brute creation.[27]

So far so good, especially when compared with Locke or Jefferson. But then Calhoun proceeds to argue that society cannot exist without government because men are basically self-centered—hardly the foundation for a noble political community. From this positing of human egoism as the basis for government, Calhoun goes on to espouse a Lockean-inspired, social-contract reading of the American Constitution and advocates reforms designed to protect southern interests which would have made the Constitution even more Newtonian and mechanistic than it already was. Calhoun too was a liberal.

Northern opinion did not take Calhoun any more seriously than it had the romantic apologists for slavery. Slavery threatened free labor, and it introduced unnecessary and undesirable structural rigidities into the liberal economic system. The debate was appealed to the jurisdiction of Mars, and the North won. After the Civil War the South was ruthlessly exploited by northern finance and industry and their southern collaborators, with racial differences being used to divide the white and black poor, a process which has continued into modern times.

The Triumph of the Vandal Ideology

Liberal society reached its political and economic apogee following the Civil War. The Republican party, beginning with Lincoln and the Free Soilers, created a democratized Whiggery, forging a durable alliance between big-business and the aspiring "little man." When the Democratic party came under the control of populist agrarian radicalism in 1896, Republican forces united with organized labor to virtually destroy the Democratic party as an alternative government for more than a generation. The Great Plains became the base of a vast capitalistic agricultural economy dependent on eastern and European markets, run by Lockean farmer-businessmen. The Indians of the West were defeated in battle by these forces. "Custer Died for Your Sins," a popular automobile bumper-strip of the late sixties, could have read, "Custer Died for Locke's Virtues."[28] The robber barons triumphed throughout the continent and the West was turned over—by stages, but inexorably—to corporation farms, railroads, and large-scale mining and grazing interests. Since no one would abandon the good old American tradition of squatting—least of all the powerful—public land laws were so written as to formalize their depredations.[29]

While the orgy of expansionism and corruption was nationwide, the development of the West wrote a new chapter in the history of American society's war with nature. Taming the Great Plains, opened to settlement through the Homestead Act, presented a special problem to immigrant easterners and Europeans because of the vast distances, harsh climate, undependable rainfall, and insect plagues. The "plow that broke the plains" brought dustbowls and floods in its wake.

In order to deal with some of these problems, science in government came into its own with the creation of the Department of Agriculture, followed by the founding of the great land-grant universities. These publicly endowed institutions, a testimony to America's faith in education as a tool for the promotion of the material success of its citizen-proprietors, were speedily converted into fortresses of agribusiness and other corporate interests.

When the tide of westward expansion reached the intermountain West and the Southwest new and even more difficult problems were encountered. Earlier cartographers had labeled most of the area "The Great American Desert," but promoters and speculators succeeded in suppressing this usage. Yet, in truth, most of the American West did consist of a few oases surrounded by deserts and mountains, a vast but uniquely fragile natural system. Even the ecological island called California was a life-system in delicate balance with its environment, and its virtual destruction at man's hands has been a vivid culminating chapter in our war with our continent.

To make the West profitable, vast amounts of labor, technology, and capital were needed. Labor was provided partly by immigrants from other continents—many of them refugees from the destruction they or their ancestors had wrought elsewhere—and partly by the overrun original Spanish-speaking inhabitants and their cousins from south of the Rio Grande. The requisite technology was devised and made available in large part by the federal government. Capital was created by giving away the public lands to the railroads and in the form of public works (highways, dams, electrical systems) paid for by federal taxes in an ingenious new "American System" whereby the citizens of the rest of the nation had money transferred from their pockets to the large corporations who were the chief beneficiaries of federally financed cheap irrigation water and cheap power, part of a general process of subsidizing growth in areas where it is ecologically dubious by levies on the general public. Shipyards on the West Coast have long enjoyed a 6 per cent differential in bidding for federal contracts, and once the military-industrial complex got into full swing during and after World War II, the West Coast and Texas were favored in contract allocation, a classic example of how the liberal state can be used to promote the growth of new groups of capitalists. It is an ironic but vivid illustration of the power of the myths upon which the liberal state rests that some of America's most militant defenders of "free enterprise" reside in these regions.[30]

The problems created by rapid and uncontrolled, though subsidized, growth of the West gave rise to some doubts about the viability of the liberal economy in its relations with nature. Even before the Civil War a few speculative souls (the novelist Cooper among them) had been concerned that the continent's resources were not infinite, but it was not until the shores of the Pacific were reached and secured for America that a psychological sense of boundaries could enter the national consciousness, creating the often subconscious realization that there was no place left to run. The modern conservation movement, spurred by President Theodore Roosevelt's interest in the West, was the result. But it was the expression of a small minority, isolated by background and education from the rest of the population, and split between those who were interested in the preservation of natural amenities for their own sake—who looked upon nature as a value in itself—and those concerned with maximally efficient use of finite resources.[31]

Conservation, whether based on respect for nature or fear of resource depletion, was doomed to be of limited appeal during the period between the Civil War and World War I. The Gospel of Wealth of Andrew Carnegie was everywhere dominant, even in pulpits putatively dedicated to other tidings. Theoretical liberalism of the baldest sort was a popular platform commodity throughout the land; the English liberal philosopher Herbert Spencer was a public personality on his visits to American shores, and Social Darwinism was the reigning American creed. Political movements such as Progressivism waged a futile war against corruption, the inevitable concomitant of the vandalization of the nation's natural and human resources, rural and urban. Economic protest was embodied in such causes as Henry George's Single Tax movement and Edward Bellamy's Nationalism, which grew spectacularly though evanescently. In large part the failure of these movements was due to the fact that they, and even the once-promising American Socialist party, were not protests against liberal society as such, insofar as their supporters were concerned, but rather the complaint of those excluded from the division of the booty—Jacksonianism in a new guise. As soon as new ships to loot hove into view and new towns were found to sack these movements faded away, though it took the repression by the Wilson Administration during World War I to finally destroy the Socialists as a significant political force. Everyone save a few isolated intellectuals—men like pioneer ecologist George Perkins Marsh, naturalist John Muir, or government scientist and explorer John Wesley Powell—unreservedly embraced liberalism and its doctrine of the acquisition of wealth through the ruthless exploitation of nature.

Liberalism at Bay

The great proof that the liberal system worked was World War I. True, conformity triumphed, and under Woodrow Wilson dissidents and radicals were persecuted as never before in American history, and government was forced to adopt a kind of national planning uncongenial to some liberal ideologues, but from American factories and fields poured the materials, from American homes the men to defeat Imperial Germany. As a result of abandoning Jefferson's dream of a republic of yeoman farmers and of allowing the robber barons to lay the foundations of an industrial colossus, we had become the world's most powerful nation.

Yet within little more than a decade pride was tinged with despair upon the onset of the Great Depression, the convulsion that marked the end of old-fashioned market capitalism throughout the world and the transition to whatever it is we live under today. The 1920s were a transitional era. The inability of liberal society to provide a sense of community during peacetime was vividly illustrated in the "Roaring Twenties," when vulgarity and corruption ruled unchecked. During these years, in order to combat the growth of the labor movement, American industry launched a systematic campaign on behalf of an "American Way" in which the interests of workers and owners were held to be the same. But most Americans needed no convincing. They were eagerly watching the rising stock-market or reveling in the wonders of the mass-produced automobile, the motion picture, and the radio. Calvin Coolidge enunciated the American creed in a matter-of-fact way: "The business of America is business."

But the Great Depression put an end to mass faith in old-style entrepreneurial capitalism. International financial turmoil combined with domestic economic imbalances to cause a collapse of business activity, catastrophic unemployment, and an end to the dream of a society in which everyone could become rich by riding the tide of speculative growth. In the meantime, floods and duststorms gave the first warnings of the over- and misutilization of the nation's physical resources. Americans were the children of Prospero, but Caliban was taking his revenge.

A bewildered nation turned to Washington for guidance. Under the pragmatic and intellectually diffuse leadership of Franklin Roosevelt, the federal government administered a variety of conflicting remedies for the nation's ills. An early attempt to create a corporate state under the NRA was succeeded by the "Second New Deal" and a flurry of antitrust activity. Attempts to stimulate economic growth through the application of such Keynesian prescriptions as increased government spending and a generally inflationary policy were carried on concurrently with attempts to create a planned economy based on the belief of some that America had become a "mature" economy. New Deal economic planning especially was the result of a variety of intellectual inspirations, but important among them was the early conservation movement and the ideas of men such as Lester F. Ward and Herbert Croly, who were essentially anti-Lockean in their premises— men committed to a more holistic, less competitive view of society than that of liberalism. Such New Deal intellectual leaders as Rexford Guy Tugwell and Henry A. Wallace were the heirs of a counterliberal tradition that could be traced back even further to the influence of men such as the pre–Civil War economist Henry Carey. The efforts of these men (embodied especially in rural resettlement, TVA, and the planned community of Greenbelt, Maryland) were submerged by other tendencies and by World War II itself. But for the first time in American history there were systematic attempts on the part of forces within the American government to seek ecological balance and social community as a matter of public policy. The men who led these attempts were optimists about nature's bounty and, even more, about the promise of technology, but they at least realized that left to themselves the forces of the liberal market place would create not order but chaos.

Neoliberalism in Power

World War II marked the end of traditional liberal society in America and its replacement by a phenomenon which has not yet been given an acceptable name. Some Marxists speak of *state capitalism*, though only a few recognize that it is a phenomenon Marx never envisaged; others still prefer the dated term *monopoly capitalism*. Some free-enterprise-oriented economists call it *syndicalism*. The ingenious Professor Galbraith has dubbed it "the new industrial state" and the "technostructure."[32] Political scientists, when talking about it in a political context, usually call it *interest-group liberalism*.

One of our difficulties as a people in coming to grips with our current problems is that, having no adequate, accepted, popularized name for our current system, we continue to mouth the ideals and ideas of the past; though knowing in our hearts they are meaningless, we are unable or fear to give the new reality a new name. Perhaps the best term for it is neoliberalism. It is liberal in that it is still based on the Lockean belief in society as a congeries of special interests all seeking their own private gain and in that it regards economic growth as the major source of human motivation. It differs from traditional liberalism, however, in that the competitors for power and profit are now no longer individuals but organized groups: industrialists, farmers, workers, professions, churches, educational and scientific communities, ethnic groups, and so on.

Old-style Lockean liberalism in its pure form had visualized the struggle for power as one waged among individuals, faceless men who—especially as far as classical economics was concerned—were without families, social status, or group ties. Even the capitalist class in the writings of Locke and his followers has this disembodied, ahistorical quality, though there was never any doubt as to the real identity of those on whose behalf Locke and his followers wrote.

By contrast, in neoliberal society, status—both ascriptive status such as race and sex, and achieved status such as possessing a college degree or a good credit rating—plays a major role in determining membership in the various and sometimes shifting and overlapping groups which now wage the war of each against all. In the political arena today organizations such as the AMA, the American Farm Bureau Federation, Standard Oil, and Lockheed are the contestants in the struggle for power and privilege.

Some political scientists describe this new society as "pluralist," since in it many forces contend for status and power.[33] But such a description contains a hidden bias: It implies that all the particular interests in society are adequately represented in the struggle for power, which is not the case, and it postulates that all the contenders for power are on an equal footing, which is clearly false. The heads of ITT obviously have greater access to the power of government and the rewards government can bestow than do ordinary citizens and taxpayers, to say nothing of welfare mothers or unemployed black teenagers. Common social interests which run counter to the liberal gospel of increase and aggrandizement, such as peace or ecological balance, have no claim to representation, since they are interests of the society as a whole, and pluralism denies the possibility that such common interests exist.[34]

Despite the fact that the new social order is one in which government plays a major role in the economy—a role which makes it simultaneously both the creator and the creature of corporate wealth—to describe this new order as "postliberal" is deceiving. For corporate liberalism or state capitalism or whatever one may wish to call it continues to be faithful not only to the ideal of struggle as opposed to that of community but to the central Lockean doctrine of increase. The collapse of the old liberal economy in the Great Depression has only served to strengthen devotion to this principle. Belief in the capitalist has been replaced by faith in the machine. Technology has become an ideology in its own right.[35]

The present structure of neoliberal society rests on the belief of the population at large that increased material wealth is the proper goal of life and that if anything is done to disturb the existing power structure, the population will suffer materially. The misnamed "free enterprise" system of the technostructure holds the key to felicity; to interfere with its workings is to court disaster. Struggle is properly limited to seeking a share in the status and wealth the system provides. As a result the system itself is unchallenged and unchallengeable.

The evolution of liberalism into neoliberalism has been going on for a long time, but the most crucial changes have occurred since the Great Depression, especially during World War II and the postwar period.[36] The New Deal began the process of openly legitimating group interests within society. The N.I.R.A. with its fascist corporate-state aspects died in court, but its spirit lingered on. The process of allowing segments of industry feudal rule over their own domain behind the façade of federal regulation goes back as far as the creation of the Interstate Commerce Commission in the nineteenth century, but it became universal in the post-Depression era. The A.A.A. organized farmers into quasigovernmental bodies, and again adverse court decisions could not stem the process, as is apparent today in the crop-support programs. The Wagner Act made labor a partner in the system. Even the blacks began to be recognized as a special group with special problems.

As government increasingly and more openly intervened in the economy, the true nature of liberal politics became apparent. The conflict was not among unorganized voters or their representatives, or between political parties, but between various "estates" of the realm—each typically composed of an interest group, a federal administrative and/or regulatory agency devoted to advancing its interests, and a standing committee of Congress which controlled legislation relating to the group and oversaw the workings of the relevant executive agencies. The legislature as a whole, the political parties, and the Presidency, were brokers among these interest groups, and by no means always honest brokers. If there were any group interests not adequately represented in this system, they were out of luck. If there was such a thing as a public interest, it had virtually no voice or leverage in the political process.

The principal difference between nineteenth-century liberalism and that of the twentieth century, or between what some would call traditional liberalism and reform liberalism, is the somewhat greater weight given human rights as opposed to property rights, a conflict sometimes described as the struggle over equality.[37] What this has meant in practice is simply that new groups, defined largely by such ascriptive characteristics as being black, female, or young, have been allowed, to some extent at least, to participate in the group struggle. But this change has not in any way altered the nature of the struggle which is still essentially Lockean—indeed Hobbesian—in character.

World War II did even more than the New Deal to create a neoliberal system in the United States. It was midwife to the birth of the contemporary military-industrial-technological complex. In the thirties the Nye Committee had held hearings on the role of the arms industry in promoting wars, especially World War I. Now all industry was arms industry. Organized science began to play a larger role in both business and government, a role which culminated in Hiroshima and the Atomic Energy Commission. The war led to the creation of a controlled economy, which could not be dismantled simply by letting wages and prices rise after 1946.

When the war ended, America was clearly a managed economy, and it was an economy based on groups rather than individuals. The technostructure came into its own simultaneously with interest-group politics. This is not to say that the technologist or the manager has replaced the capitalist in the "managerial revolution" visualized by some social scientists.[38] Bourgeois values and bourgeois men still call the tune. But a new amalgam has been created, just as in Europe and Japan the old aristocracy and the new bourgeoisie merged when modern industrial capitalism was created. America is today a syndicalist, bureaucratized economy and society in which the individual is at the mercy of large organizations. It is a state capitalism, if one likes, but one based not on socialist norms of equality but on bourgeois norms of inequality, where the prime function of the state is clearly to advance economic interests, to spur economic growth and technological advance, and to promote new forms of inequality. Locke might find it strange, but Hamilton would feel quite at home.

World War II did more than solidify, expand, and legitimize group liberalism. It gave new meaning to the Lockean dogma of increase. Productivity was the key to victory; all available resources of money, labor, and physical nature were mobilized to win the war. The curbs on expansion which had resulted from the limitations upon effective demand inherent in the old market system were removed as the war machine became the insatiable consumer. The notion of some New Deal planners that the United States was a mature economy was swept aside by a new burst of growth as it was proved that America could have guns and butter too. Some previously favored groups found themselves less well off under wartime rationing, but in general there was more meat being eaten—sometimes by draftees and war workers who had rarely seen it before—and more clothing worn than ever before. Government and the war industry consumed the vehicles and gasoline and energy that civilian sources had previously consumed, but several times over.

World War II left America with a vastly expanded level of economic production and increased the hopes of millions for the material good life. The neoliberal structure of interest-group liberalism was combined with a restored belief in America as a land of milk and honey. We had fought the war for "God, Mother, and apple pie," but not necessarily in that order. Not only was there at war's end a vast productive plant which had to be used, and new mass appetites to be satisfied, but there were new aspirations to be fulfilled. Store clerks and filling-station attendants who had become officers thought of themselves as future executives. The G.I. Bill of Rights sent millions of veterans to college, many of whom might not otherwise have gone, and sleepy little normal schools became multiversities almost overnight. Through this process a whole generation with humble social beginnings rose to positions of leadership—men who had grown up in a world of dogtags, chowlines and mass organization. Social mobility, affluence, and bureaucratization triumphed together.

Liberalism's Indian Summer

To describe America of the early 1970s as the heir to all that has gone before is to appear to award the palm of moral victory to Lockean liberalism. Who could argue against the desirability—in the simplest human terms—of a society in which more people—absolutely and in proportion to any large national population—enjoy generally higher standards of diet, housing, health, education, and recreational opportunities than ever before in human history? If Nixon represents the rearguard of traditional American liberalism as one of his biographers contends,[39] who could say that the American people were wrong in rejecting the wave of protest of the late sixties and passing an overwhelmingly favorable verdict on him in 1972 and, in so doing, on the works of Hamilton, Jefferson, Locke, and even Hobbes?

Yet there are dissenters—not only "pointy-headed intellectuals," but masses of people who voted for his opponent or chose Nixon only as a lesser evil or refused to vote at all. The presidential election of 1972 was described by *Time* magazine as a struggle between two Americas, but it was, as *Time* also recognized, more than that.[40] It was a struggle between two perceptions of reality. In the last analysis it was another confused battle between the two primal concepts of the meaning of America: the dream of the saints and the dream of the pirates.

For there is another side to the coin, and, when all the returns, not only of elections but from social and physical dynamics, are in, the liberal society will be proved to be unworkable. Because political conflict focuses on popular perceptions based on the immediate and the particular, the obvious rather than the hidden, the results do not always accurately reflect the state of the nation. Often they reflect the hopes and fears of the electorate rather than the underlying reality, just as a dying patient may sometimes elect to retain the services of an incompetent physician. When the American people chose "four more years" in 1972, it looked like the beginning of a long Indian summer for liberalism, a last reprise of the "great barbecue" of the late nineteenth century. But what started out as the "Satisfied Seventies" is fast becoming the "Sour Seventies," a decade of scandal and scarcity. Even Americans who see the Watergate investigations as partisanly motivated or much ado over little cannot escape the unpleasantness of having to cope with closed gas-stations and empty butchers' cases. But even before these cracks began to appear in the American economy—the fragile picture window through which Americans view reality—it was becoming apparent that liberalism's legacy to America was far from being a wholly happy one.

Americans are increasingly fearful, lonely, and alienated. Income is still maldistributed, as is political power.[41] Social conflict is endemic and rising, despite cyclic trends of expression and repression: black versus white, parents versus children, labor versus management, workers versus union leadership, teachers versus students, intellectuals versus nonintellectuals, even women versus men. We have a growing sense of being controlled by faceless social machines—government agencies, private corporations, labor unions, even schools and churches—which not only ignore but actively seek to obliterate our identities as individual human beings. Crime reaches new heights, and the abuse of drugs—both alcoholic and nonalcoholic—remains a major problem among all age groups and at all economic levels.[42] Boredom and a feeling that their allotted tasks are meaningless as well as tedious spreads among workers, especially the young. Corruption plagues government and business at all levels.[43] Some of the astronauts, with their hidden postage-stamp caper, even carried corruption to the moon.[44] Proposals for reform are legion, but legislatures usually manage to balk at the last minute.

The ordinary citizen is increasingly alienated from and cynical about the workings of the political and economic system, though he can conceive of no acceptable alternative.[45] Undeclared war has become a norm of American foreign policy, and Lieutenant Calley became a popular hero. Even an enormity such as Watergate barely shocks and does not surprise. Increasingly the norms of what is organizationally convenient or technologically possible or economically profitable determine how we live in a regimented and automated society, and individuals—and society as well—suffer from "future shock" as a result of being forced to adjust to rapid, technologically conditioned social changes they neither welcome nor comprehend.

Millions of Americans, faithful acolytes of the new national religion, are glued to their TV sets weekend after weekend watching the often-drugged gladiators of pro football,[46] a spectacle encouraged by the nation's political leaders in a perversion of the Puritan and liberal ethic of struggle. They respond (though with decreasing enthusiasm) to the small-boy mentality which demands that America be "Number One" and tell those who cannot love what America has become to leave.[47]

The private environment becomes more luxurious while the public environment deteriorates.[48] "The public domain has ceased to be civilized territory."[49] Homes become arsenals; streets are deserted at night, even within blocks of the White House.

This picture of degradation and unhappiness is obviously overdrawn, since it fails to take adequate account of countertrends and of the millions of Americans who still happily enjoy their private lives, asking little save a modicum of security and subsistence[50] as most humans have throughout the ages. Yet this picture is easily documented. Although most Americans still see themselves as happy and hopeful, doubt grows.

But whatever happiness exists (if we accept John Adams's equation of happiness with wealth and ease) has been bought at a terrible price in environmental degradation. Floods are worse than ever; they have been getting worse ever since the flood-control act of 1936 which instituted a vast program of federal expenditure designed to encourage and subsidize settlement where people had no business living in the first place.[51] Rivers and lakes are polluted, and it has even been seriously suggested that the only way of dealing with the most famous example, Lake Erie, is literally to fill it in. The continental shelf and the oceans themselves are in danger of becoming lifeless sump-pits.[52] The growing "energy crisis" threatens both the Alaskan environment and the beaches and mountains of America, while nuclear power, the touted alternative to fossil fuels, threatens the health and the very lives of future generations through the production and accumulation of radioactive wastes.[53]

Cars are unsafe, and food is contaminated as well as increasingly expensive.[54] Consumerism joins environmentalism as a protest against the world liberalism has made. Commuting to work is a nervewracking chore, and the social evils from which the commuters have fled have begun to spread to the hitherto favored suburbs. Air is dirty and smog-alerts a common event in many cities. Pesticides threaten the survival of the American eagle, the nation's once-proud symbol,[55] while the Lincoln Memorial begins to crumble because of air pollution.[56] Recreational facilities are crowded; "getting away from it all" is increasingly difficult.[57] Ironically all these things are happening just as—and in large measure because—more Americans have the money, leisure, and taste to care about their air, water, beaches, safety, long-range health, and cultural amenities. Everyone has become rich, or at least almost everyone, and being rich means less and less.

Just as the good life seemed within the reach of most Americans, the economy has begun to falter and even the vaunted American standard of living upon which so much of our national identity rests is beginning to decline. Food prices rise, and Americans are told by government leaders that the era of cheap food is over for good. The energy crisis erupts, and Americans learn the meaning of brownouts in summer, fuel shortages in winter, and unavailability of gasoline for the automobiles upon which our way of life depends.

American liberal society has, above all, purchased its currently faltering levels of economic life at a terrible cost to others. American prosperity has been based on using the resources of the total planet in a way which could not continue even if the distribution of military and political power in the future permitted it.[58] The developed nations—33 per cent of the world—use 75 per cent of its nonrenewable resources. Each child born in the United States consumes five hundred times as much of the resources of the world as does a child born in India.[59] As for the future, a sober study prepared under United Nations auspices argues:

> No imaginable disposition of the planet's resources can give the 1000 million Americans who could arrive, via the three-child family, within a hundred years, conditions as spacious and promising as the standards enjoyed by three-quarters of America's two hundred million inhabitants today.[60]

In other words, we cannot go on this way. We limit our births or our standard of living, or both—or else we and the world perish.

We may perish before ecological disaster strikes, however. Despite the détentes achieved with the Soviet Union and with China, nuclear war is still possible. In years past, imperialism has been the handmaiden of liberalism, and our present economic status as a nation owes much to our political and ultimately our military power in the world. In the future we will necessarily be tempted to maintain the American standard of living by using subversion or force against rival economic interests, so that local wars will remain a constant possibility, and there is always the danger that a devastatingly destructive world war could grow out of them.

As the warmth of liberalism's Indian summer fades, the winter storms will break, whether the liberal power structure and the population which supports it can see them or not. Their inevitability is a fact of life.

The ghost is coming to the banquet.

5

Technology and the End
of Liberalism

LIBERALISM has failed. It has failed because it is historically outmoded. It has failed because technological change and population growth have created a world in which its fundamental premises of individualism, competition, and unlimited growth are incompatible with social reality. Its problems and shortcomings are not the result of temporary or accidental historical circumstances, but are the inevitable outgrowth of its basic philosophical inadequacies. It is therefore beyond reform.

Liberalism is an unworkable philosophy for a technological society. All human societies are of course based on technology, but the society into which we are rapidly moving—the coming society of the twenty-first century—differs radically from societies of the past in several key respects.

Contemporary technological society is based on sources of energy which are nonrenewable, which are being rapidly depleted, and the use of which produces increasingly insupportable side-effects in the form of pollution. Our present industrial technology is based on mass production, organized by and through highly centralized bureaucratic systems. Our technology involves a high degree of physical interaction—represented by the increasing urbanization of the human population—and of spiritual interaction—represented by the increasing extent to which not physical "work" in the traditional engineer's sense but "communication" is the end product of technology. The planet as a whole is fast becoming a single dense network of complex physical and mental interrelationships and dependencies, making the individual more than ever one point of self-consciousness within a larger body. We are creating a planetary system in which a high degree of coordination and planning is required for individual and social functioning and survival. All of these developments make it less and less possible for society to operate according to the liberal norms of individualism, competition, and unlimited growth.

Most important of all, presently emergent technology has a capability for controlling human beings and their physical environment which is virtually unlimited, subject only to the existence of a social desire to use it, the marshaling of adequate economic resources, and the laws of nature themselves. We can move mountains (with or without faith) and divert rivers into new channels, if we have the will to do so. We are becoming increasingly able to modify the weather, and we can turn night into day if we choose. We can join the Atlantic to the Pacific; we can "kill" the Great Lakes, the Baltic, the Mediterranean, or the wide oceans themselves, or we can save them. We can—for a period of a generation or two at least—feed all the world's existing peoples with a minimum of labor, or we can let famine destroy vast numbers. We can bombard any place on the face of the globe with advertising, propaganda, or nuclear bombs. We can have sex without procreation and procreation without sex. Through psychological and pharmacological means we can change perceptions, opinions, behaviors, and personality. We can allow economic and technological expansion to destroy the last areas of wilderness—even the last open spaces—and create an entirely artificial world, if that is what we want. The world can support billions in comfort and dignity or more billions in poverty and squalor, as we choose. We will soon be able to create new forms of human beings and other animals through genetic manipulation, or forms of man-machine symbiosis should that be our will.

In short, there is virtually nothing about the world that can any longer be taken for granted, nothing that cannot be altered or controlled by humanity—or individuals or groups—as it or they choose. Certain choices, of course, logically preclude others, and we cannot do everything we might wish to do all at the same time. But subject to these limitations we are moving into an era of history—or posthistory—in which man can (by most traditional measures) play at being God. Unless one postulates a direct intervention into the workings of the world as it now exists by superior extraterrestrial forces, divine or otherwise, the only limits on humanity's powers are the amounts of matter and energy within our reach and the inherent laws governing their interaction.

The philosophy of liberalism is clearly inadequate to deal with the problems presented by this new world of unlimited powers and virtually total interdependence. This is especially true given certain specific characteristics of humanity's cultural state—its value-pluralism and lack of central institutions—and of its physical state—the unequal distribution of wealth and of resources throughout the globe and the onset of ecological crisis.

The social nature of man was obvious even to the ancients but the liberal denial of human interdependence is even more absurd today. On a planetary as well as a local scale human beings can only function as elements of total systems. To pretend otherwise—as liberals have—is an outmoded intellectual aberration.

Technological society intensifies the reality of the human as a social being and underscores the fact that the individual exists only within society. But to acknowledge this fact means to acknowledge that the social contract is just as dangerous and misleading a myth for human beings as it would be for the cells of our bodies. If we cannot choose to live alone, it is obviously pointless to speak of choosing to live together. Related to this liberal misrepresentation of the relationship of the individual to society is the liberal penchant for seeing society as an arena for the clash of the preexisting ideas held by its members, since such ideas (like the individual members themselves) are a product of society and a result as well as a cause of the political and social process.

It is the final measure of the inadequacy of liberalism that not only are its individualistic premises clearly meaningless in an interdependent technological society, but under modern conditions they are perverted and turned against themselves. The dynamics of the economic and political order created by liberal individualism have resulted in the undermining of both individual autonomy and human equality. The economic free market has degenerated into a new corporate feudalism without chivalry, and the political marketplace has become a bargain basement for overpriced, shoddy goods in which the individual is increasingly at the mercy of the media and large campaign contributors and is at best a pawn in the struggle for power among interest groups, political parties, and bureaucratic elites.[1] Only through the destruction of the myths of liberalism and the creation of a political philosophy which can adequately describe technological society and can prescribe the values necessary to sustain it, can the present combination of economic and political power be broken up or humanized in such a way as to make room for whatever measure of individual freedom and equality the nature of man, society, and the physical world allows.

The End of Social Stability

Liberalism, a force for unlimited change, came into being within, and assumed the continued existence of, a stable social and cultural context. The liberal political and economic order assumed the unalterability of certain forms of social life—among them the landed proprietor, the private industrialist, and the nuclear family. It also assumed the permanence of certain values—a kind of civic religion consisting of elements of Christianity, deference to authority, classical culture, the work ethic, and the sanctity of contracts.

Liberalism—especially the liberal economic market system—then proceeded to destroy much of the social and cultural stability on which it rested, like a parasite destroying its host.[2] Because liberalism promoted the ideal of technological and economic growth, the nature of agricultural and industrial production and ownership has changed drastically, the family and the community have been gravely weakened, and traditional codes of behavior have been undermined.

Winning—liberalism's prime goal—has come to overshadow the rules of how the game should be played. The late pro-football coach Vince Lombardi, with his creed of victory at all costs, was the ultimate liberal, and was appropriately admired by men such as Richard Nixon. In the world of liberalism in extremis, opposing players are trampled on, umpires spat upon, and the goalposts of the stadium itself consumed in the bonfires of victory. Lockheed and other defense contractors are rewarded for incompetence; duplicity and theft are defended as necessary to national security. Increasingly people doubt the personal utility of hard work and fair dealing, and less and less fear the judgment of God or the efficacy of the police. Unable to provide a concept of community higher than shared material gain or a concept of morality higher than adherence to procedural rules, liberalism has turned western and, increasingly, world society back toward not merely the state of nature which Locke postulated, but the state of war of each against all which haunted Hobbes. The myth of the beginning is converted into the reality of the end.

Our new technological society makes possible, indeed requires, a system of human interaction through which *substantive* as well as procedural decisions can be made on the basis of norms above and beyond the liberal rules of fair competition, which have lost both their meaning and their credibility. The powers which technology affords are too awesome to become toys for men with the mentality of small-boy playground bullies. The value-systems which have disappeared since (and largely as a result of) the triumph of liberalism must be replaced by new value-systems.

Public policy decisions can be based in large part on scientific knowledge, spread as widely as possible throughout the society in order to avoid gnosticism and elitism and a consequent lack of public understanding and support. But the mere spread of scientific knowledge is not enough to restore legitimacy to political institutions and policies. A system of values based on nature and man's knowledge of nature—the nature of the world, including the nature of mankind—must be the foundation of the basic moral consensus on which society rests.

Liberalism is intrinsically hostile to values other than those embodied in the political process itself. Just as it destroyed the old consensus of society, it seeks to undermine the possibility of any new substantive value-systems.[3] Yet there are only two alternatives to a socially accepted value-system based on knowledge of and respect for nature. One is the dominance of the most simplistic technological value of maximum efficiency (Ellul's *bête noire*), a standard which is meaningless anyway without reference to goals. The other is the breakdown of any social order except that based on coercion by whatever forces are strongest at any given moment in time.

The End of the Outside

Central to liberalism has been the belief that politics could be divorced from all other areas of life, especially the economic. It was assumed that life went on regardless of what happened within the political order, and that the political system could take many things for granted: the economic, social, and physical characteristics of its participants, the shape and quality of the lands and the seas, and the contours of society and culture. For liberalism, politics constituted a subsystem within a larger system which existed independently and beyond the reach of the political process. Forces from outside politics could impinge on the political process, but the political order was not supposed to affect these outside forces in any substantial way; political action was essentially limited to the political system. Certain things were outside of and beyond the reach of politics. Politics might help keep things as they were, as in protecting the existing division of property, but it could not change their character.

But today we live in an unbroken web of life and action. Political decisions can determine what we do about pollution, population, resource depletion, atomic energy, weather modification, education, genetic engineering, and a host of other problems that affect the shape and quality of the economy, society, culture, and even the nature of the human species. Not taking action on these matters is itself a political decision, since we now have the ability to affect them. Political decisions not only reflect but also condition the political forces generated by the "outside"; not only are political decisions influenced by "external" events, but these events themselves are, or can be, directly affected by the conscious decision-making of the community—that is, by politics.

Most critics of liberalism have focused on its failure to recognize the obvious interrelationship between politics and economics. Tax laws, we have long realized, not only are influenced by who has economic power now but are influential in determining who will have economic power in the future. But if we face up to the realities of life in a technological society we will recognize that not only the distribution of economic power but the relationships between human beings and machines, the characteristics of the environment, and the shape of man himself are now capable—unlike the situation when liberalism was born—of being consciously altered by collective human action, that is to say by politics. Thus in the new world of technological society there is no longer anything "outside" the political system: everything is necessarily political, everything is necessarily subject to conscious collective human choice.

The Limits to Growth

The heart of Lockean liberalism is the dogma of increase. Increase not only benefits mankind in a gross material sense, it short-circuits what has traditionally been the central problem of political life: the attainment of justice. If the good things of life are known and finite, society must divide them among its members in accordance with some system of social ethics. But if the machine can produce ever more things without destroying those already in existence, if the increase of material wealth is potentially unlimited, the problem of distribution is no longer a problem.

During the nineteenth century especially, some English liberals became concerned about where liberalism was taking society, and skeptical about the human value of indefinitely increasing production and population. John Stuart Mill, for instance, deplored what man was doing to nature and began, as his thought matured through various editions of his *Principles of Political Economy*, to think in terms of a "steady-state" economy.[4] But American liberalism, making common cause with America's optimistic democratic traditions, never admitted his insights. Nor could it realistically do so, so politically necessary was the vandal ideology as a banner under which all the forces of American society could march. In contemporary post–World War II western society it has been the domestic economic policy of capitalist and socialist nations alike to still any mass complaints about relative shares of economic goods by following expansionist economic policies which enable the poorer to become better off absolutely even though the capitalists or bureaucrats may be becoming better off not only absolutely but even relatively to the people as a whole. The rich get richer, but if the poor are somewhat better off complaints are stifled.

But continuous growth cannot solve the problem of distribution for two reasons. One is psychological. The more people get the more people want, as long as there are others who seem to have even more. This phenomenon, which is sometimes described as "relative deprivation," takes place both within and among nations, and as a consequence no unequal allocation of economic goods, no matter how generous the smaller portions, can ever make everyone completely happy, especially if the possibility of further growth of available goods is believed to exist.

The second reason why growth cannot solve the problem of distribution is physical: The production of additional material goods simply cannot go on forever, because of the finiteness of the earth and its resources. As the limits to growth are approached, some decision about redistribution must be reached. Ethical norms of distribution which go beyond liberalism's moral agnosticism must be devised and mechanisms of social control adopted which cannot easily be accommodated to liberal norms. The vandal ideology of increase-for-its-own-sake will have to be superseded by a new political and social philosophy.

The Decline of Compromise and the Restoration of Justice

Creating a human future within technological society will demand comprehensive, decisive, and relatively rapid collective action based on clear-cut choices. But political liberalism places its highest value not on the output or results of the political process, but on the process itself. Deadlock often means that no action is taken. This in turn usually means that the status quo is maintained or that those with the economic power to do so are left free to alter the status quo in whatever way they consider to be to their own benefit. Not to decide is to decide. If nothing is done to provide medical care for the indigent poor, a decision has been made: The poor shall suffer. Deadlock also usually means minority rule, as even Hamilton noted.[5] If a majority prefers Situation A, involving a change, and a minority prefers Situation B, involving no change, then leaving things as they are does not mean that no decision has been made but rather that the minority has been given the power to make the decision. Thus liberalism—because it places the highest priority on procedural rules, built-in minority rights, and similar checks on governmental action—has an inherent propensity toward deadlock and inactivity and thus is an apt philosophy for those who like things the way they are and wish to keep them that way.

But liberalism can also be a powerful agent for change. If political action is not taken to affect events, other nonpolitical forces can and will step in to effect change—change without any social check, an abdication of collective human control over the future. If the mining companies are destroying the countryside through stripmining, not to prohibit stripmining is a decision to let private power change the landscape. If the medical profession decides to engage in certain kinds of genetic manipulations and the law does not regulate it, then this constitutes a decision to change man's nature, a change left to a very few.

Throughout history liberalism has presided over the remolding of western society and has furnished the context for the future shock which is now upon us. It allows the old to be destroyed without giving humanity any control over what is to take its place. Contemporary technology has increased both the scope and the intensity of the problem. The more technology makes it possible for small segments of mankind to effect changes (in man, in society, and in man's relationship with nature) which affect the future of humanity as a whole, the more a political philosophy such as liberalism, which institutionalizes political inaction and minority desires, deprives humanity of control over its destiny.

Compromise, the principal positive product of political liberalism, is simply a more complex form of deadlock, whereby future events are shaped but not enough to make any crucial difference. Many important issues are intrinsically unamenable to successful compromise. In some cases this is because they concern values. If the liberal political process produces an abortion law which permits abortion only prior to the third month of pregnancy, jurists and politicians may breathe easier, thinking they have produced a statesmanlike compromise. But neither those who believe that a woman's rights over her own body mandate abortion on demand, nor those who believe that the fetus is a human being with an inherent right to life, will be satisfied. Other issues are difficult to compromise because they are crucial watersheds in natural life which are either crossed or not. If a lake cannot renew itself after 40 percent of its waters have been subject to eutrophaction, keeping 50 percent of it free of detergents does no one any good. Half a loaf is often no better than none, just as a bridge which only goes three quarters of the way across a river is totally useless.

Because of the priority which liberal society historically gave to economic issues—and to economic solutions to noneconomic issues—compromise was often a useful outcome of the political process. Most people regard ten dollars as better than five dollars: Any amount of money is good for something in a market economy, and money is easily divisible. But if political conflict is focused on things that money cannot buy or which are not divisible into parts in such a way that a little money can buy a little, compromise solutions cannot successfully resolve the conflict. If I bribe three of my five children to sleep late on a Sunday morning I am little better off as far as quiet is concerned than if I bribe none; analogous situations exist relative to industries which pollute the air and water. Even some differences in degree, though important in themselves, are of only minor or temporary effect in arriving at desired objectives: If each mother bears only three children instead of four, this still does not solve the problem of constant population growth.

Many issues raised by man's relationship to technology and nature are compromisable; most probably are not. In addition, the increasing prominence in politics of such emotionally charged issues as peace, racial equality, the death penalty, and the use of drugs—issues involving not so much clashes of interest as clashes of worldviews and lifestyles—has already severely strained the liberal political process in the modern world. Demonstrating youths and irate housewives are a lot harder for traditional liberal politicians to deal with than corporations or labor unions. And really coming to grips with the problems of preserving human life within technological society is certain to strain any political system based on compromise to the breaking point.

In the history of western political thought, liberalism represents a retreat from the pursuit of justice—of giving each man or god or thing its due—and it exalts struggle—albeit structured struggle—in its stead. Liberalism has substituted the pursuit of the possible for the pursuit of the good. But we are entering a world in which the good is all that will really be possible, because some of the new possibilities will be unbearable. Liberalism was an unsatisfactory political philosophy before the present technological revolution; it is now a completely unviable one.

6

Roads to Nowhere: Conservatism, Socialism, Anarchism, and the Antipolitics of Despair

IF liberalism has failed as a political philosophy for the contemporary world, what will take its place? Or is it possible that it can be reformed?

The so-called "reform" liberals (or "radical" liberals), a group which includes not only intellectuals but a number of practicing politicians, would argue that liberalism is not hopeless as a creed, that with a little tinkering liberalism can still provide acceptable solutions for our problems. Reform liberalism has been defined as starting from essentially the same principles as liberalism but placing a greater emphasis on equality and a somewhat reduced emphasis on "property rights, materialism, and individual self-interest."[1] Radical or reform liberalism would meet the currently popular criticism of liberalism—the argument that it leaves some elements (the nation's poor and racial minorities, for instance) out of real participation in the struggle for power—by providing increased political and economic power for such groups through state intervention if necessary. Such measures as electoral reform, changes in the tax structure, extensions of the welfare state, compensatory education and job-hiring policies would be instituted in an attempt to ensure that everyone would arrive more equally armed on the liberal political and economic battlefield. This was the inspiration behind the liberalism of the supporters of Senator McGovern, whose ill-fated presidential campaign was a nostalgic rerun of the themes of Jeffersonian and Jacksonian democracy, an essentially reactionary attempt to call

America "home" again to the era which preceded the triumph of neoliberal corporate capitalism. For reform liberalism the "revitalization of the basic values (with some changes in emphasis) and reopening of the political processes (with some modernizing improvements) are . . . the keys to reform."[2]

But reform liberalism is doomed to failure. Since it shares liberalism's fear of absolutes and the premise that the workings of the liberal political process are more important than its substantive results, it must accept the verdict of that process. If the workings of the process produce results through the outcome of free elections which the reformers dislike, they are helpless, unless they seek refuge in the power of the courts,[3] a democratically dubious and, in the long run, highly undependable escape route.

The basic impulses of reform liberalism are correct. It *is* necessary to put better men in office. It *is* necessary to alert the public to the dangers confronting it. It *is* necessary to avoid violence as the normal method of change, since it is likely to foster repressive measures which only strengthen the status quo. It is therefore necessary to count on effecting change by means of an elite (defined as those regardless of race, age, sex, class, or education who are aware of what is wrong with the present state of affairs) who spread their views to the rest of the populace through an educational process as a prelude to a democratically based legal change in political power and priorities. Such a process of change is both feasible and ethically defensible. But reform liberalism fails to recognize the extent of the change in consciousness which must occur—a change which must run counter to the whole intellectual basis of liberal society—before civilization can be reformed and restructured. The basic problems are not questions of economic or social equality, or particular issues of pollution or urban congestion or bad schools, but the whole relationship of human beings with nature, the machine, and their fellow humans.

Incremental change can lead to total structural transformation if it is guided by a vision of what the good society should be, but without such a vision changes either will not take place or else will simply cancel each other out, thus preserving the status quo. Reform liberalism today lacks such a vision of the good society because it is liberal and focuses on means rather than ends. While reform liberals can be useful allies in the cause of social regeneration, as long as they remain liberals they cannot become its leaders. To lead you have to have some idea of where it is you want to go.

Conservatism

Conservatism is the most ancient and in many ways the most noble of political creeds. Unlike the various brands of liberalism, it is not so focused on the political process as to lose sight of substantive standards of right and wrong. Ancient philosophers were all in some sense conservatives, because they sought a standard for human conduct rooted in the nature of things to which political and social institutions must conform in order to be deemed just. But what calls itself, and is popularly called, conservatism in America today is a futile and tawdry political movement, a congeries of intellectual dandies and political cynics fronting for a gang of used-car salesmen and real-estate developers—conservatives who don't even believe in conservation. To understand how this has come about is to realize why conservatism is not a solution to our problems but is itself one of the more serious of those problems.

Conservatism in America[4] is a failure at two levels: Intellectually conservatives have failed to understand the nature of the human person, and historically America has failed to provide the social base for an intellectually and morally principled conservatism. The intellectual failure of conservatism stems from the central error of classical philosophy, the inability to recognize the progressive nature of the human animal. The key tenet of Aristotle's *Ethics* is the belief that the essence of man is fixed and immutable, when in fact humans are evolving beings.[5] Human nature is real, but an essential part of human nature is this capacity for change. Humanity evolves. As a result, what is proper to mankind in one time and place, as a legitimate expression of human nature, will not be universally so.

Of course, not all evolution—much less all change—is progressive. Some change is dysfunctional and perverted. Not everything which is possible is desirable. It has historically been the chief glory of conservatism to insist on this fact, and to oppose change simply for the sake of change. Thus when liberalism and modern industrialism appeared on the scene in England and Europe at the beginning of the modern era, conservative thinkers and social forces sought—in the long run unsuccessfully—to place a brake on the often antisocial side effects of these movements. Retaining the ancient notion of the primacy of community and properly skeptical of social-contract theories of society, venerating throne and altar, conservatives often fought to protect the interests of the peasants and the poor against the rising bourgeoisie, and were the original proponents of what Marx in the Communist Manifesto was to call "anti-socialist socialism"— the struggle against capitalism on behalf of precapitalist values and institutions.

Seeing the fabric of life ripped apart by industrialism, conservatives often tried to curb the forces of antinatural technological change let loose by capitalism. They recognized that the capitalist philosophers "dissolved all ideas of the sacred as standing in the way of the emancipation of greed."[6] Believing in hierarchy and order, they rejected the liberal doctrine that "human excellence is promoted by the homogenizing and universalizing power of technology."[7] They were faithful to the teaching of the ancient philosophers who, as Leo Strauss notes, "in contradistinction to many present day conservatives . . . knew that one cannot be distrustful of political or social change without being distrustful of technological change," and they "demanded the strict moral-political supervision of inventions . . . [to] determine which . . . are to be made use of and which . . . suppressed."[8] Conservatives often made the mistake of confusing the accidental with the essential, of mistaking transient customs for what is fundamental to human nature. But nonetheless they clung to the belief that there are some things which are proper to humanity and some things which are not, that there is meaningful order in the universe.

Today in America most of those who call themselves conservatives—and virtually all who are politically active and identified as such—are not conservatives at all, but out-and-out liberals in flimsy disguise. They accept the fundamental liberal premises of an intrinsically necessary social struggle and of the ethical primacy of growth and change. In part this is due to the fact that they cannot escape being Americans and America is a society "which has no history before the age of progress";[9] there has never been the social basis for a true conservatism in America. But in large measure conservatism's failure is due to the fact that even some otherwise authentically conservative thinkers have adopted the liberal capitalist myths about human nature. Burke could write eloquently about society being a compact among the living, the dead, and the unborn and about the "unbought grace of life," but he could also, as Marx disgustedly notes, hold that the laws of commerce were part of the laws of nature and therefore of the laws of God.[10]

Contemporary American conservatives have identified the fundamentals of human nature with the accidental characteristics of Lockean man. What passes for conservatism in America is "a peculiarly American inversion of liberalism, a caricature of liberalism" which shares the liberals' optimism and liberalism's "ultimately materialist values."[11]

This can be easily documented by a perusal of contemporary conservative writers. There is much fervent rhetoric about tradition, but what is embraced is the liberal capitalist tradition. Inevitably so, since what is enshrined is the American constitutional system, which was created by liberals.

The leading conservative publicist, William Buckley, tells us that "American conservatives are opposed to state control of the economy . . . whether direct or indirect. They stand for a free economic system."[12] Buckley quotes with favor the program of the American Conservative Union, which endorses "free enterprise" not only in the name of individual liberty but in the name of material increase, asserting that "no other system can assure comparable living standards and growth."[13] Buckley sometimes seems uneasy about the philosophical mésalliance he has helped create—at least subconsciously aware that conservatism and capitalism may not be the most compatible of bedfellows. He bitterly and at length condemns Ayn Rand because she is avowedly and consistently atheist in her free-enterprise philosophy of selfishness, but the vehemence of his condemnation is that of one who does protest too much.[14] For the late Willmore Kendall, Buckley's former mentor and long-time collaborator, the most important task of the American conservative was to keep the struggle for wealth and power free of governmental interference, and to defend the inegalitarianism enshrined in the Constitution as originally written in 1789.[15]

Despite all the ritual bowing to Burkean ideas of community, contemporary American conservatism denies the fundamental Aristotelian premise that the political community is capable of helping people to lead the good life, of helping them to become more human. Conservatism's primary touchstone, according to the late Frank S. Meyer, is the "individual person," not the community.[16] Journalist Garry Wills, during his early conservative period, condemned both Aristotle and the medieval and modern followers of Thomas Aquinas for asking too much of the state, which, he argued, cannot create justice but at best can only take care of man's needs for material goods and social order.[17] Buckley has endorsed the political agnosticism of the British political theorist Michael Oakeshott, who doubts that the state can devise any strategic policies at all "pursuant to an act of will."[18] What community ends up meaning for most American conservatives is not women and men working together for the common good but shared prejudice and jingoism, the "love it or leave it" solidarity of mass ignorance and complacency.

Traditionally conservatives have sought standards in nature and revered the texture and shape of humanity's relationship to the beauties of creation, but there is none of this in American conservatism. The only "natural" things it reveres are the misnamed natural laws of the marketplace. One conservative writer has deplored the fact that the major American conservative organ, *National Review*, gives such scant attention to man's relationship to nature and the machine: "One would like to see in a conservative magazine, a continuous concern for the conservation of the American countryside . . . Conservatives will also have to devote more thought to such things as city planning and the problems inherent in automation."[19] But, as a perusal of "conservative" literature attests, he is a voice crying in a neon wilderness. Whatever the capitalist economy wants to do—tear down noble old buildings, pollute the air and water, destroy the wilderness—is all right. Growth is sacred, interference with the workings of corporate greed blasphemous socialism, impious meddling in the affairs of free men. Occasionally common sense overcomes dogma. William Buckley's brother James, the conservative United States senator from New York, an amateur ecologist of long standing, swallowed the supersonic transport, though with some distaste, but did work against the highway lobby's recent gutting of the environmental controls designed to slow down the paving over of America.[20] But such behavior by so-called conservatives is rare. The major thrust of conservatism is clear. Economic growth through technological innovation is the supreme good of humanity.

The extent to which conservatives are enamored of growth, even when it destroys the physical and social past they claim to revere, is illustrated by their attitudes toward population. An economist respected by many conservatives—Wilhelm Ropke—shared Mill's doubts about the benefits of a world of infinitely expanding population,[21] and occasionally a voice like that of the politically conservative demographer William Petersen raises the obvious question of how jarring change can be averted in a world of unbridled breeding.[22] But for most conservatives any thought of a governmental population policy is simply another example of antihuman tyranny over the individual and his sacred absolute rights (a position bolstered by the large number of Catholics in the American conservative movement).

Thus, although William Buckley can describe conservatism as a "spirit of resistance to the twentieth century"[23] and has expressed the hope that our time will be known as the era in which "the individual overtook technology,"[24] American conservatives are faithful servants of the spirit of the age, lackeys of the machine; and there is no gainsaying the judgment of the Canadian philosopher George Grant that while those of the "right" may "seem to have some hesitation about some of the consequences of modernity . . . they do not doubt the central fact of the North American dream—progress through technological advance."[25]

Ironically the most severe indictment that can be made of American conservatism is that it is not conservative enough. As the late Paul Goodman wrote in his last major work, "Conservatives at present seem to want to go back to conditions that obtained in the administration of McKinley. But when people are subject to universal social engineering and the biosphere itself is in danger, we need a more neolithic conservatism."[26]

Conservatism in America is a philosophy of futility. What has been described as the classical conservative dream—rejection of mass industrial society, in favor of an ordered hierarchical society not centered on the city[27]—is meaningless for the America of today or tomorrow. Not only can we not return to such a past, it was never our past anyway. The traditional conservative veneration for continuity in human life and hostility to change is pointless in America because the most salient institution which American conservatives can find to venerate—the American capitalist system and the Constitution designed to bolster it—is the greatest force for change—save for technology itself—that the world has ever known, and to embrace this system is to find oneself in the arms of Lockean liberalism. Also, the conservatives' suspicion of the intelligence and good will of the masses, and its yearning for an elite of learning and power, is futile in a society where the elite outside of the business world generally reject the conservative creed.[28] The spectacle of so-called conservative intellectuals leading a populist mob in an attack upon the universities and the courts is as pitiful as it is absurd.

But conservatism is dangerous as well as futile. The betrayal by American conservatives of truly conservative ideals threatens us all. Despite the objection of critics that a conservative mass movement is almost a contradiction in terms, it is far from impossible. In Europe and elsewhere large numbers of ordinary people have been organized to struggle against change throughout history: peasants in the French Vendée, orthodox Hindus in contemporary India. But could Americans, schooled in the inevitability of change from childhood, be organized in a conservative movement? The answer is, unfortunately, Yes. Recent experience shows that mass fears can be played on and people rallied under the banners of conservatism, not against the technological and demographic forces leading to undesired change but, ironically, on behalf of the liberalism that legitimizes these forces.

Today in America we find people being mobilized to fight any attempts on the part of the community to alter the liberal economic system so as to aid its victims, mobilized to support liberal imperialism abroad, and mobilized to oppose community attempts to rationally influence the environment in order to preserve the patrimony of the past and the health and dignity of the unborn. People are being mobilized in these essentially individualistic, technocratic, liberal causes by leaders who call themselves conservatives and exhort the masses to fight "liberal" "social engineering."

Conservatives are not merely helpless in being unable to restore and protect what they claim to revere—this would make them nostalgic figures, worthy at least of pity and admiration, gallant soldiers in a lost cause. Rather, American conservatism is a sinister force which in the name of tradition is collaborating with the destroyers of our physical and social heritage, and in the name of patriotism and loyalty is committing treason against the American earth and the American community.

Socialism

Socialism is as protean a term as conservatism, and far more in vogue throughout the contemporary world, especially outside the United States. Adolf Hitler called himself a "National Socialist," and one American socialist theorist has even claimed George Meany for socialism's legions.[29] But the mainstream of socialist thought derives from the work of Marx and Engels, and socialism can fairly be judged on the basis of their philosophy and the attempts made to implement it in nations which claim to be inspired by their teachings.

Marxian socialism fails to offer an acceptable philosophical alternative to liberalism because it shares the Lockean belief in the possibility of unlimited growth and in the unconditional triumph of man over nature through the use of technology. Despite passages in Marx in which he complains that capitalism alienates man not only from his own nature but from nature itself, for socialism man is the measure of all things and nature is an enemy to be conquered.[30] The conquest of nature is a precondition for entering the kingdom of freedom. Science and technology are intrinsically good because they enable humanity to bend nature to its will. Capitalism is (despite Marx's many nostalgic backward glances at feudalism) a progressive force, and one of the basic socialist indictments of capitalism is that, in the priority it gives to private profit, it sometimes inhibits technological advance rather than allowing it free rein.

Not only is economic growth good in itself, but so is population growth. Central both to the origin and logic of Marx's economic theory is his repudiation of the early nineteenth-century economist David Ricardo's "iron law of wages," which was derived from Malthusian thinking.[31] For Marx, as far as one can tell, under socialism the population of the world could increase without limit, avoiding poverty through the benign influence of a fully realized technology. Nature sets no limits to what man can and should do. For Leon Trotsky, socialist man would become a "superman," moving mountains and altering his own nature as his fancy dictated.[32] One searches in vain in Marx, despite his allusions to man's projected harmony with nature under socialism, for any feeling for nature at the concrete existential level. Rural life is "idiocy" and is so, one suspects, not only because of its frequent cultural and social poverty.[33] Man is a maker, a doer, a conqueror. The worker with his hammer is a fit Communist symbol, Prometheus a fit Marxian myth. Man is in revolt against all limitations and all gods, the lares and penates and the spirits of the forest and field as well as the God of the churches.

Engels, who knew far more about science than Marx, had a more realistic view of man's relationship with nature. "Let us not, however, flatter ourselves overmuch on account of our human conquest over nature. For each such conquest takes its revenge on us," he wrote in words which are echoed by ecologists today.[34] But his reservations have been ignored by virtually all socialist thinkers.

It is no wonder, therefore, that in so-called socialist nations—where a minority of bureaucrats and technicians rule in the name of socialism— technological change is embraced with the same uncritical fervor it is by most American political and corporate leaders. That automation and cybernation may lead to alienation and loss of human identity, or that scientific advance presents basic ecological problems, are heretical notions rarely and only circumspectly raised within the socialist camp.[35] The record of the Soviet Union and Eastern Europe in the field of environmental pollution is as bad as or worse than that of the capitalist industrial nations in part for technical reasons, since socialist accounting makes calculation of negative externalities even more difficult than it is under capitalism, but primarily because growth has become the great socialist god. The fetishism of commodities which Marx condemned as a feature of capitalism is just as strong a force in the socialist world as in the West.[36]

The growing warnings of Soviet scientists about the true nature of the environmental crisis, and some initial perception of the real economic costs of pollution and reckless industrialization, have recently led to policy vacillations within the Soviet Union and the assertion that socialism can deal with environmental problems better than capitalism.[37] But these problems are still regarded as marginal side effects of bad housekeeping rather than harbingers of a basic imbalance between man and nature and of a possible barrier to a victory over scarcity through the technological conquest of nature. The Soviet Union has also blown hot and cold on the subject of population control, sometimes following the orthodox anti-Malthusian Marxist line—especially when doing so can reap propaganda benefits among developing nations suspicious of Western-inspired birth-control programs—but tolerating and in the past promoting birth control domestically.

Some reports suggest that the People's Republic of China is more concerned with humanity's relationship to nature and the machine than other socialist nations.[38] To attribute this to a different view of the world, however, as by arguing that Chinese lack the western desire to conquer nature, is of course nonsense.[39] Traditional Chinese culture did not share the messianic progressivism of the West and did not seek to conquer nature as an end in itself, but the Yellow River floods are eloquent testimony to centuries of heedless deforestation.[40] Small-scale decentralized industry, pushed largely unsuccessfully, especially during the "Great Leap Forward," and the recycling of waste materials seem to represent making virtues of necessity.[41] Current Chinese poverty makes such policies highly functional, just as it does the regime's attempts to control births through late marriage and contraception. But the gleam in the eyes of Chinese visitors at the sight of large-scale technologically sophisticated western industrial processes and public works, and the heightening pall of smoke over Chinese cities, suggest that salvation from the problems of technological society is not likely to be borne on the east wind.

In capitalist nations where socialist parties have achieved a share of power, the traditional liberal emphasis on growth for growth's sake has been maintained. Since the capitalist power structure is too strong in nations such as Britain or Sweden for a frontal assault, the problem of the distribution of goods is sidestepped by an emphasis on producing more, and socialism's greatest claim to political support rests on its alleged ability to cut through certain limitations on faster economic growth caused by the capitalist emphasis on profit. Indeed, economic expansion plus the welfare state has become the working definition of postwar European socialism.[42] Only the faintest beginnings are being made to integrate the reality of the ecological limits to growth into socialist thinking.[43]

One looks in vain in the literature of American socialism for any significant discussion of the problems which economic growth and technological change pose for man's relationship with nature, his cultural integrity, or the world's ecological future. The revisionist historian William Appleton Williams begins a book designed to demonstrate the relevance of Marxism for America by telling us that "America's great evasion lies in its manipulation of Nature to avoid a confrontation with the human condition and with the challenge of building a true community,"[44] but never follows through on the implications of his theme. While he believes that cybernation has created the material foundation for an abundance which can be the basis for a "vision of community,"[45] he does not spell out how this is to come about, and though he argues that technology must be socially controlled he does not discuss the means or standards for such control.

In their important book *Monopoly Capital*, a classic orthodox Marxist interpretation of contemporary American capitalism, the economists Paul A. Baran and Paul M. Sweezy devote the overwhelming portion of their work to stressing the mechanics by which corporate giants appropriate the surplus created by modern industrial processes, the way in which this surplus has been used by business to dominate government, and how such a system has led to imperialism abroad and to failure to provide for economic needs at home. But while they attack the evils which they claim characterize modern American industrial society—high crime rates, the low quality of family and social life, and technological unemployment—they clearly imply that these would vanish if the industrial system was state-owned, rather than viewing them as possibly inevitable byproducts of technological society itself.[46]

The political and intellectual weakness of contemporary American socialism is dramatically illustrated by the career of its leading intellectual, Michael Harrington. In his recent massive work *Socialism*, Harrington seeks simultaneously to emphasize the humanism of the early Marx of the "economic and political manuscripts" of 1844, to relate socialism to the mainstream of American political and social life, and to deal with the challenges which the ecological crisis poses for orthodox socialist thought. Harrington recognizes technology as a basic challenge to human freedom and identity, yet at the same time he is bound to the Marxist belief that mankind can only enter the kingdom of freedom if scarcity is first abolished and "man's battle with nature has been completely won."[47] But while he states that technology must be brought under social control if man is to remain free, he provides no new criteria for its development and use and no new insights into how it can be controlled, and indeed has less to say on the subject than such humanist socialists as Marcuse, Fromm,

Lucian Goldman, or Eugene Kamenka.[48] In dealing with the problem of scarcity and ecological balance, he tells us that "Marx's categories . . . can be used to analyze automation, which he anticipated . . . the crisis of the environment and so on,"[49] and then suggests remedies—worldwide birth-control and a less materialistic and more austere standard of the good life—which are as close to the thinking of the ecological movement as they are alien to that of Marx and the socialist tradition.[50]

Further complicating Harrington's task of reconciling socialist ideol-ogy and the demands of the current planetary crisis is his recognition that socialism is above all a theory of how change in society is to be brought about and that, as another American socialist intellectual puts it, Marxism is "unthinkable apart from its premise of proletarian victory and its liberat-ing effect."[51] Yet Harrington knows (though he fudges the issue) that the traditional American working class is more reactionary on most issues than any other group in society—the war in Vietnam being a prime example—and is also declining in political strength. He therefore postulates that the laboring class will be joined in the revolutionary effort to change society by a "new working class" of proletarianized professionals and technicians, who he assumes will be content to remain the industrial workers' followers and auxiliaries, and will also be joined by many of the alienated young, who though selfish and flighty have some of the right instincts about the inade-quacy of life under capitalism.[52]

But how viable would such a coalition be? We already have some clues. Alienated youth and middle-class individuals constituted the core of the "new politics" movement which nominated George McGovern in 1972 and were crushed because they could not gain the support of either the leaders or the rank-and-file of organized labor. As a final ironic refutation of Harrington's thesis, agents of his erstwhile hero George Meany have captured control of the American socialist movement itself, resulting in Harrington's personal political isolation and in his resignation from his own position of leadership.[53]

The lesson seems clear that any forces concerned with the future of humanity can expect neither political nor intellectual leadership from the ranks of socialism. If we must wait for the end of scarcity before we can enter the kingdom of freedom, we will wait in vain. If we expect socialism to provide us with a blueprint for how human beings should relate to their technologies, we are going to be disappointed. If we look to socialism for a framework for restoring man's shattered harmony with his natural environ-ment, we are looking in the wrong place. It is not inconceivable that groups which call themselves socialist could, though somewhat inconsis-tently with their Marxian premises, play a useful political role in the regen-eration of society. But they cannot lead it because they do not know the way, and in the short run they are just as much the enemies of future freedom as conservatives are of the preservation of the past.

The New Left

Orthodox socialism has never made important inroads into American politics. Despite gross inequalities in wealth and living standards, greater than most Americans are actively conscious of, Americans—including the working class—have been generally content with their material lot in life. But in the sixties an increasing number of Americans became critical of their culture as a whole. For many intellectuals and important segments of youth and the middle class, the real crime of liberal capitalism was not economic oppression but spiritual disfigurement. Out of this consciousness, fostered by impatience at the refractory nature of the problems of racial discrimination, poverty, and the Vietnamese war, the short-lived New Left movement was born.

The *New Left* is an imprecise term for a varied, shifting, and intellectually and organizationally chaotic group of radical critics of contemporary American society, possessed of a bewildering variety of ideological perspectives.[54] But, at the risk of gross oversimplification, its intellectual origins can be traced to the Marxist concept of alienation, as interpreted by such writers as Erich Fromm and Herbert Marcuse to mean not only the absorption of man's identity through the process of production but the destruction of his identity in the process of consumption. Man is naturally good, stated the New Left's seminal document, the Port Huron Statement of the Students for a Democratic Society in 1962; humans are "infinitely precious and possessed of unfulfilled capacities for reason, freedom, and love."[55] Yet "loneliness, estrangement, isolation describe the vast distance between man and man today" in a degraded and degrading society.[56]

The New Left was "left" in that it presented a root-and-branch challenge to the institutions of liberal capitalist society, a challenge flung down in the name of, if not actually by, "the People." It was "new" in that it rejected several of the most sacred doctrines of orthodox Marxism: It generally repudiated the dogma of the increasing economic misery of the masses and the liberating role of the industrial proletariat by making the doctrine of universal alienation the key concept of political analysis. What was really going on, according to the central theme of many of the new leftists, was that capitalism was working only too well: Man was losing his soul in the consumer society. Liberation could only come from those who recognized this, generally speaking members of the upper economic strata who had had riches enough to know their hollowness and their cost in human values.

As the Port Huron statement illustrated, its authors were the children of alienated affluence: "We are people of this generation, bred in at least modest comfort, housed now in universities, looking uncomfortably to the world we inherit," it began.[57] New Left theorist Gregory Calvert argued that the difference between the leaders of reform liberalism and real radicals was that the latter, even though middle-class, could see themselves to be just as directly oppressed by society as the poor.[58] Capitalism was a universal oppressor and young radicals were often wont to declare that being brought up in Scarsdale was an even worse fate than being brought up in the ghetto.

In opposition to classical Marxism, the New Left took a dim view of technology, especially organizational technology. Participation in large-scale, rationally organized, centrally directed organizational structures is dehumanizing. Such organizations, usually designated by the term *bureaucracy* used always in a pejorative sense, destroyed the freedom and integrity of both their members and all those subject to their power. The technology of large-scale organizations is an intrinsic source of alienation. "I am a human being. Do not fold, spindle or mutilate," was the battlecry of students in large, computerized universities. The power of such organizations over human beings is the result of their key position in modern technological society and of the general belief that they are, through the technology they marshal, the architects of popular affluence. In order to destroy these organizations' hold over humanity, what is needed is not only structural change but renunciation of the values of consumer society by their members and clientele.

An important New Left insight was their emphasis on the extent to which industrial technology is losing social saliency to the new technology of information, communication, and control. This emphasis enabled New Left theorists to attack a badly exposed flank of liberal society, its implicit theory of public opinion. Liberalism had assumed that public opinion was in some sense a given, an exogenous force which could be used to attack and control the economic and technological structures of society. Even orthodox Marxist critics of society, for all their exposés of capitalist control of the press, had always in practice assumed the continuing existence of working-class culture and the growing self-consciousness of the proletariat. But the New Left criticism, strongly indebted to earlier sociological theories of mass society and mass culture,[59] assumed the existence of a common culture created by the total "system," a culture which rested on foundations far broader than capitalist control of particular channels of expression. If public opinion and culture itself were products of the existing power structure of the society, the claims of liberal democracy to represent the public will could be dismissed. Existing social structures could not be legitimated by reference to a process of "free" choice which was in fact under the control of these structures and hence not free. Democracy could therefore be redefined as government not by the public but by those who somehow were outside and above the system of manipulated communications, claiming to act on behalf of the public interest.

The New Left failed both intellectually and politically. Intellectually, despite its criticism of the mechanization of society it never, any more than its sometime mentor, Herbert Marcuse, distinguished systematically and comprehensively among particular technologies and their potentialities and uses.[60] Participatory democracy—its solution for the technologically conditioned evils of bureaucracy—was hardly an all-purpose answer to the problems of human living on a crowded, interdependent earth. Despite its increasing overlap with elements in the counterculture, the New Left never faced the problems inherent in man's relationship with nature, focused as it was on man's injustices to his fellow men.[61]

The political failure of the New Left was even more clear-cut. Because of its apocalyptic mood, it wanted change in a hurry—a mood reinforced by the ongoing horror of the war in Vietnam. But since it could not deny that the liberal political system worked, in the sense of representing what people thought they wanted, it had limited strategic options. In order to achieve change the New Left increasingly resorted to extralegal acts, including acts of violence, designed to awaken public consciousness, but such actions more and more turned the general public not only against the New Left but against all reform, while it forced potential allies to disassociate themselves from the New Left.[62]

As the uselessness of direct political action as a tool for revolutionizing society became apparent, members of the New Left increasingly deserted its ranks for those of the counterculture, looking to personal development in isolation from society as the basis for a revolution in consciousness which would provide the precondition for social revolution, or a substitute for it, or a means of individual salvation in a situation in which general revolution was impossible.[63]

Despite its failure and swift disappearance from the American political scent, the New Left did leave some significant legacies to American political thought. Its theorists performed a useful service in pointing up the extent to which American society was permeated not only with political and economic inequality but with alienation, mechanization, and bureaucratization, and to the fact that these were all related phenomena. Its political failure illustrates the difficulty which any reform movement based on upper-middle-class idealism faces in a political arena in which the old Left is powerless and obsolete, organized labor part of the problem rather than the solution, and oppressed racial minorities—however understandably—almost exclusively concerned with their specific problems rather than interested in those of technological society as a whole. Both the limited theoretical successes of the New Left in penetrating toward the deeper cultural roots of the ills of modern society, and its practical failures in creating a political force capable of restructuring society, stand as important lessons for any postliberal political movement aiming at social regeneration.

Anarchism

Anarchism is the belief that since man is naturally good, there is no need for a state, or at least for any large-scale, centralized political organization of society exercising coercion.[64] Most anarchists are also socialists in that they believe that private property, like the state, is an unnecessary institution which deforms human existence.

Anarchism has never received significant public support in the English-speaking world, but in the late nineteenth and early twentieth centuries it was an important political force in Southern Europe, where it built on the traditions of peasant resentment of centralized national power and capitalist institutions. Despite its lack of popular appeal, anarchism in various forms and to various degrees has been embraced by many intellectuals in Britain and the United States because of its consonance with the romantic reaction against industrial society and against centralization, bureaucratization, exploitation, and war.

In recent years there has grown up a small school of what might be called "ecoanarchists," who see a parallel between the ability of the natural biological world to achieve order and apparent harmony without centralized direction and the anarchist concept of a human social world capable of existing without state power.[65] As the leading ecoanarchist Murray Bookchin puts it: "An anarchist community would approximate a clearly defined ecosystem; it would be diversified, balanced, and harmonious."[66] The naturalness of harmony in both nature, as described by ecologists, and society, as conceived of by anarchism, is central to ecoanarchism. "The most impressive aspect of the law of the jungle is not ruthless competition and destruction, but rather interdependence and coexistence," another ecoanarchist quotes approvingly from biologist Rene Dubos.[67]

Ecoanarchists reject two of the central assumptions of their Marxist-socialist cousins: the desirability of conquering nature and the belief that unlimited economic growth is possible. "The concept of 'dominating' nature emerged from the domination of man by man," Bookchin writes; "both men and nature have always been the common victims of hierarchial society."[68] Environmental concern is not a spurious, counterrevolutionary diversion, he argues against the mainstream of leftwing opinion. We can do all sorts of things to man and nature, but "the *totality* of the natural world . . . cancels out all human pretensions to mastery over the planet."[69] Unless the liberal order is destroyed, mankind is doomed. "Either revolution will create an ecological society . . . or humanity and the natural world as we know it will perish."[70]

The ecoanarchists even recognize the gap in values which will necessarily separate any viable future social order from the antinature bias bequeathed us by Calvin, Locke, and Marx. The outcome of an anarchist revolution would be a society which would enable men to regain their "sense of oneness with nature," a society in which "culture and the human psyche will be thoroughly suffused with a new animism."[71]

The ecoanarchists realize that the path to a decent human future does not run through a Luddite attack on technology as such. Technology, the ecoanarchists agree with their socialist cousins, is the necessary key to opening the doors of the kingdom of freedom to mankind. Technologically conditioned abundance alone can make possible the end of the repressive institutions of the state and of private property, which are necessary in an era of scarcity. But ecoanarchist spokesmen such as Bookchin go further than traditional socialists such as Harrington in distinguishing among technologies, and asking which ones contribute to centralization or decentralization, alienation or community, human enslavement or human liberation; they set up some standards for the social and moral assessment of technology rather than embracing the uncritical technophilia which most Marxists share with the heirs of Locke.[72] Thinkers who have been strongly influenced by anarchism such as Mulford Sibley and the late Paul Goodman share this critical but responsible attitude toward technology, which distinguishes ecoanarchism from socialism, liberalism, and simple romantic reaction against the machine.[73]

Despite its many insights into the nature of the problems facing humanity today, ecoanarchism suffers from some fatal inadequacies. There is a basic error in the premise that the ecological order of nature is directly comparable to that of a healthy human society. Mankind cannot fully accept an order which is based on the destruction and suffering of others, as the order of nature is. Balance in nature when it exists is the result of much destruction of life: Big fish eat little fish, constantly and necessarily; cougars protect the vegetation by killing the deer who would otherwise overgraze it; the deer die in pain, their mates mourn. Human beings prefer to avoid such suffering. Classical liberals claimed that poverty and unemployment were necessary to the automatic workings of capitalism, that these were the dictates of the laws of nature operating within a self-regulatory economic system—a creed which was in many respects an illegitimate union between anarchism and private ownership.[74] But the market system was so contrary to human instincts and needs that it was only imposed on modern western man by the force of state power. Later generations of men have increasingly modified it. Normal human beings will reject any system which eliminates conscious human governance in favor of a self-regulatory society. The laws of nature are too harsh. We hear the screams of the wounded deer because they are akin to our own.

A further reason why a humanly ungoverned society is an impossible ideal is that, while animals other than men are conscious—even self-conscious—to varying degrees, they are rarely subject to "false consciousness," to appropriate Marx's phrase. They perceive reality without cultural or personal blinkers; instinct guides them to their goals more or less accurately. Hens are incapable of electing foxes to rule over them; young sheep rarely mistake wolves for their mothers. Our animal cousins don't try to second-guess, much less dominate, nature. But human society—unlike ecological systems in physical nature—has as its elements people, who have differing and sometimes erroneous perceptions of their interactive roles. As a result they foul up the workings of any system designed to be automatic. We reason that if we leave our homes fifteen minutes earlier we can avoid the worst of the rush-hour traffic. Many of us try it, resulting in an earlier rush-hour and fifteen minutes less sleep. Human systems cannot operate on the automatic meshing of desires alone. Our desires are mediated through culture which distorts them. For human systems to operate effectively, desire and innate drives must be supplemented by law, custom, and conscious direction. A laissez-faire society which would be tolerable to ordinary human beings is just as chimerical as a laissez-faire economy. Anarchists are right to strive for a society which maximizes freedom and self-fulfillment, but such a society could never exist without some degree of overall coordination and control over individual passions and follies.

What is true in general—the inescapability of some form of government—will be especially true in the coming decades, as the worldwide ecological crisis worsens. There is a curious parallel between the practical estimation of the world's problems made by many of the ecoanarchists and by such orthodox but sophisticated socialists as Harrington. Both accept the fundamental Marxist premise that abundance is the key to the kingdom of freedom which they both seek. Both recognize that to some extent advanced technology can free mankind from the restrictions of the past, and therefore should be used rather than rejected. But both also realize that this ability of technology is not (as the liberals assume) unlimited; even the most advanced technology possible and humanly acceptable cannot, they both recognize, do away with global poverty unless the size of the human population is stabilized by birth control and unless human beings, especially in the currently developed nations, are willing to accept a redefinition of what the material good life consists of and a more austere standard of living. They necessarily disagree, however, as to how this reduction of population and demand is to be brought about. On this issue, socialism makes more sense than anarchism. Considerable large-scale centralized direction, based in the last analysis on coercive power, will be necessary to institute and maintain the control over human social and economic life required to create global ecological balance, something socialism can admit but which is unacceptable to anarchism. Government, always a requirement to some degree in human society, will be especially necessary in the years to come.

If anarchists are less realistic than socialists about the need for the state in the emerging world of the twenty-first century, they are both more and less realistic about the means by which the new world of the future should be built. Bookchin recognizes that the traditional working class is too reactionary to be a useful revolutionary vehicle and that Marxists who cling to their faith in the proletariat are intellectual troglodytes who have lost touch with reality. Anarchists share the New Left perspective that everyone is alienated and oppressed by current society. This being the case, they proceed to argue, it will be from among the most alienated that the mass revolt against society will come. For the anarchists, there is nothing wrong with "mobs" and "riots"; they are in fact the first step toward "individuation."[75] "Ultimately," Bookchin asserts, "it is in the streets that power must be dissolved." All revolution comes "spontaneously" and from "below," and the abortive May-June 1968 uprising in France is a prototype of international importance. The lumpenproletariat, the drifters and cast-offs of society, are to be the initiators of the new order brought about through disorder, and "the spirit of negativity must extend to all areas of life."[76]

Such a revolution is as unattractive as it is implausible. The most alienated are unlikely to be the best at curing alienation. Woodstock was followed by Altamont, the Diggers by Charles Manson's "family." An anarchism dependent upon the outcasts of society is, as Paul Goodman has observed, likely to end in dictatorship and authoritarianism.[77] Many people used to argue that the western world would be saved from Communism by ex-Communists, as if having been an asylum inmate was a precondition for being a psychotherapist. That a community of human affirmation can be built by means of a process of absolute negation seems just as unlikely. People will probably be different after the revolution, but mob rule based on hatred will almost certainly make them worse rather than better. All violence, whatever its goals or necessity, brutalizes rather than ennobles.

Nor would such a revolution be likely to permit, much less expedite, entrance into the kingdom of freedom. If scarcity is to be done away with, or at least tamed, by technology, then maintenance of the technostructure, physically and socially, is absolutely essential. Yet mob action led by the lumpenproletariat and based on universal negation can only lead to breakdown, disorder, and poverty, in ever-descending spirals. In the case of many spontaneous local uprisings of workers and students, both social property and even work activities have been preserved, but this could not be the case in a universal uprising. As Goodman has commented on many occasions, technological society can only be redeemed and reformed by people with skills, discipline, learning, and willingness to engage in sustained work.[78] A negative upsurge of the masses would destroy all possibility of redemption.

Happily, this is not likely. Structures of social control in advanced industrial societies are more than adequate to the task of repressing any uprising by the groups and for the ends contemporary anarchists envision. The 1968 revolution in France failed. The activities of the New Left in the United States have led not to revolution but to reaction. However attractive many elements of the anarchists' vision of the good society may be, its concept of political strategy and tactics is barely a step above that of the infantile tantrum on a mass scale, and it is as doomed as it is self-defeating. Political power may grow out of the barrel of a gun, as Mao claimed, but the liberation of humanity is unlikely to be the result of burning down buildings and throwing rocks at the police. Anarchists may have some idea of where to go, but they have little idea of how to get there.

Dropping Out and Copping Out: The Antipolitics of Presumption and Despair

Some of the most trenchant critics of contemporary society deny that any political solution to our problems is possible or necessary. Some among these assume that the redemption of humanity can come painlessly, even automatically. In classical Christian theology this is known as presumption: the belief that salvation is assured without any need to struggle against evil. Lewis Mumford, a man to whom every American concerned with the social implications of technology owes an immense intellectual debt, direct or indirect, ends his latest work, The Myth of the Machine, by telling us that, if we do not like the kind of prison which he contends that science and technology are creating for us, we can leave any time; "the gates will open automatically . . . as soon as we choose to walk out."[79]

Not bloody likely. As Mumford himself points out, the present "megamachine" which is responsible for war, pollution, and dehumanization could not exist unless it was in the interest of powerful individuals and groups that it be what it is. Obviously then, if there is a door leading to escape the keys will have to be wrenched from the hands of humanity's jailers. This means that some sort of political action to free man from the megamachine will be necessary.

During the late sixties, Yale Law professor Charles Reich caused considerable controversy with his thesis that America was undergoing a process of "greening,"[80] led by its young people with their new, freer lifestyles based on rock and folk music, drugs, new forms of dress and hairstyles, and freer sex lives. Americans were entering upon a state he called Consciousness III, a cultural change which would be marked by rejection of the aggressive, nature-denying authoritarian norms of liberalism—classical or reformed—and by the embracing of a new lifestyle based on leisure, communal living, and communion with self and nature. While clarity of thought is not one of Reich's strong points, what he seemed to be saying was that a precondition for Consciousness III was the basic change in cultural possibilities and probabilities which resulted from general affluence, a condition which in turn owed its existence to technologically created abundance and leisure.

Many social critics such as Bookchin and the French observer Jean-François Revel[81] are in substantial agreement with the Reich thesis of redemption by the dropout young.[82] But, like other prophets of liberation through abundance, Reich ignored the fact that entrance into what is in essence another version of the Marxian kingdom of freedom might not be possible on a worldwide basis, as well as the more obvious fact that the very abundance which made the new lifestyles possible for a minority of the affluent young in the United States and other developed nations was predicated upon the inexhaustibility of the cornucopia produced by the very technostructure which the young were rebelling against. The rebellion against technology (from which stereos, motorcycles, contraceptives, and other favored artifacts of the young are, of course, exempted) would triumph by a process of cultural transformation, in which everyone would in time come to see the virtue of "tuning in, turning on, and dropping out," in the famous phrase of the psychologist-turned-psychedelic-prophet Timothy Leary. Power over the future of society was to be achieved by fleeing from the economic and political arenas in which power is normally held to reside.

Reich's mistake is much like Mumford's. He assumes that the existing interlocked complex of cultural attitudes and political and economic interests which he and his confrères oppose will simply roll over and play dead. For the Sorelian myth of the apocalyptic general strike, which lies at the root of the fantasies of the anarchists, he has substituted the even more apolitical myth of the apocalyptic cultural withdrawal.

While most, if not all, of the attitudes which Reich celebrates are in themselves healthy reactions to the dehumanizing and denaturalizing trends of contemporary technological liberalism, his program for redemption through inaction was fatally flawed. The established order was able to fight back and did. New recruits from young people still on the make were available as necessary to fill the positions of leadership which the dropout elite young were abandoning, the process predicted and described by sociologists Peter and Mary Berger as the "blueing of America."[83] The fact that only a minority even of the young were really "into" Con III was illustrated by the support which youth, especially working-class youth, gave to political leaders such as Nixon and Wallace at the height of the "youth culture."

From the instant the first scouts of the mass media made contact with Haight-Ashbury, the counterculture became another item of mass cultural consumption. American commercial society took over many of the overt symbols of the rebellious young as profitable novelty items, while deflating their revolutionary significance in a textbook example of the "repressive tolerance" which Marcuse has described the system as capable of, exploiting even the struggle against exploitation.[84] Long hair, psychedelic art, sexual freedom—even, to some extent, the new drugs—became acceptable to wide segments of the more sophisticated part of the Establishment. By the early seventies it became apparent that while cultural civil war might still smolder between those in positions of power in the technostructure, who decided that some aspects of the counterculture could be assimilated into the system as safety valves, and the more moralistic and inflexible who failed to appreciate or accept this fact,[85] the system itself ground on—with war, economic growth, and destruction of the manmade and natural environment continuing unabated. Dropping out had meant—it could only mean—relinquishing power to those who stayed in.

Nor did relinquishing power mean escaping from its effects. Radioactivity from atomic testing—still carried on by China and France—was just as much present in the milk that mothers breastfed their babies in remote rural communes as it would have been had they remained in the cities, while the creeping tide of urban sprawl and atmospheric pollution threatened even the trees under which they sought shelter. What supporters of the "greening" analysis of American culture failed to realize was that, while ideas have consequences and culture influences politics and economics, this is a two-way street. Politics and economics also influence culture, and the only way to achieve the revolution of lifegiving forces is to infiltrate and transform the structures of economic and political power, not to walk away from them. The open door, insofar as it existed, was a gateway to irrelevancy.

The work of the popular humanist critic of technology, Theodore Roszak, is permeated with the same self-defeating, antipolitical futility. Roszak, who coined the term *counterculture*, has presented us with telling indictments of what reliance on technology as an end in itself has done to mankind.[86] But being against technology is simply not enough. To seek more human meanings and goals than those embodied in rationalistic and quantifiable science is a useful and necessary task, but it is not in itself a program for social change. If technology is to be mastered, it must be by people who understand it and know how to use it and judge it.

Cultural revolution must become incarnate in social action. For this to occur the machinery of the technostructure must be grasped and bent to mankind's will. To talk about changing norms and attitudes without discussing how power is to be achieved is to drop out just as surely as to seek salvation through personal nonparticipation in the system. Hermits and monks—and the communes are the monasteries of the new generation—may save their own souls. They may even provide useful examples of how to deal with small-scale social problems, serving as pilot plants and sources of information for a reformed society, in much the same way that the

medieval monasteries served Europe through their dissemination of learning and new agricultural practices and early technology. But eventually the
real world of action and power must be tackled, and the dual rejection (in
fact if not in theory) of technology and politics by the alienated intellectuals and the dropout young is an added problem rather than a contribution to a solution. "The abolition of power is not a political program but
antipolitics, an affirmation of nihilism."[87]

Another form of copping out is religiously inspired, though it has had a
more widespread influence than its origin would suggest. Many religions,
confronted with the problem of evil and the difficulty of explaining how
pain and sin can exist in a world created by a beneficent diety, have fallen
into the trap of dualism and postulated a universe in which the powers of
light and darkness contend on virtually equal terms. The Zoroastrian religion of the Persians was such a creed and through Manicheanism has
strongly influenced western Christian thought. The usual form this dualism
takes is one in which matter is the home ground of evil and good triumphs
only after matter is transcended by spirit, often in some kind of afterlife.
Traditional Christian interpretations of scriptural warnings about "the
world, the flesh, and the Devil" have led to tacit acceptance by many
Christians of the idea that Satan is the prince not of darkness but of the
natural world. Much of the impulse behind western technological development derived from a desire to conquer the world of matter in the name of
man and God—a war in which the conqueror may be about to become the
conquered.

If the world belongs naturally to the forces of evil, this suggests the
impossibility of winning on the enemy's home territory, and while some
Christian theologians are now arguing that the redemptive work of Christ
extends to the world itself—to matter and society as well as to individual
souls—the central tradition of Christian thought has been pessimistic and
antiutopian. Modern liberal western society is primarily Protestant in its
religious origins, and the especially pessimistic attitude toward the world
which Luther[88] and other Protestant leaders derived from Augustine has
been the dominant one, even among Catholics.

Thus Jacques Ellul is so convinced that technology controls all of
modern society and will inevitably dominate any future civilization that his
message becomes one of despairing quietism. His Calvinist background
seems to predispose him to see humanity as preordained to collective sin
and its punishment, and vitiates any message of warning which a work
such as *The Technological Society* contains.[89] For if a man is inevitably
corrupt and incapable of salvation, why bother to tell him about it, except
to give him an intellectual foretaste of the suffering which awaits him after
Judgment Day? Classically, despair was regarded by theologians as the
complementary sin to presumption, one of the tools by which the Devil
secured his domination over his victims. If technology is as evil a force as

Ellul seems to believe, then to try to convince us that its victory over man is inevitable is to do the Devil's work.

Even leaving the particular question of technology aside, a profound pessimism about the possibility of creating the good society on earth pervades Christian political thought. A leading exemplar of this trend in modern American political philosophy is Reinhold Niebuhr.[90] Though Niebuhr was a prolific writer and a complex thinker, the thrust of his work was always defeatist in relation to the problem of the possibility of creating the good society. Christian political philosophy, in his interpretation, meant the rejection of the Greek conviction that the good life could be found in and through the political community. In his major work, *Moral Man and Immoral Society*, Niebuhr's theme was the "inferiority of the morality of groups to that of individuals,"[91] an inferiority based on less rationality, less capacity for self-transcendence, and less ability to comprehend the needs of others. The logical consequences of such a position is obvious: Do not try to make things very much better, certainly not in concert with others, the attempt is doomed to failure. Paradoxically, despite the essentially conservative cast of his thought, Niebuhr himself was a political and social activist, one of the original founders of Americans for Democratic Action, but his activism stemmed from his belief that one had to move fast in order to stand still and also from a desire to keep the field of social action from being preempted by overly optimistic, utopia-oriented reformers. But if things did not go well, if reforms were not adopted or failed, a Niebuhrian would be grieved but not surprised. To act in society is necessarily to become implicated in evil.

This position leads to certain difficulties, however. It is not easy to act when filled with grief and guilt. Luther—believing sin was inevitable—held that one should sin joyfully. The next step is to turn sin into a species of virtue. Niebuhr's thought greatly influenced the so-called realist school of theorists of American foreign policy during the period after World War II, and much of the support for the American misadventure in Vietnam came from those who, believing it necessary, began to deem it good. Once one denies the hope that action can make things better, it is all too easy to embrace as good the evil that one must necessarily do.

Any attempt to restore humanity will have to struggle against this widespread belief that to try to create the good society is presumptuous, foolish, and even dangerous. Walter Lippmann, a fellow-traveler of Niebuhr on these issues, says of reformers: "The harder they try to make the earth into heaven, the more they make it into a hell."[92] This aphorism, an echo of Pascal, is the major stock in trade of the conservative defenders of things as they are when they make a pretense of philosophical profundity.

But such pessimism is meaningless in an era in which the future of humanity is at stake. It is one thing to say that we should not seek perfection but simply muddle through as best we can. But suppose the times are such that muddling through is equivalent to doing nothing at all? Humanity cannot muddle through the problems of mechanization, environmental degradation, and population growth. Utopia is the new name for survival. There will be no more deals with history where we split the difference between perfection and disaster. To argue that any attempt to do better means doing worse is a copout. If we don't do better—and soon—we will do very much worse, inevitably and irrevocably.

The more one examines the problems which confront humanity, the more one recognizes that to drop out, either into a private heaven in this world or a hoped-for heaven in the next, is to collaborate in dooming the earth to becoming a hell. Human beings are called on to act and to act hopefully. The task of political philosophy is to inspire, validate, and guide that action, not to deny its possibility, either directly or by default.

II

Ecological Humanism

7

The Real and the Ideal:
Philosophy and Politics

WHO needs theory? Why not take the world as it comes and act in accordance with whatever limited information we have, using common sense as our guide? Surely most of the world's problems are due to attempts to force men and societies into the Procrustean bed of somebody's grand theory.

To the complaint that theory is unnecessary and dangerous, there are two obvious answers. The first is that everyone is a theorist perforce. To deny metaphysics is to make a metaphysical statement, and the argument that all is flux and mere appearance is just as much a philosophical position, requiring a philosophical justification, as is the contention that there are underlying realities in nature and history which we are capable of comprehending and which have certain definite, specifiable characteristics.

But there is a practical objection to using pragmatism as a norm as well. It doesn't work. What seems practical is very often impractical. For decades now American national policy has been in the hands of those whom the late C. Wright Mills labeled the "crackpot realists," and we all know the results.[1] President John F. Kennedy brought to Washington scores of hard-headed, technologically oriented young men from universities and foundations and law offices, and elevated pragmatism to a dogma in his famous speech at Yale in which we were told that America's problems were no longer ideological but technical.[2] Our legacy from the Kennedy Administration was the bungled, shameful war in Vietnam—liberalism's war, but the realists' war also. Nor has corporate liberalism, as embodied in the Johnson and Nixon administrations, been any more adept at solving the problems of poverty, environmental degradation, racial violence, drug addiction, crime, and cultural alienation.

Common sense itself would seem to dictate that we ought to go beyond common sense and try to understand the broader currents of history, both national and international, and recognize that history unfolds within the framework of the given nature of man and his world. To believe that the cures for our social ills will be produced by spontaneous generation, as some critics of the present disorder do, or that we should practice a kind of homeopathic social medicine and look in the same quarters—intellectual, economic, and political—which have produced these ills for their solution, is to act counter to all that reason and experience tell us about the world in which we live.

We are in difficulty not because of mechanical malfunctions in our social machine but because it is badly designed. We have not simply lost our way, we have been going in the wrong direction and toward the wrong destination. Before we can extricate ourselves from our difficulties, we are going to have to stop and reorient ourselves. We are going to have to take time out to ask the most basic questions about the kind of beings we are, what we want, and what life means.

The Creation of Technological Man

In a previous book I argued that the greatest problem facing humanity is the need to gain control over technology.[3] Man has, I asserted, achieved virtually godlike powers over himself, his society, and his physical environment. As a result of his scientific and technological achievements, he has the power to alter or destroy both the human race and its physical habitat. Man is thus on the threshold of a further evolutionary step of almost unimaginable importance. He now has the potential for transforming himself into a new man, one who is able to understand his powers and is willing to use them responsibly to control himself and his world in order to create the first truly humanized physical and social environment. This human being I call "technological man," the creature who both creates and controls his technology. Alternatively, man now has the capacity of degrading himself into a mere object, a physical cog in a less than human society, or of allowing society to dissolve into primitive poverty and chaos. Humanity, in Buckminster Fuller's phrase, now faces a choice between utopia and oblivion.[4]

The new technological man, who seeks to control the world of which he is the potential master for humanistic purposes, must necessarily have a very different cultural and philosophical outlook from the bourgeois man who has created liberal society. The bases for this new outlook are three overall synthesizing principles: naturalism, holism, and immanentism.[5] The new philosophy is naturalistic in that it is rooted in the assumption that man is part of nature and his salvation lies in acting in accordance with this fact. The new philosophy is holistic in that it is based on the realization that everything in man's world—the physical planet he lives on, the society he lives in, and himself—is closely interrelated in a single system, and that any descriptive or prescriptive principles will have to take into account this entire universe. Finally, the new philosophy is immanentist in recognizing that the reordering of human society and man's nature can never come from outside or "above," nor can it be blueprinted in advance; it can only grow out of whatever already exists. The form of the new society will only be determined in the course of the process of interaction among individuals and groups and society as a whole as they strive to achieve a greater sense of identity and purpose and a renewed planetary order. Technological man and the new humanist society are "emergent properties" of this interaction.

What do these high-sounding generalities mean theoretically and programmatically? What specific ideas about man and society are required or implied by these principles? What ideas are incompatible and excluded? How is this philosophy to be operationalized and what will a new world-order created in accordance with these principles look like? In short, what will the philosophy and politics of technological man, of postliberal society, be?

Such an analysis must necessarily be carried on on two levels. On the theoretical level, I will attempt to set forth certain basic principles which appear to be scientifically tenable and logically necessary. On the more practical level, I will try to show how these principles can be used as guidelines toward the creation of new structures in the real world around us and how they can be used to help shape the course of events. In accordance with the principle of immanence, this second level cannot be expected to have the same logical necessity as the first. On this level, it will only be possible to indicate parameters and general courses of action rather than to specify all necessary details; it will only be possible to suggest tentatively rather than to prescribe definitively how order can be achieved. But to reject dogmatism and a superimposed blueprint is not the same thing as to embrace the complacent stand-pattism or "muddling through" philosophy of the reigning liberal establishment, the philosophical nihilism of most of the counterculture and New Left, or the optimistic meliorism of the prophets of the new "greening." One need not prescribe what color eyes and what shape nose an unborn child must have in order to be able to distinguish the birth of a human being from the emergence of a monster, and one need not have planned the interior decoration of a house in order to begin to dig its foundations.

Political Philosophy and the Facts of Life

Human beings are part of the world of nature, of the world of natural living things; they share the characteristics common to other elements of the physical and biological universe and are affected by the characteristics of the whole natural system of which they are members. From this fact follows the first general requirement for any philosophy of politics designed to meet mankind's needs: *Political philosophy must take into account and be in conformity with the objective nature of the universe*, insofar as science can ascertain that nature.

This is not as self-evident as it may seem. Some political philosophers would argue that although we can no longer accept the scientific worldview of the ancients, their political philosophy alone deserves to be called truly human, and that, while we are forced to accept the modern view of the physical world, we must in the name of humanity reject any political philosophy derived from it.[6] They would hold that although we cannot any longer give credence to Aristotle's description of how the solar system is constituted and operates, as human beings we also cannot live in the atomistic social world of Hobbes, Locke, and their successors, which these political theorists still assume to be consonant with contemporary scientific thought.

Dualism of this sort seems to have a built-in appeal in times of major intellectual discovery and change; witness the Averroistic reaction to the introduction of Aristotelian science into the medieval Islamic and Christian worlds. But such dualism runs counter to our most fundamental intellectual and psychological instincts; we are all monists at heart. And sooner or later, one of the elements in a dualistic worldview will triumph. We must try at the outset therefore to look steadily at reality and see it whole, albeit recognizing the possibility that human uniqueness and indeterminacy may itself be part of the basic order of things.

This is not to say that it is necessary to construct a theory of society which is strictly analogous to the picture of the physical universe established by the physical sciences, or even less a theory in which social laws are reduced to physical laws. The scientific view of the universe, including the world of living things, has changed over time and will probably continue to change. Simply to extrapolate from it would be to doom any political theory to becoming dated as soon as scientific theories are revised.[7] On the other hand, no philosophy of society which runs counter to currently established scientific facts and laws can expect to long survive, nor should it. Societies based on illusion cannot expect to prosper or even endure: "Ethics, sociology, politics are ultimately subject to infestation by the germ that is born when a discovery in pure science is made."[8]

A major objection to liberalism is that it is based on an outdated view of the universe. In referring to the American Constitution, Jefferson said, "All its authority rests upon the harmonizing sentiments of the day,"[9] that is, the Lockean worldview. But these harmonizing sentiments have evaporated (although this is not yet popularly realized), and with them its authority must also wane. Political theories must, like scientific theories, "save the appearances," that is, they must be consistent with the world as we perceive it. This requires that they be consistent with the data of the physical as well as the social sciences.[10]

But not only is it important to recognize that theory must be consistent with, though not overcommitted to, existing structures of knowledge in the physical sciences; it is also necessary to keep in mind the fact that although the universe is a self-consistent whole, different principles of order operate at different levels. Political theory deals with the activities of human beings, who are, as individuals, midway in size between the nebulae and the electrons, the largest and smallest "objects" in the universe.[11] Theories about human behavior must be compatible with those laws of nature appropriate to human beings in this middle kingdom.

Facts and Values

But if political philosophy must be in accord with the facts of life, these facts are not what it is about. Political philosophy is above all a normative enterprise. Its norms must ultimately derive from the nature of things; political philosophy cannot set goals and standards contrary to the order inherent in nature. But it is by no means simply a reflection of how things are.

Perhaps this is as good a place as any to set to rest one of the greatest bogeymen of modern ethical and social philosophy, at least since Kant: the supposed inherent dichotomy between facts and values, the alleged human inability to derive the "ought" from the "is." Forests have been devoured and rivers of ink poured out in the technical discussion of this question by philosophers.[12] There is therefore no possibility of recapitulating, ordering, and resolving the argument here in technical terms. In part it is a quarrel over words. We use specialized language for various purposes and sometimes fail to recognize that words are tools, not full embodiments of reality. Descriptive and prescriptive statements are of different kinds in and of themselves simply as a matter of linguistic usage. But this is not really saying anything more than that "is" has two letters in English and "ought" five, or that some languages have more than two words to express the concepts involved in these words while others make no such distinction. The real issue is at once both more complex and simpler than our convenient linguistic distinctions.

Aristotle said all that needs to be said on the subject two thousand years ago when he defined the end of man as happiness.[13] This is both a descriptive "is" statement and a normative "ought" statement. He meant both that men normally seek happiness and that men should seek happiness. If anyone seeks to be unhappy we can say one of two things: that for him being unhappy constitutes his happiness (as in some forms of psychological disorder, such as masochism), thus validating our proposition formally though weakening it substantively, or that such a person is so seriously deranged as to be no longer human, not worth arguing with, and hardly worth worrying about. Men are uniquely free in being able to choose to act contrary to their nature. Or, to put it in the most mechanistic terms, their behavior can be determined in such a way that they will act contrary to their own objective good, just as animals without the human being's facility for choice sometimes end up being miscued by nature and eating things which poison them. But we do not feel it necessary to consider such arbitrary choices in establishing our norms. A person who is a masochist and enjoys being tortured is considered to be perverted, and we usually can find ways of accounting for his aberrations. We do not, because of his existence, consider it impossible to make the statement that it is the nature of man to avoid pain, and that he therefore should avoid certain actions which will cause him pain in order to be happy. Pain is basically a mechanism by which our body tells us that something wrong is happening, something which under normal circumstances we should avoid or correct.

The matter becomes clearer if we think in terms of health. All living creatures seek life and power, the full exercise of their potentialities. Animals sometimes choose death in accordance with instinctual drives to preserve their young or their herd; also, under extreme conditions such as overcrowding, their internal controls may break down and they may engage in behavior which is group-destructive as well as individually harmful. But life—the fullness of life—is the norm.

Health is a word we use to describe the maximization of life, both in terms of longevity and of scope of activity. If our physician tells us that research indicates that the intake of certain substances *is* conducive to illness and that we *should* therefore avoid them, he is making a simple statement at once both descriptive and normative, the legitimacy and cogency of which we easily recognize. We may, of course, act contrary to his recommendations. We can decide that we prefer other things to personal health, either for personal or altruistic reasons, but if we do so we usually recognize that our choice is one which involves setting one value (wealth, let us say) above another (continued health). We would never think of arguing that our medical adviser's statement is philosophically illegitimate or illogical. We may as a result of compulsive behavior patterns continue a course of action such as cigarette-smoking—or, on a national level, polluting the environment—which we know or have been told is dangerous, because we refuse to break these patterns, but we would never deny that doing what harms us is wrong in some meaningful sense. Continued life and health are regarded as good, and what can be shown to imperil them as evil. Even suicides, in so far as they act out of rational choice rather than compulsion, are saying that the quality of their life—or its effects on the lives of others they cherish—has become so negative as to negate the value of life itself. And it is obvious that a terminal-cancer patient, living on in pain with no hope of resuming normal activity, a burden to the lives of others, who chooses death rather than a continuation of his situation, can hardly be said to be denying the intrinsic value of life and health.

What is true of the choices of individuals applies to social choice as well, even though social choice, like individual choice, may not be rational but rather the result of compulsive behavior-patterns. Few societies seek to be unhappy or unhealthy according to their perceptions of what constitutes health and happiness, even though to the outside observer it appears that they may have chosen a path bound to lead to disruption and decay. No social system seeks extinction, although political leaders and their followers sometimes pursue policies which can only lead to destruction. The problem is not solely—or even primarily—one of overcoming an inherent philosophical inability to derive meaningful prescriptive statements from descriptive ones—the "ought" from the "is"—but in convincing people, including ourselves, to do what we know we ought to do.

Much of the professional ethician's confusion and that of many social scientists who should know better stems from a failure to recognize that the problem derives from the fact that "is" statements can be made universally and "ought" statements are made most usefully only about human beings. To say that a rock is subject to the laws of gravity is a complete statement to which nothing useful is added by saying that therefore rocks "ought" to fall, though we often say so colloquially. Rocks have no choice or apparent awareness in the matter; they will follow whatever physical laws apply to their situation. In the case of animals the matter is more complex, depending on how much volition we believe we can ascribe to them. The fact that the Hound of the Baskervilles is *not* barking is significant because we think he *ought* to be. His actions reflect a kind of choice, and because *is* and *ought* are not completely congruent in his behavior, it is necessary to account for his behavior.

In the case of human beings, we know both from introspection and observation that their behavior is purposeful. Even the supporters of the most behavioralist schools of philosophy or psychology, even those who in the abstract deny the existence of free will, in practice respond to the behavior of their wives, children, colleagues, and students *as though* the actions of these others were more than determined behavior over which they had no control. Professor Skinner feels—or at least manifests—annoyance with those of his colleagues who "choose" to prefer what he considers to be ineffective methods of social control. The human being is above all a goal-setting animal, and therefore for him the "is" and the "ought" 'are necessarily conjoined.[14]

Some scientists try to avoid the implications of man's purposiveness by observing and describing him in terms of his external behavior (including his verbalization of norms) as if he were a rock or a dog. Whatever satisfaction it may give to the observers' own sense of "oughtness," such a method seems foolish as a universal procedure. However useful it may be in some cases to suspend judgment about motivations until after behavior has been observed, sooner or later any description of human behavior to be at all interesting as an explanation or useful as a predictor of actions must come to grips with the "causes" of behavior, among which are motivations, conscious or unconscious. Even a complete determinist would have to deal with the obvious fact that overt human behavior-patterns are normally preceded by and correlated with internal mental states which must be assumed to be of some potential explanatory significance. It is impossible to describe human beings without discussing their purposes. We are human, and the more we react as humans the more we are able to understand how men and women in fact behave.

Today some social scientists have moved all the way to the opposite end of the spectrum and insist that there can be no understanding without complete empathy, sympathy, or participation—that only the poor can understand the poor, only blacks can adequately describe black society, etc.[15] This obviously overstates the case, but no matter how detached we try to be, we face a choice between increased comprehension based on some sharing accompanied by possible loss of perspective on the one hand, and a lesser degree of comprehension as a result of detachment together with a possible gain of perspective on the other.

Even to perceive is to judge. One cannot describe without using language or without imposing some framework and some criterion of choice on a myriad of details. One cannot talk about human affairs without using valuational terms. The legal difference between homicide and murder is a descriptive matter normally dependent on antecedent factors such as premeditation, but the difference has obvious valuational connotations and consequences. There is a difference, as Americans (especially southerners) know, between a "rebellion," a "civil war," and a "war between the states." We can invent supposedly value-neutral terms to replace obviously value-laden ones—we can call actions "dysfunctional" rather than "bad," and "operationally inappropriate" rather than "futile." We can talk about "instability" rather than disorder, or call statements "inoperative" rather than lies. But we still use such language in order to evaluate. For in political discourse the descriptive is necessarily normative.[16]

Because there is no real difference between descriptive and normative statements, and because all language is evaluational, it is absurd to argue that science cannot be a source or standard of values. Once we have made the basic decision to survive or function effectively as individuals or as societies, *science can tell us what we should or should not do*, what actions are capable of leading us to our goals and are therefore good and what actions will frustrate the achievement of our goals and are therefore evil. Insofar as science describes what in fact hurts or heals us, it provides us with substantive norms. However much philosophers may quibble, science can furnish us with norms of behavior appropriate to the achievement of basic human purposes.

Actually, as the sociologist Ernest Becker notes, the "separation of fact and value is an historical anomaly that has no place in contemporary science."[17] Not only do any descriptive statements reflect our own perspectives and concerns, including our universal characteristics as human beings, —not only do we necessarily create nature in our own image in a process sometimes called the "social construction of reality"[18]—but, unlike the situation in the nineteenth century when science conceived of man as a detached spectator of an objective universe, we now, as a result of Heisenberg's work in quantum mechanics, must accept "a partial fusion of the knower and the known."[19] Even assuming that there is some "objective" nature "out there" which is independent of human perception, it is in itself purposeful, as I shall emphasize later. Nature is not the mere fortuitous outcome of the random jumbling of subatomic particles, as the early modern physics which developed along with liberalism held, but a hierarchically organized structure in which no constituent element can be described except in terms of its function in relationship to other elements in creating meaningful wholes.[20]

As a simple operational matter, political philosophy must be normative because it is by definition prescriptive. But if we are going to make statements about what should be we must have some general definition of the good. This is the Achilles' heel of pragmatism, no matter how it masquerades as moderation, compromise, or common sense. We cannot act to avoid present ills, unless we can define the probable outcome of our actions as something better. We cannot say a thing is good simply because it works. It must be good *for* something, it must work in terms of some end which is sought. Any set of statements about what we *should* do to surmount our current crisis must ultimately be based on some clearly conceived idea about what man *is* and what therefore is good for him. We can no longer act as if we could get out of the maze simply by running faster. We cannot defeat the alien invasion of earth unless we can tell what is human from what is inhuman, unless we have a banner under which we can rally.

The Tasks of Political Philosophy

There are three basic questions with which any philosophy of society or politics must deal: (1) What is man? (2) What is the nature of the universe as it affects man? (3) What is the relationship between man's values and the way in which the universe works? Whether these questions are answered explicitly or implicitly they cannot be ignored. One of the things which separates the great political philosophers such as Plato, Hobbes, and Marx from mere ideologues is that the former deal with these questions directly and attempt to relate subsidiary propositions about the nature of justice to their answers to these basic questions about human nature and destiny.

Unless we know what man is, we have no way of knowing what the possibilities or limits of his actions are, nor can we know what will make him healthy and/or happy. Since society consists of men, we cannot discuss society without discussing its human constituents. Nor can we understand man or society without reference to the physical universe within which they both exist. This universe has certain regularities which we call laws, and man—composed of atoms and subject to the laws of physics and chemistry (or statistics)—is bound by them. To understand what man and his society must be, and what they can aspire to be, we must first understand this context. Does the universe have human meaning? Is it moving in any direction that man can perceive? Is man part of a "great chain of being," as the central philosophical tradition of the West once averred,[21] or is he, as Lycurgus and Hobbes would have us believe, simply the product of a fortuitous concourse of atoms in the void?

To discuss man's nature intelligently we need not penetrate (even if this were possible) to the ultimate meaning of the concepts of time and space. Aristotle got along quite well as a philosopher without knowing whether or not the world was eternal, and we need not solve the conundrum of infinity in order to save our planet from disaster. Whether entropy rules universally or whether the universe is involved in a process of continuous creation need not be definitively answered here, but to know that we live in a small subsystem where that antientropic phenomenon called life flourishes, does matter, and the important question for us is what we can learn from this fact.

Political Philosophy and Theology

At this point it might be objected by some that all political questions, all questions of the nature of man are at bottom theological, and that therefore we cannot discuss the meaning of the universe without dealing with the question of the possible existence and nature of God. In a sense this is true, but this does not mean that the traditional problems of theology as such need concern us. Increasingly theologians have come to recognize that the full reality of God is as hidden from us as He was from the

Hebrews. God may not be "totally other" as Barthian neoorthodoxy holds, but the relationship between his nature and his manifestations in the universe is not a simple one-to-one correspondence as medieval piety supposed.[22]

In the broadest sense, however, the revolt against liberalism is implicitly theistic. Theologians are coming to recognize that whatever provides the source of our value-orientation is our god. The revolt of life-affirming forces in the modern world against the idolization of technology, the implicit appreciation by the ecological movement that we live in a numinous universe which must be respected if we are to respect ourselves, is a religious movement. Nor is it an accident or total misconception that has led many commentators to refer to the recent youth culture—despite its distrust and in large measure abandonment of formal religion—as being the expression of the most religious generation in our nation's history. The spiritual vacuum left by liberalism is being filled.

There is one theological point on which postliberal political philosophy must take a stand, however it eschews traditional theological disputes. Any religious doctrine which holds that the world is fundamentally evil or at best a neutral, valueless stage on which individual men and women work out their individual salvation, or any doctrine which radically separates man's earthly and eternal destinies, is not only contrary to the most fundamental insights of the traditions of most of the great religions, which have traditionally held the divine to be immanent as well as transcendent. Such religious beliefs engender human alienation of the most profound sort, serve the cause of the forces of evil in the world, and imperil the future of humanity.

In dealing with political matters, we need not utilize the special language nor speak to the special concerns of the theologians. We can simply talk of human beings and the universe in terms of what we know and feel, through our minds and our bodies. Unlike Molière's character, if we speak the prose of theology as well as the language of science and philosophy, we need not be surprised or concerned. Insofar as many contemporary theologians seek to be social philosophers and prophets rather than the more otherworldly specialists in a God who exists outside his creation that most of their predecessors were, our concerns may overlap, but that is their problem rather than ours.

Man's Values and Nature's Norms

Throughout the history of the West there have been conflicts between society and society's laws and those who claimed that there were laws of God superior to the laws made by men. Recent struggles over the principle of conscientious objection or resistance to "unjust" wars are a contemporary expression of a long tradition of belief in some kind of "higher law." It is relatively easy to speak of the relationship between the "laws of God" and the "laws of man" when one has an anthropomorphic image of God as a conscious, willing being, much like ourselves except for being omniscient and omnipotent. It is more difficult to relate man's values to the norms of the universe if the universe is thought of in nonpersonal terms.

The classic natural-law tradition of Plato, Aristotle, Cicero, and the Stoics and the Scholastics held essentially that there is an intelligible order in the universe and that man is a part of this order. This order is eternal and unchanging, and it is man's moral duty to conform to its dictates.[23] For other philosophers, man's will is set over against the universe's nonwill, its simple existence. What if man is subject to the laws of physics and biology; why should they invalidate his desires? Man alone can will and therefore man alone can be a bearer of morality. Man's role is therefore to struggle against nature and the barriers it poses to his will.

Any philosophy of politics must take a position on this central issue. Does the universe provide norms of behavior for men in society? Are there laws of nature, and if so, are they of moral significance or simply physical constraints which we are free to circumvent rather than obey? The issue of the extent to which the universe can be a source of value for humanity must be met squarely by any political philosophy worthy of the name.

Equally basic to the task of reconciling our view of man and of nature is the need to logically relate our view of human nature to our view of society. Historically, the idea that man is "intrinsically" evil has been used to justify both strong and weak governments, and the same is true of the opposing position that man is inherently good. Clearly there is no simple one-to-one relationship between our concept of human nature and precise forms of the social and political order. But if there are no simple means by which political and social forms may be univocally deduced from premises about human nature, there are, as in the case of physical nature, parameters for man and his behavior which are set by his natural constitution. Only totalitarians dream that everything is possible. Any coherent political philosophy must therefore be clear about what it holds to be the irreducible essentials of human life and how it relates its ideas about human nature to its social and political goals.

Utopia Without Utopianism

Any political philosophy meaningful for our time must meet two further criteria: It must be universally relevant, and it must be capable of being incarnated in political and social reality within the relatively near future.

Any new political philosophy must be universal because our problems are planetary; the oneness of the world, the crisis of the world ecosystem, and the need of the human species as a whole to regain mastery over its technology are the factual assumptions underlying our normative inquiry. This does not mean that a new political philosophy must postulate a world in which all cultural and social subsystems have disappeared—only that the basic premises for the solution of our planetary problems must be universally adopted and implemented. A political philosophy for the next century must be equally acceptable and intelligible to East and West, to developed and less developed nations alike. We are entering an age of worldwide community, and therefore an age in which universal political and philosophical consensus is both possible and essential.

Any new political philosophy must eschew traditional utopianism. Utopianism refers to the tendency of social philosophers to create ideal states without telling us in any useful fashion how they are to be achieved. Our utopias must be relevant utopias. Ideals are important as standards of what should be, but we must have guidance as to how to achieve them as well.

Some utopian thought (Plato's *Republic* is an example) premises an all-powerful ruler, acting as a *deus ex machina* for the society, creating the good order by force of decree, independently of the normal dynamics of social life. Rousseau's "legislator" used persuasion but struck while the iron was hot, and could therefore operate successfully only in a few special moments in history, the existence of which the legislator could not influence. Other utopians have depended on some kind of fortunate isolation from the complexity of ordinary social processes. Thomas More's *Utopia* was an island, and from the nineteenth century to the present America has seen the birth—and death—of utopian colonies based on the withdrawal from the world of a select and usually economically independent few. None of these philosophies or movements comes to grips with the problem of how one moves a large, powerful, already existing social and economic system from peril to utopia. The great appeal of Marxism has been that Marx, in direct contrast to earlier socialists whom he derided as "utopian," attempted to provide a scientific explanation of the dynamics which made possible (and inevitable) change from an unsatisfactory present to a desirable future.

Any serious political philosophy must have its own theory of political dynamics, based on its assumptions about the nature of man and the universe. In addition to defining the good society or denouncing the existing one, a political philosophy must be able to tell us how the new society is to be brought into being, and it must postulate means which are compatible with the ends it seeks to achieve, because, as Marx also recognized, means inevitably shape ends.

Environment and Social Philosophy

All of these requirements have always held for any serious political philosophy. They are in effect the philosophical and methodological criteria for determining whether we are talking seriously about man and society, or whether we are spinning fantasies. What is unique in our own time is that there is a further overarching, substantive requirement which political philosophy must meet. The nature of man and society must be considered in terms of our relationship to physical nature, including our own material artifacts, since the central issue of today is how man should deal with technology—how he can control the machines he has created so that they do not destroy him and how he can control his machines and techniques so that they do not destroy the ability of the planet to support decent human life.

It is interesting to note that classical political philosophers were always conscious of the relationship of the physical environment to human society. Plato and, to a lesser extent, Aristotle were aware of the impact and conditioning force of geography, economics, and biology. So too were their modern heirs, Rousseau and Montesquieu. But the most influential modern political thinkers—the founders of liberalism, Hobbes and Locke—ignored the physical bases of human existence and spoke in terms of abstract universals independent of geography and history. Despite their rootedness in the study of history, and Marx's special concern with economics, both Hegel and Marx ignored or were unaware of the biological and physical constraints under which human society on earth exists. They therefore contributed to a politics of unreality paralleling the economics of unreality of their age and—till recently—of our own, in which man chooses to forget he is part of the kingdom of nature, akin to other children of Mother Earth.

But even the ancients, despite their concern with the size and location of cities, the effects of climate on peoples, and so on, did not deal with the central issues of man's relation to physical nature and his tools. How could they? Nature was an enemy far from conquered. The size of a city might condition the nature of its political system, but the earth as a whole was an unknown and unnecessary concept in their calculations. For Plato, Aristotle, and Thomas Aquinas, man was dominated by his own nature (however conceived), while physical nature was an external reality so boundless as both to invite and defy conquest. Machiavelli's heroes fought against nature in striving for glory, yet they were the playthings of the external world as personified in Fortuna, the goddess of chance; both man and nature were capricious for men of the Renaissance and the early modern rulers. For Hobbes and Locke, man had the task of dominating a nature which was humanly meaningless; for Hegel and Marx, nature had meaning, but only because it produced men who would increasingly manipulate it.

It must not be supposed that modern western man has been unique in abstracting himself from and ravaging nature. Many other civilizations have apparently destroyed their resource bases. Neither the ancient Greeks nor the Romans, the Mayans nor the Plains Indians, ancient China nor ancient India have always, despite their philosophies, treated nature with the respect it demands.[24] Our primitive ancestors were responsible for the extermination of many species of animals through overhunting and the results of their agricultural practices.[25] We are not the first generation of men to destroy some of our animal cousins forever. It may not be specific cultural traditions only, but instead deep-rooted psychological impulses toward aggression that shape our attitudes toward other species and the earth.

But we of the last quarter of the twentieth century and the dying decades of bourgeois industrial society inherit not only the age-old proclivities of most human groups to foul their own nests and destroy the basis for their subsistence, but also the intellectual rationalizations for such action developed by modern western man in liberal society. Thus the prime requirement for a political philosophy for technological man, a philosophy adequate for dealing with the social and physical ills created by the unbridled growth of population and technology, is a new theory of man's relation to nature which will tell us both what the good life and society are and how we can attain them. Such a philosophy is essential if we are to survive as human beings within a human society. The only alternative to discovering such a philosophy and creating a new order based upon it, is the destruction of any human civilization worthy of the name, and possibly even the destruction of the human species and its mother planet as well.

The outlines of this new philosophy of the proper relationship between humanity and nature are beginning to become apparent in the work of scientists, philosophers, theologians, and poets and also, increasingly, in the attitudes and behavior of the educated young in industrial societies.[26] Ecological humanism is emerging as the basis of a new political and social philosophy because existing philosophies fail to meet the challenge of our time. What is ecological humanism? To answer this question we must begin at the beginning, and ask ourselves what it means to be human, what the place of our species is in the pageant of the universe, and what the nature of the universe itself is.

8

Man in Nature and the
Nature of Man

THE RELATIONSHIP of man to nature is the central political issue of our time.

This statement runs counter both to common sense and learned consensus. Surely the subject of politics is the relationship of men to one another. Of course it is. But today it is impossible to answer any of the fundamental questions about how men do or should relate to one another unless we first deal with the issues arising out of the man-nature relationship. What men are and how they sustain their lives necessarily determines the problems and possibilities of their lives together.

Two basic questions about man's place in nature must be answered by any consistent political philosophy: Does man have a relatively fixed nature which can be the source of the constraints or norms which ought to govern man-man relationships, and to what extent can physical nature itself be a source of moral prescriptions? Both of these questions have long been staples of dispute among philosophers, as the history of the controversy over the existence and possible implications of "natural law" testifies. Both questions must be faced together. They are intertwined at the theoretical level because one cannot hold that man's nature is the source of norms for his conduct without first establishing that nature itself can be a source of norms. At the practical level the interweaving of these questions is clearly apparent in the central political issues of our day: the need for the human race to cope with the related problems of overpopulation, scarcity of resources, environmental pollution, and the mechanization of human relationships, and the need to create a just and humane political order.

The Separation of Man and Nature

Western culture is in large measure based on a radical divorce of man from nature. The moral norms of Plato, the Stoics, and even Aristotle, though all were phrased in terms of man's nature, were derived more from metaphysical propositions about man's allegedly spiritual essence than from any observations about man as a creature rooted in the flesh and blood of physical nature, an animal among animals. The earlier books of the Old Testament made no distinction between body and soul or matter and spirit, but these distinctions, drawn from the philosophical thought of the Babylonians and Greeks, begin to appear in the "Sapiental" books of the Old Covenant, as the Jewish tradition had to combat the nature-worship of the surrounding pagans.[1] This later Hebrew tradition—in combination with the dualism of Near Eastern religions and with neo-Platonism—made Christianity a radically dualistic religion which set the things of the spirit against "the world, the flesh, and the Devil." Nature had value only insofar as it served the needs of man.[2] Man was superior to nature because of his special "nature" and was destined to rule it.

A philosophical countertradition can be found in the teachings of St. Francis of Assisi and others, in which nature is regarded as good in itself and man is not simply the master but the steward of God's creation. But the main thrust of Christianity, above all of popular Christianity (despite the animistic pagan underground which long persisted in the tradition of witchcraft), was that there was a radical dichotomy between nature and man and that man, who was superior to the rest of nature, had the right to exploit and manipulate that which was of little intrinsic importance. "Go forth and multiply, and subdue the ends of the earth," man is enjoined in Genesis; and for most men this meant that conquest and growth were divine injunctions.

The Enlightenment introduced somewhat different ideas, and though the Christian emphasis on the practical exploitation of nature was maintained, the Enlightenment tended to reject dualism and to make nature its standard of value. Humanity was urged to find rules for its conduct not in the prescriptions of otherworldly religion but in the way things supposedly were, to turn for guidance from the priest to the natural philosopher. Natural law, as the term was used by such as Hobbes and Locke, tried to take observed human behavior as the basis for political norms. A disdain for the artificial was a component of Enlightenment ideals, and eventually gave birth to Romanticism, with Rousseau and others opposing the "natural" to the artificial as a standard of life.

But the new-found interest in what was natural, in unspoiled physical nature, was not dominant long, despite the Romantic counterattack against the rising forces of rationalism and industrial civilization.[3] The power over nature which Enlightenment-encouraged science and technology gave to man was too potent to resist, the lessons of man's omnipotence too easy to draw, and humanism began to glorify man to the exclusion of nature, as well as of God. Man was increasingly the measure of all things and the master of his fate, so that a twentieth-century humanist, Ortega y Gasset, could write, "Man has no nature, he has only a history."[4] Man's nature was not a derivative of his relationship to the natural world; man's nature was whatever he could make of it.

This statement of Ortega is so palpably absurd that it is difficult to see how it could possibly be taken seriously. Obviously men and women have natures. They are erect bipeds with certain genetically derived physical characteristics. They must ingest nutrients, eliminate waste products, avoid extremes of heat, cold, pressure, and so forth. Ortega, of course, knew this too. What he was saying was not that humans were endlessly protean but that what we are physically and biologically is irrelevant to the values we choose to live by. Of course, our physical characteristics limit our ability to act, but there is no reason that they should determine our desires as well. In short, nature is a constraint but not a source of values. Human beings and their wishes are autonomous, and exist for their own sake and in their own right. We have no set mission to fulfill within the world, no role in any prescribed drama; we are free to improvise our own parts and even to desert the stage if we choose. Values, as Nietzsche insisted, are something men impose on a meaningless universe.

This philosophy contrasts starkly, of course, with the alternative belief that nature is value-laden and should be an object of piety and a source of guidance. Primitive peoples have usually worshipped nature, seeing in conformity to what is (or what seems to be) the basis for an ethical existence. Eastern religions and philosophies have found value in conformity with or even absorption of the individual into the whole. Even Christianity, especially the Aristotelian-Thomistic tradition of Roman Catholicism, has used aspects of man's physical nature as the source of moral norms. In our own day renewed recognition of the ethical importance of man's place in nature is reflected in such poetic expressions as Robinson Jeffers's injunction, "the greatest beauty is organic wholeness, the Divine beauty of the Universe, love that, not men apart from that."[5] At a more mundane level the development of systems theory enables a contemporary philosopher of science to argue that "objective value norms can be deduced directly from the contemporary scientific understanding of natural systems." "The objective basic values of man are those which he shares with all natural systems."[6]

Is man the creation of nature or of his own history? The only sensible answer is that he is both. In man the indeterminism inherent in nature achieves conscious expression. It is man's singular ability to choose consciously, and thus create a history: "History does not unfold of itself, but evolves through man's evolving."[7] Man is by nature somewhat above the rest of nature; he is by nature "unnatural." It is this paradox which makes any discussion of nature as a source of human value so difficult. *Nature has created man as a transcending animal.*

But before we can discuss the thorny problem of how and to what extent nature can serve as a source of human values, we must first be clear about what we mean by nature and which characteristics of nature are relevant to the human condition.

The Nature of Nature

What is nature? Throughout our discussion we have been talking about the relationship of humanity and nature, but what does the term *nature* really mean? Like many of the basic words on which we daily depend, nature has more than one meaning.

When the Greeks looked about them they saw a world of physical objects, including their own bodies. They sought to understand this world, and, as humans always have, they projected their ideas of themselves upon the world. Men had customs and laws, then so must things. Every object in the universe had a principle—a way of behaving—that was its nature. Nature was that principle which made things what they were. Nature also had a second, less usual, meaning for the Greeks. Nature also meant the sum total of beings and things in the world.

In English and related languages one word, *nature*, has both these meanings. Nature means the way things are ("John is surly by nature") as well as the world as a whole, the "natural" universe of animals, plants, and stones. This dual meaning has resulted in a good deal of confusion, since it makes nature both a principle and an object. Political philosophy in the past has been much concerned with "natural law"—with nature as a principle: how man should act if he acts in accordance with his own essence. Political philosophy has been little concerned (as we are here) with how man should relate to nature, to the preexistent universe around him.

A further confusion arises because of the distinction often made between the natural—the preexistent, uncreated by man—and the artificial—that which is manmade. This confusion can be found in virtually all discussions of technology. For everything in the universe is necessarily natural, even manmade things, in that they must follow the laws of nature. The artificial is natural also in another, less generally recognized sense. It is of the nature of man to make things, to create the artificial.[8]

The two different meanings of the concept *nature* caused less confusion for the Greeks than they do for us. Since law was equated with reason, the universe, participating in regular, lawful activity, was conceived of as being rational, as being an organism just like man himself.[9] The distinction between spirit and matter, mind and body, was unnatural to the Greeks, in every sense of the word, despite the origins of dualism in Platonic philosophy. Socrates takes their union to be a common assumption of his hearers, and is apologetic when he has to treat them separately. Because the world was in ceaseless motion, it had to be alive. Only living things moved. Machines, though they existed, had such a marginal role in Greek culture that they played no part in shaping the Greek view of the universe. The universe had a mind and a soul.

Medieval philosophers regarded the universe as having been created by God, who was the personification of the prime, unmoved mover of Aristotle, but they still retained a sense of the universe as being in some sense alive. Only slowly, albeit inexorably, did the emphasis which Christianity placed on man's spirituality lead to a dualism which denied spiritual qualities to the nonhuman.[10] The heavenly bodies were celestial intelligences, ranged in hierarchies like the nine choirs of angels. The universe was a "great chain of being" in ordered rank from the highest to the lowliest thing, all manifesting the glory of God.

The Renaissance brought a radical change in man's concept of the universe, a change which provided the foundation for the worldview of Locke and his contemporaries. Humanity became fascinated with machines. The universe was no longer thought of as a living organism, a relationship among intelligences, but as a vast machine, an automaton composed of lifeless parts. This machine—the universe—was assumed by most philosophers to have had a creator—God the Divine watchmaker—but He was distant from and outside of it. His reason had established the laws which governed the universe but, like a wind-up toy, once set in motion it ran of its own accord, following the immutable laws which he had built into it. Man could achieve wealth and power, and come closer to God in the process, by learning these laws and taking advantage of them. From this time on science and technology marched forward, eventually hand in hand, tearing the mechanistic universe apart to see what made it tick, and striving to order the parts into new combinations more amenable to man's will.

The Origin of the Universe

But what is the universe really like? Where, in the first place, did it come from? All cultures have had creation myths which attempt to answer this question. Human beings have always been able to look around them and see that everything living is subject to generation and decay, that manmade artifacts are created at some point in time prior to which they did not exist. Where did nature come from? Did it have a beginning or not? Some myths answer both Yes and No, positing a cyclical theory of creation.[11] The Greek philosophers were divided. Plato toyed with creation myths. Aristotle believed the world was eternal. Philosophy could provide no acceptable answer. Either a Yes or a No led to difficulties, and this is still the case. Most religions have posited a God as the uncreated Creator. "In the beginning," says the Judeo-Christian scripture, "He created heaven and earth."

Twentieth-century astronomers still argue the question of whether the universe had an origin in time or is the result of a process of continuous creation. For a while the "steady-state" hypothesis of Fred Hoyle was widely accepted; it held that new matter is constantly being created throughout the universe. Today, however, the "big-bang" theory is once again dominant, and the universe is generally assumed to have originated in an explosion of a primal mass of energy/matter, the debris of which—the galaxies—is still racing apart. The astronomer Martin Ryle places the origin of the universe at about ten billion years ago, and many astronomers argue that our telescopes reach as far now as they ever will, since they see stars some nine billion lightyears away; and there can be none older. Where the mass which exploded came from is an unanswerable, and to most scientists, a meaningless question. Whether the process has happened before and could happen again, whether the universe at some future instant might collapse in upon itself and start all over, is a more meaningful though no less difficult question. Some astronomers, such as Allan Sandage and Robert Dirke, believe in an "oscillating universe," in which creation and destruction could occur again and again.[12]

Which of these basic theories about the origin of the universe is correct cannot be positively determined at present, though some observers of the current revolution in astronomy believe that a consensus may be arrived at in a matter of decades. These theories are important to a picture of the place of man in the universe not so much in themselves but because of their implications.

Modern physics has long postulated what has come to be called the second law of thermodynamics: entropy. According to this law, in a closed system, energy—and therefore organization—runs down. Put cold water in a vessel filled with hot and you get lukewarm water. Philosophers of science do not agree, however, about whether or not it is proper to call life—including man's ability to decrease the randomization in the universe through the creation of information—an antientropic (or "negentropic") process. Some philosophers would argue that life is simply a special case (like filling an inside straight in poker) where random events lead to greater rather than less organization of the materials of the universe; others assert that life obeys different laws from those governing nonliving matter. A "big bang" theory of the origin of the universe is more consistent with the second law of thermodynamics than is a steady-state theory which seems to contradict it or an oscillating theory which reduces its "universality." A "big bang" theory also makes it easier to think of time as an objective constant—the beginning, nine or more billion years ago, gives us a benchmark—and increase in entropy becomes a measure of duration, as when we use carbon-dating in archeology to establish the age of objects. If the decay of radioactive materials was not a unidirectional process it could not be used as a measure of time. Despite Einstein's theory of relativity, which states that space and time have no reality save as we perceive them, entropy still "points the direction of time. . . . Nature moves just one way."[13]

If the "big bang" theory is correct, there is a "direction" to time—and even if the oscillation theory is accepted there is a temporal direction within *our* universe—a passage of time between *our* cosmos's creation and destruction. Within our universe, differences in levels of energy are decreasing and organization is declining. Yet despite this, life—for some period in the span of cosmic time—is working in a different direction, toward heterogeneity and increased organization. Whatever our eventual fate in a universe which is running down, we living creatures are engaged in a struggle to create order and complexity against the tide of time, a struggle whose ultimate outcome we can hardly imagine.

Paradoxically, it is only this cooling down of the universe that makes life possible. Anything we normally think of as life could not exist except at levels of temperature which only come into being in the course of the "running down" of the universe. Actually, life as we know it is possible only in the presence of a large number of contingent circumstances, and the extent to which these conditions might be present elsewhere in the universe is a matter of dispute among scientists. Most would argue that given the vast number of stars in the universe there must, simply on a statistical basis, be other planetary systems which could support life.[14] Whether it would ever be possible to communicate with other life (even if such a feat were desirable) is doubtful. No form of communication imaginable to man can travel faster than the speed of light, so that no meaningful dialogue can take place between individuals lightyears away. Communication with other life-forms could occur only if they were able to leave their place of origin and come into closer physical contact with us, or vice versa. For practical purposes, therefore, man is probably alone in the universe,[15] regardless of what other life-forms may have evolved in other galaxies.

The Constitution of the Universe

While it taxes our imaginations and reasoning powers to deal either with a universe which has no beginning or one which begins out of nothingness, none of the theories of the origin of the universe impinge directly on our common sense view of the world. But modern physics does. Samuel Johnson refuted Bishop Berkeley by kicking a stone and proclaiming it real, and in so doing spoke for most men throughout western history. But modern physics assures us that he was wrong.

Our senses tell us that the book we are holding, the chair on which we are sitting, the pen we write with are things, bodies composed of parts, occupying space. Since Newton, the development of modern physical science has been a constant attempt to determine the composition of these parts, based on the belief that by understanding the constituents we can understand the whole. Physicists have searched for smaller and smaller and therefore presumably more fundamental elements of matter, the "stuff" of ultimate reality. By the turn of the century ultimate reality was found in the atom. The picture of the universe of early atomic physics—perhaps the last attempt to make the world of physical reality visualizable and verbalizable to ordinary men and women—itself did violence to common sense. The stone which Johnson kicked was declared to be composed of minute atoms, each of which was mostly nothing, a miniature solar system composed of electrons in "orbit" around a nucleus. But the electrons and their brothers the protons were at least real. They had electric charges and positions, if nothing else. But as time passed, more and more subatomic "particles" were discovered—or hypothesized in order to account for scientific observations—and reality became ever more unreal.

The electron—difficult as this is for us to believe—does not have a unique position at every instant in time; it is in a sense nowhere part of the time, yet it exists.[16] "Matter is no longer material."[17] The hierarchy of matter "trails away without hitting rock bottom, until matter dissolves into patterns of energy-concentration, and then perhaps into tensions in space."[18] If this is true, we are forced to refashion our whole view of the universe and man. For, as the naturalist Joseph Wood Krutch points out, if matter need no longer occupy space and have weight, there is no difference between a man who says that he is a materialist because everything is material and one who says he is not because nothing is.[19]

In contemporary science the controversy between materialism and idealism which exercised Dr. Johnson loses its meaning. *Reality is a relationship; substance is dissolved into function.* What distinguishes things is their structures; but *structures are only processes which last over a relatively long period of time.*[20]

Prior to modern physics and ever since Aristotle, it was assumed that motion was something (an "accident" in Aristotelian terms) that happened to a body.[21] Peter kicks the football; but the body—the football—has its own nature regardless of such accidents, regardless of whether it is kicked or not. Now we can only say that matter is what it does; its motion is not something added to it but constitutes its very existence. Sometimes electrons behave like waves, sometimes like particles, and sometimes like both, but they are neither.[22] What they are depends on their context; they exist only in and through their action. In Whitehead's words, "Process, change, activity are the only matters of fact. At an instant there is nothing."[23] Or, as a contemporary commentator puts it, "The object (or elementary particle) is nothing but the nexus of the various relations in which it participates."[24]

Since the abandonment of the concept of the "ether," generations ago, it has been accepted that forces can act upon each other across "empty" space, without "physical contact." Now, as Collingwood expresses it, "The modern physicist recognizes no empty space. Matter is activity, and therefore a body is wherever it acts, and because every particle of matter acts all over the universe, every body is everywhere."[25]

The implications for human life of this new worldview are of course vast and will be spelled out later. Not only does it necessarily change our conception of the physical universe as such, but if physical bodies can be said to be merely functions of their environment, what does this imply for human existence and identity? If every body attracts every other through gravity—which to some attenuated degree each does—is not every living human being a real force in the life of every other? If we perceive light from stars long dead as being as "real" as that from those still existing—as we do—how is the effect on our lives of friends and leaders long dead any less real?

But even if there is no difference between matter and nonmatter, things do happen in the universe as we observe it, and they happen in regular patterns. How can this be explained? Obviously there are real patterns even if objects as such have only a derived existence. The patterns, the relationships, are the irreducible components of reality. Changes in the world derive not from the motion of entities but from changes of form.[26] The whole—the system or field of which a "body" is a part—defines the nature of the part, just as you or I are defined in some measure by our education, life-history, and social and economic rôles. And as relationships change, the universe itself changes. The nature of reality is what traditional Chinese philosophy has held it to be, "dynamic and not static, relational and not absolute."[27]

The modern atomic theory of matter takes its name from the atomism of the Greek philosopher Democritus, but it is actually radically different from Greek atomism—the inspiration of Hobbes—which was still seeking the infinitesimal pebbles with which the house of reality was built. Atoms by themselves mean little: The smallest units of matter which have significant characteristics as far as their quality is concerned are molecules. Some molecules are composed of only one kind of atom; most are combinations. Sodium and chloride combine to make ordinary salt, a substance which has characteristics radically different from its constituents, characteristics which could not be predicted in advance from the characteristics of its components or from their atomic structure. The characteristics of molecules are the simplest examples of what has come to be called "*emergent properties,"* that is, cases where the behavior of the whole is different from the sum of the behavior of its parts.[28] Just as subatomic particles make up atoms and atoms make up molecules, molecules grouped together make up compounds, which in turn behave differently from their constituent molecules taken in isolation—just as men and women often behave differently in groups or crowds than when they are alone. *The structure of a relationship determines the characteristics of the objects which constitute the relationship. Changing relationships are what we call processes. Nature consists of processes, the character of which is determined by emergent properties which come into existence when certain things are combined in certain ways.*

These relationships which define the characteristics of processes are hierarchically organized; that is, each level of combination and complexity presupposes certain characteristics of the elements of which it is composed yet goes beyond them, and this level in its turn becomes part of a larger relationship which presupposes yet goes beyond the characteristics of its constituent parts.[29] The post-Newtonian attempt to take the watch of the universe apart is of limited value because a chemical analysis of the parts will never tell us how, taken together, they can constitute a watch which tells time. The parts can never be fully described in terms of their internal composition—steel or gold or whatever—but only in terms of their relationship to other parts, their function in the whole. In the case of subatomic particles the "parts" only "exist" in relation to the whole. Thus one can say that molecules are emergent properties of atoms; more complex organic substances emergent properties of molecules; life an emergent property of matter; and consciousness an emergent property of life.

But if material things do not exist in themselves but only by virtue of relationships, our view of the position of man in the universe is radically altered. Locke and the men of his time sought to reduce sensible qualities —color, sound, heat, and cold—to basic physical quantities. In the process they defined these qualities as in some sense illusory, as nonessential aspects of the object manifesting them, a subjective human reaction to basic physical properties. Light of a certain wavelength is perceived by normal human beings as blue, light of a higher wavelength as red. Quality was considered to be a function of quantity, the ultimate reality. But human beings are real, so the nature of whatever objects they have a relationship to is determined by that relationship just as certainly as the "nature" of an electron is determined by the field in which it exists.

Is there sound when a tree falls in an empty forest? No, sound waves exist, but there is no sound unless there are creatures to hear it. Is the sound that is heard real if there is a hearer? Yes, it is just as real as the soundwaves are. Secondary qualities, says the philosopher Samuel Alexander, are objectively real; they are "functions of patterns themselves composed of material elements."[30] Is a rose red? Yes, answers Whitehead, since human beings exist equally with roses; redness is a function of man-rose society.[31] Thus leading modern philosophers of science agree with the common-sense observation of the poet W. H. Auden, that if the word *real* can be used at all it must be in reference to the only world which is real to us, the world "in which all of us, including scientists, are born, work, love, hate and die . . . the primary phenomenological world as it is and always has been presented to us through our senses . . .",[32] this is the world in which "finite man is incarcerated by his essential nature."[33]

Nature is part of the reality that is socially constructed by man. Many scientists have over the years accused naturalists of committing the "pathetic fallacy," of attributing human qualities to natural things, of saying that storms "rage" when we know that storms are systems of energy incapable of emotion or saying that animals "smile" when they move their facial muscles in ways analogous to a human smile.[34] But if men perceive animals as smiling, they do in fact smile. Just as a rose is red if we see it as such, animals feel pain if we perceive them as feeling pain. At one time these qualities may not have existed in nature, but with the advent of brains—animal and especially human—then nature added color and sound, pain and pleasure to the universe.[35] They are as real as the brain is real; and the brain is as real as the subatomic particles which compose it. It may be that much of the value found in nature is value which we ourselves have put into it, but we have a perfect right to do so because what we think and feel about nature is just as real as any other properties it possesses.

The Nature of Life

Some philosophers would go even further and argue that the universe is alive in itself (inorganic as well as organic matter), not simply as a result of a relationship with living humans. Life, as men as diverse in their orientations as Teilhard de Chardin and Paul Weiss would hold, is movement, a striving to function, especially any movement in the direction of greater organization and complexity. For Teilhard there is a continuum of consciousness linking rocks, human beings, and God.[36] Nonliving beings and animals are free, Weiss holds, to become different in nature; evolution is "change of concern" on their part, as new conditions emerge which present challenges to which a substance must react.[37]

The distinctions between life and nonlife, matter and consciousness, are issues I will examine later in more detail. But whatever can be said about these differences, the distinction is not a dichotomy between matter and spirit. The universe is not composed of matter in aimless motion; it is a structure of order and what can only be described as will. It is a system, which to some extent and in some way is aware of its own existence.

Human beings are part of that universe not only in the "material" aspects of their bodies but in their total being.

Man is part of nature in that there is nothing about the human body and its observable characteristics and activities that cannot, at least in principle, be described in the language of the physical sciences. This does not mean, however, that all of biology can be reduced to physics and chemistry. While all activities of living things can be described in physicochemical terms, biological laws are autonomous in that they are necessary if we are going to understand what activity is going on in living creatures. Living molecules are wholes which are greater than the sums of their parts (just as all molecules are more than the sum of the atoms which constitute them), and they act according to regulatory principles of their own. As Arthur Koestler writes, the organism "is not a mosaic aggregate of elementary physicochemical processes, but a hierarchy in which each member, from the subcellular level upward, is a closely integrated structure, equipped with self-regulatory devices, and enjoys an advanced form of self-government."[38] Chemical reactions are involved in the process by means of which DNA determines our hereditary characteristics, but the laws of genetics are not chemical laws. The dianucleotic code gives us the vocabulary of heredity, it does not provide its grammar.[39]

In addition, something which amounts in practice to teleological explanation is necessary for describing living things. They and their functions can only be understood in terms of ends or purposes, as even Darwin admitted.[40] The lungs of animals exist in order to help oxidize sources of energy. But the belief of Darwinians—classical and neo-Darwinian alike— that the structural characteristics of living creatures can be fully explained by a process of evolutionary development—itself consisting of random mutations preserved by natural selection according to statistical laws—is increasingly under fire.[41] To say this does not of itself require (though of course it does not preclude) the acceptance of a concept of a predesigned natural order. But that phylogeny (the development of species) is the result of some basic and as yet unknown purposive force of living things— or life as a whole—just as much as is ontogeny (the development of individuals), appears increasingly probable.[42] Reduce embryonic tissue or adult flatworms to pulp, and they reconstitute themselves.[43] Single-celled animals—necessarily without nervous systems—construct microscopic dwellings of great complexity.[44] Any purely physicochemical explanation of how they do so would lead us into a realm as unreal in traditional mechanistic terms as that of the positionless electron.

The fact that man is a living creature does not separate him radically from the natural world around him. Different scientists and philosophers of science define life differently. Some would contend that the terms *life* and *nonlife* do not define mutually exclusive categories of existence but simply, like *hot* and *cold*, indicate extremes of a spectrum.[45] Others attach such descriptive characteristics to life as the ability of a system to sustain itself by using materials from its environment and to grow, or the ability to move and to reproduce. But crystals and fires sustain themselves and grow; machines are mobile and capable of being designed to reproduce themselves—even in the form of improved models. Just as the nature of the fundamental elements of matter is a question of relationship, so the differences between life and nonlife seem to be structural in origin and nature. Experiments with freezing simple living creatures indicate that certain arrangements of molecules are alive when energy is present and not alive when it is not. Life, it would appear, is an emergent property of nonliving matter. Matter arranged in certain relationships is alive.

Some philosophers refuse to accept the concepts of process and form as the basic explanation of reality and insist on the pre-Hegelian distinctions between being and becoming, but there is a developing consensus that the universe is a continuum in which some phenomena are more highly organized than others, a hierarchy of systems, and that this higher organization yields the characteristics we call life. "A living organism," writes von Bertalanffy, "is a hierarchical order of open systems which maintains itself in the exchange of components by virtue of its system properties."[46] The difference between life and nonlife, Bohn holds, lies "in the degree and kind of intrinsic order which has thus far resulted from the process of evolution."[47] Living and nonliving things are distinguished only by the arrangements of their parts; their difference is "not one of ultimate nature but of scheme and degree of complexity, nothing more," says Sherrington.[48]

This complexity which distinguishes life from nonlife, and higher forms of life from lower, is the product of evolution, an evolution which is proving to be a more subtle and complex process than the Darwinians envisaged, one with a forward movement directed somehow by overall patterns of order, the origins and nature of which we cannot at present comprehend. "The universe," in Dobzhansky's words, "is not an accident; it is an enterprise."[49]

Man is a part of nature—created through the evolutionary process. His blood has a salt content which reflects not the oceans of today but those of millions of years ago.[50] His hormonal activities vary with diurnal and seasonal changes based on the motions of the moon and the earth and the sun and perhaps other rhythms of the cosmos which we do not yet know about.[51] He was made out of dust: His whole body is "redolent of earth, whence it was dug."[52] But this earthliness of man's nature need not be thought of as materialistic in the old-fashioned usage of the word. Mechanistic materialism in the nineteenth-century sense is dead. The increasing emphasis given to the importance of form and with it emergent properties, in which life itself is regarded as an emergent property of "dead" matter properly arranged, results in a mysterious and numinous universe. All nature becomes in a sense alive, and we seem to be moving not into the age of the robot but into the morning of the magicians.

Mind and Matter

Even if the human body, including the physical apparatus we call the brain, evolved from the bodies of other living creatures, which in turn emerged from nonliving matter, can we accept this as being true of the human mind? Are not human beings possessed of a unique property called self-consciousness which differentiates them from all other creatures and cannot be accounted for by the workings of physical forces?

The only possible answer is that consciousness does exist; it is a primary datum which cannot be denied in practice any more than can our own existence. Because it is primary, it cannot be adequately defined.[53] Animals, even the most primitive, possess consciousness to some degree. What is unique about human beings is the degree to which they are self-conscious—which is what we usually mean when we speak of consciousness in a human context—a degree manifested by our use of language and especially of abstract concepts. Consciousness—the distinguishing aspect of what we sometimes call mind— is an emergent property of matter.

Most of the difficulty people have in accepting such a proposition stems from the dualistic heritage of western civilization with its dichotomy between body and soul, spirit and matter. The Platonic tradition led philosophers to think of the soul as something separate from the body, imprisoned in matter, and "mind" is largely a modern version of the Platonic concept of soul. Aristotle made no such distinction. The soul, in what is usually called the hylomorphic doctrine, was simply the form of the body, no more separate from it than the shape of a statue is separate from the clay of which it is composed. Thus every living thing has a soul. Medieval philosophers juggled the Platonic and Aristotelian traditions; the latter made more sense, but the former could provide a more secure basis for the theological concept of individual immortality, which had no place in Aristotle's scheme.[54] The triumph of the mechanistic view of the universe in the early modern period only made matters worse. Unable to give up the self-evident fact that the human mind existed—"I think, therefore I am"—dualists such as Descartes and Spinoza were faced with the insoluble conundrum of how an immaterial soul could affect a material body and vice versa.

The only solution lies in a return to a monistic position, which recognizes that "matter" and "spirit" are the same reality looked at from different perspectives, a position for which contemporary physics lays the groundwork. "Fundamental in the evolutionary view," as embryologist Robert

Francoeur points out, is the "rejection of matter and spirit as fixed categories."[55] Body and soul, the distinguished French scientist Alexis Carrell wrote a generation ago (essentially reverting back to Aristotle), are the same thing, and dualism simply engenders false problems.[56] "Mind is not a thing or an aspect of a thing; it is a process that takes place in the body," says biologist George Gaylord Simpson.[57] "Mind and consciousness are dynamic, emergent (pattern or configurational) properties of the living brain in action," according to neuropsychologist R. W. Sperry.[58] They arose out of the evolutionary process,[59] and to deny their evolutionary origin, as the philosopher Mortimer J. Adler does, is to deny the whole possibility of evolution, and to reject process in favor of a world of fixed, changeless entities.[60]

But simply because mind and consciousness are products of evolution does not lessen their reality nor the extent to which they are uniquely (as far as we know) developed in humanity. This something which differentiates humans from other animals is radically different from anything that previously existed in nature. "Something about our abilities and behaviors is different and was once, even if not suddenly, new," Simpson contends.[61] "Self-awareness . . . possibly the most fundamental . . . characteristic of the human species . . . is an evolutionary novelty," writes Dobzhansky.[62]

How far we have escaped from our earlier, apparently primarily reflexive, animal brains is a matter of dispute. Koestler and some others have argued that the neocortex, physical seat of our higher mental powers, exists in uneasy domination over our older mammalian and reptilian brains, and that this is the cause of many of our individual and social-psychological problems.[63] But there is no question that in creating humanity, as in creating life, nature transcended its previous self. Evolution has worked not simply toward survival—as classical Darwinism would force us to hold—but toward something new, called "awareness." In man the universe has become capable of self-consciousness.[64]

Free Will and Determinism

How is human freedom possible if the human mind is not separate from evolution and physical nature? Is it perhaps not simply another machine like the computer? Many computer scientists argue that eventually computers will become truly intelligent, by which they often mean capable of self-awareness, and some look forward to the day when men will be forced to abandon their claim to being more than machines themselves.[65]

How we will know when that day arrives, will of course be difficult to determine. If we ask a computer if it is self-conscious, it may say Yes but not be telling the truth. We can directly perceive the self-awareness and freedom of choice of no one but ourselves, though we assume our fellow humans possess both these qualities at least some of the time. A sophisticated mathematical-logical argument holds that no computer could possibly be effectively programmed with the knowledge that it was being programmed and therefore could never be truly aware of itself.[66] But could not self-awareness arise spontaneously, as an emergent property of the computer's internal communication processes? Whatever the theoretical possibilities, true artificial intelligence probably remains further away than most of its proponents hold, and possibly by the time it exists human beings will have developed even higher forms of consciousness than they now normally possess. Whatever the future of artificial intelligence and whatever analogies exist between intelligence in man and in machines, the human biocomputer is of a different order of complexity and efficiency and versatility than any computer ever constructed.[67]

Yet even if manmade machines cannot think, this does not prove that we ourselves are not thinking machines, subject to that determinism which traditionally has been considered an inevitable consequence of being a machine. To make the physical basis of human mental activity the underpinning of an argument for determinism requires, however, that we assume that the physical universe can be adequately described in purely mechanistic terms, and this we have seen is false. "One day we shall certainly 'reduce' thought experimentally to molecular and chemical motions in the brain; but does that exhaust the essence of thought?" asked Frederick Engels. Though he is often thought of as a materialist, the way Engels asked this question indicates that his own answer would be No, and we can agree.[68]

It would appear that the observable physical activity of the brain leaves little room for mind as more than determined motions of matter. "Between the individual human mind and the individual human forebrain considered in time and space there is a 100 per cent correlation," wrote the distinguished physiologist Sir Charles Sherrington a generation ago in a classic study;[69] yet he continued to hold that mind was something more than brain, that the brain was simply the "organ of liaison between energy and mind," and that nature had evolved humans as compounds of energy and mind.[70] This position creates a dualism within nature, since it postulates that there is something other than energy—the mind—which is natural. More reasonably, Julian Huxley has argued that evolution as a philosophy of human development forces us to think in terms not of dualism but monism, "of an enduring world-stuff, an X which possesses both material and mental . . . attributes, according as it is experienced objectively or subjectively," which makes it possible to reconcile necessity and free will.[71]

Since Heisenberg's principle of uncertainty became a basic foundation of quantum mechanics in physics, many philosophers and scientists have tried to use it as a means of specifying how Huxley's combination of monism and free will could be made more credible. For some, events in the mind can be considered as potentially indeterminate because they are based on events in the physical world of the infinitesimal; that is, there are so many neurons in the brain that their individual activity falls within the statistical range in which indeterminacy operates, yet the brain is so organized that the effects of their activity can be "amplified" so as to affect events in the larger world of ordinary causality.[72] Others argue in favor of a relationship between mind and body analogous to the general principle of "complementarity" derived from quantum mechanics, holding that mind and brain may be the same thing behaving according to different "rules," just as electrons sometimes behave like waves and sometimes like particles. The physicist Niels Bohr, father of modern atomic theory, held that humans could thus be thought of as free or determined depending on the viewpoint of the observer,[73] and J. Robert Oppenheimer embraced similar views.[74]

Henry Margenau cogently sums up the current state of the question: "Physics has little to say about freedom of the will," yet we are all aware in our own experience of conscious decision-making as a reality. "The latitude needed for action consequent on decision is guaranteed by the probabilistic features of physical reality. This is as far as we need to go at this point."[75] Thus contemporary physical science's picture of a nondeterministic universe does not guarantee human freedom—we may still be slaves of various social and psychological forces, as behaviorists such as B. F. Skinner assert —but it does affirm that we are not denied freedom simply because we are wholly part and product of physical nature. As the distinguished philosopher Karl Popper wrote recently, "Man is certainly a part of nature, but . . . has transcended himself and nature, as it existed before him. And human freedom is indeed part of nature, but it transcends nature—at least as it existed before the emergence of human language and of critical thought, and of human knowledge."[76]

The relationship between the human mind and nature has other aspects besides the simple fact of our materiality. Earlier I referred to the extent to which reality is social, constituted by our acts of perception, and I argued that, whatever the physicists who claim to have a more "objective" view of nature may say, roses are in fact red. The mind which sees and in seeing "makes" roses red, the mind which constructs reality, is a special kind of mind, formed by nature in a special way. Our mental capacities were evolved on earth for earthly tasks, primarily but not exclusively for survival. "Ours is an earthly mind which fits our earthly body. It produces percepts of earthly things from an earthly viewpoint. . . . Earth's nature is our nature. We owe to earth the entire gamut of our mind's wonders, whether of joy or pain."[77] Our concepts of time, space, extension and form and all that derive from them intellectually are rooted in our sensory apparatus, and the extent to which abstract symbolization, as in mathematics, can make it possible for us to break free of these—and how far it should—

is a moot question. We relate to nature because we are part of it, and to forget this and try to exist in a world of scientific symbolization alone is to court alienation and self-destruction.

It is becoming increasingly apparent that in our dualistic, antiearthly desire to relate to the world only as an objective, "outside" physical presence rather than as a system of which we are part as both product and creator, we have impoverished ourselves as individuals and as a society. Both mystical and psychedelic experience would indicate that there are realms of reality—at once "spiritual" and "physical"—with which we have lost contact.[78] We as yet know very little about the physicochemical correlates of dream or drug experiences, and less about those of mystical states. But there is little question that a whole borderland of experience and phenomena lie in a parapsychological world which is real and is currently becoming the subject of disciplined investigation.[79] At the same time we are becoming more aware of our own bodies, and of the possibility that we can to some degree go back in our developmental history to a period when we knew certain aspects of natural reality as directly, as connaturally as other living creatures do.[80] The cultural significance of these trends for the future of mankind will be discussed later, but it needs to be stressed here that our increasing concern with these aspects of our existence in large part reflects a rejection of the dualistic worldview of liberalism and a recognition of the fact that while the human mind is the highest form of self-consciousness the world has yet produced, it does not set man apart from the natural universe. Matter and spirit are aspects of the same reality—the moebius strip of the universe that has two sides and yet only one.

9

Nature and Human Values

THE need to make value-judgments is an intrinsic, necessary part of human nature. So basic and universal is it that we can only conclude that ethics itself must be a product of the evolutionary process. Humanity has learned to make moral judgments just as it has learned to think, verbalize, and fantasize.[1]

Important consequences follow. If morality arose out of the evolutionary process, it must be functional for us, if not as individuals then as a species; it must have survival value in the struggle for existence and the drive toward development. As Bentley Glass, former president of the American Association for the Advancement of Science (AAAS), has remarked, "Man's own values grew out of his evolutionary origins and in his struggle against a hostile environment for survival."[2] Other animals survive without making judgments about right and wrong. They too have their own structures of right and wrong behavior, but these patterns are fulfilled as the result of a complex process of stimulus and response, which presumably never reaches quite the level of consciousness and choice acquired by humanity. Animals almost always do what is right from the point of view of species survival, barring physiological accidents and lack of experience in novel situations.[3] They cannot do right or wrong as we use the words, since their actions are not mediated through conscious choice. Human beings alone have evolved a moral sense which is problematic, which allows us to fail to do right because doing right involves general and sometimes complex judgments, dealing with things not immediately present to the senses.

Why does the human being—uniquely as far as we know—have this moral sense? It is not because he is a social animal. Most creatures are. Morality arises basically from two factors. One is the long period of infant dependence and maturation, which necessitates the maintenance of a particular, often extremely complicated social structure that will permit nurturance and protection of the immature offspring, and requires (as Freud long ago surmised) the repression of "instinctual" behavior which might be functional in other contexts.[4] But added to this physiological imperative at the species level is the fact of humanity's unique degree of self-consciousness and its use of language. Socially functional behavior-patterns are not transmitted to the immature simply as part of their physicochemical, stimulus-response patterns, nor only by example, but through a learning process in which symbols play an important role. Once reality is mediated through a complex symbol system, the possibilities of erroneous perception of reality increase, and the symbolic system itself supplements directives which otherwise would be "instinctually" given upon direct contact with the environment. Ability to understand moral symbolism becomes necessary both for the functioning of the human individual in society and for the functioning of society as a whole. Waddington has even theorized that the existence of authority is an essential element in morality, since so much necessary knowledge is taught and must be accepted and used before it can be understood.[5]

In any event, humans can choose to be immoral in so far as any particular moral code is concerned, but they cannot choose to be amoral, to be nonmoral. Sociopaths, those who cannot respond to moral imperatives, are rightly judged to be mentally defective. Mankind as a whole cannot survive unless those moral specifics which are functional to species survival in a given context are accepted, and it is the role of the social philosopher to discover—not to create but to discover—what those principles are, based on an analysis of the needs which are functional for human survival and development. In short, man is naturally a moral being.

Are Values Natural or Conventional?

But to say that morality arose out of nature through the evolutionary process is not quite the same as saying that nature generally and evolution specifically can provide substantive moral guidance. Yet if moral principles do not come from nature, where else can they come from? One possible alternative source is convention. Seeing that some patterns of action were universal—all humans must eat and sleep—and that some were not—different societies have differing rules for political, economic, and family life— the Greek sophists distinguished between nature (physics) and convention (nomos). But Greek thought as exemplified in Plato and Aristotle also recognized that some of these alternative patterns were more universal than others, and some more in conformity with human biological nature and the constraints it places on human choice than others. With the absorption of Greece into the cosmopolitan Roman empire, greater emphasis was given by classical thinkers to the conventional aspects of moral codes; and the struggle between the partisans of universal norms and of conventionalism has continued into our own time.

Liberal society, with its emphasis on the notion of the social contract as the basis for public order and its willingness to allow public policy to reflect the balance of power among competing interest groups, is essentially conventional in its moral premises. But as has often been pointed out, the ability of liberalism to survive has rested on the existence in western society of a substratum of commonly held principles which liberalism has undermined but never completely destroyed. Also, some conventions are more equal (less conventional) than others even in liberal society. The Anglo-American legal system originally presupposed a natural law, and in such concepts as "due process" the United States Supreme Court seeks to elevate certain principles of natural morality—especially in its definition of the rights protected by the Fourteenth Amendment—above the everyday political process.[6] Even Hobbes was forced to fall back upon moral universals rooted in the human condition—the belief that men will honor their contracts and not cause pain to others gratuitously—in order to make his system work.

If we hold that every psychically healthy human individual must have a moral code and that no human society can survive unless it has a structure of rules governing the interaction of its members, we are saying that morality is as basic to human life as metabolism, and just as there are patterns of action based on the characteristics of our physical being which are conducive to physical health and survival, there must be similar and related moral patterns. As a contemporary systems-theorist argues, moral imperatives arise out of the fact that societies are natural systems: "Each of us must (in the sense that he cannot help but) commit himself to survival, creativity, and mutual adaptation within a society of his peers; the alternative to these is death."

Convention cannot be the source of basic moral principles because some arrangements do not work and the agreement of men by itself cannot make them work. As the anthropologist David Bidney points out, nature sets limits, even if it does not determine which of the options it offers are taken.[7] Cultures differ, and some serve human needs better than others: "No one cultural system completely satisfies all human needs and potentialities; each system has defects corresponding to its virtues."[8] But, as the sociologist Morris Ginsburg argues, it is difficult for anyone to "seriously believe that all cultures are equally valid."[9]

"Despite the diversity of moral codes general principles are discoverable which are implicit in all of them and which come to be recognized as universally binding in the course of development."[10] Human beings require access to material goods in order to survive; and unless some system of property relationships (however broadly defined) exists, men will be in constant chaotic scramble over these goods, and the goods created by human action rather than the automatic workings of nature might not be produced. Children are dependent on adults for physical and emotional sustenance; unless responsibility for them is taken by someone, they will perish. Various emergencies such as illness and natural catastrophe threaten individual and group survival; unless certain coordinated tasks are undertaken, these emergencies cannot be met. Whatever is done, something must be done; and what is done by differing cultures tends to be quite similar. Amid variations in detail, moral codes everywhere exhibit striking similarities in essentials.

The complete elimination of scarcity and a resulting increase in altruism among human beings could mitigate many of the problems of human interaction now regulated by moral codes but could never completely obviate the need for moral prescriptions. In unstructured social situations even today many problems of interaction are taken care of by simple rules of thumb: "Finders keepers," "First come first served," "Charity begins at home." International maritime law is still to a large extent a compendium of such "natural" rules of human conduct. But even these conventions reflect underlying regularities. Convention can prescribe how certain human problems are taken care of in certain situations, but the problems are defined by nature. Some solutions work better than others, and all are workable only within the limits dictated by the nature of man and the universe.

The conventionalist argument in modern times takes the form of cultural relativism. The knowledge of the Greeks that others behaved differently—though in their eyes less well (the reason the term *barbarian* developed its pejorative connotation)—has in contemporary times been multiplied a thousandfold by the research of anthropologists and cultural historians, which has led to the unwarranted conclusion that there is no such thing as a human nature capable of definition save in terms of truisms or obvious physical needs which can be satisfied by different cultures in ways so diverse as to defy the construction of general social norms of behavior.

Is man not simply what his own cultural history has made him, as the cultural relativists hold, and is it not therefore nonsense to talk of man as such apart from culture? Certainly human beings do not—and cannot—exist in a cultural vacuum, any more than an abstract Platonic tree can exist apart from individual trees. But the mere fact that human beings deal with their common problems in a variety of ways—ways which may even reflect possible genetic differences among peoples[11]—does not make all these ways equally useful or even useful at all. Thus, while the anthropologist Clifford Geertz argues that one cannot find cultural universals which are substantial and grounded in biological, psychological, or social processes[12] and that "there is no such thing as human nature independent of culture,"[13] even he denies the position sometimes taken by such extreme cultural relativists as Ruth Benedict that anything which one group does is worthy of respect by another.[14]

To observe the various customs of primitive peoples and then say that all cultures are equally human, is as foolish as saying that all animal species ever created are equally able to survive under all circumstances. In some sense all cultures are equally human since all are cultures created by women and men, but some have perished and others have become frozen in isolated ecological niches.

Even survival is not the only test. Because of the plasticity of human nature, human beings are able to survive under widely varying circumstances. But survival is not happiness. Why we must necessarily judge the Dobu with their nasty warlike culture to be the equal of the Hopi or even ourselves is difficult to fathom.[15] We would not judge Adolf Eichmann to be of equal human worth with Albert Schweitzer, despite our fundamental belief that all human beings are equal and have the right to self-development according to their potentialities, simply because each of them manifested a particular human adaptation to his society of origin. Many of the primitive cultures cited as examples of human variety are clearly incapable of adapting to even the most minute changes in their environment, and thus deprive their definition of what being human means of any universal significance. Some cultures are dysfunctional even in their own surroundings, as the incredible behavior of the Soriano of the Amazon indicates, and are dying out even without cultural contact with modern western society, peaceful or malevolent.[16] Just as there are many varieties of individual human personalities, some beautiful and some monstrous, so also there are various societies, but their mere existence does not make them all equally central in defining what is human.

Cultural relativism—long, paradoxically, the governing element in the supposedly value-free science of anthropology—is an intellectual copout. All cultures are not equal in their ability to give expression to basic human values, and all are attempts to cope not only with given historical situations but also with the needs of basic human nature. We do not have to choose between nature and history as definitive of what it means to be human; man is the creature of both. But, as Bidney notes, human nature is the root factor; it is "metacultural in the sense that it is an ontological postulate which transcends cultural experience."[17] The similarities among human beings far outweigh the differences, and the values which are derivable from nature far outweigh those which are attributable to convention.

Theology and Human Values

But if convention cannot furnish an adequate basis for the values individuals and societies need in order to survive and function, is nature the only alternative? Can it not be argued that moral principles are divine in origin, the commands of a sovereign deity, and sometimes at variance with the dictates of nature? Although this is not a theological treatise, it is impossible to discuss man's relationship to nature without some reference to the problem of God. The current situation in political science, in which religion and political thought are totally separated, is an abnormal one, as any student of its history can attest; and, as one political scientist has recently noted, this becomes especially obvious when one deals with humanity's relationship to the environment, which is in the truest sense a religious issue.[18]

The relationship of theology to political and social values is a practical as well as a theoretical question. Despite the assumptions of most intellectuals, religion is a real force in the lives of most people in the world, including the peoples of western nations, above all perhaps that of the United States.[19] For most Americans rights are not conventional but "God-given," and for them the "Creator," who, in Jefferson's words, endows men with inalienable rights, is not the impersonal being of deism but the personal God of scripture, and most Americans feel very comfortable indeed pledging allegiance to "one nation, under God." To be politically effective, a philosophy which accepts nature as the source of value must be compatible with popular religion, even though it will necessarily run counter to the operational civil religion of most Americans, in which Locke takes the place of Christ, and will necessarily go beyond popular religion in stressing the immanence as well as transcendence of God, an immanence manifest both in humanity and the physical universe.[20]

Happily it is not necessary to resolve any of the problems relating to the existence and nature of God which trouble theologians in order to deal with the role of value in the universe. If one assumes that there is meaning in the universe which can in some sense be personalized, two possible alternatives present themselves. One can posit the existence of a Supreme Being who is radically "other" (in Barthian terms) than the created universe, or the universe as it is perceived by man, a Being whose ways are not our ways nor those of his creation. In this case nature is not a touchstone of value, its values are not those of the Supreme Source of values or are in conflict with his. At best nature is a neutral, valueless stage on which the drama of salvation is played out. At worst, one is left with the problem of why a presumably omnipotent deity allows a world which is not his own, which is no reflection of him, to flourish, and one feels under a compulsion to join him in hostility to it.

Alternatively, one can argue that the universe in some sense *is* God or that it is an emanation of "Him"—or "Her"—which to some extent reflects His/Her nature and will. In this case either God is immanent in the universe and it itself is holy, or else God, even if totally transcendent, created the universe as an expression of his goodness. Such religious formulations may have to ask to what degree the universe has been tampered with by the forces of evil—including perhaps man—but they make it possible to look to the universe as basically exemplifying God's will and nature.

In recent years two trends have grown up within organized religion in the West which make it less difficult to seek a source of value within the created universe. One is the popularity in some circles of theologies of "liberation" and "revolution."[21] While these theologies are the special province of social activists who are usually little concerned with nature, they necessarily imply that God is equally as immanent in the process of historical development as he is transcendent to it. More directly useful in breaking through the rigid separation between the natural world and transcendent values are the various ecological theologies which have burgeoned—especially in the United States—in the wake of growing popular concern with environmental problems.[22] Increasingly it is possible to argue, even within religious circles, that to seek value in nature, including the evolutionary process, is only to seek to peer behind one of the masks of God, and even orthodox Christians can see some validity in what a western interpreter describes as the basic Hindu belief that "the world is God playing hide-and-seek with Himself."[23]

From a strictly scientific point of view, the image of the universe which makes the most sense today is that of the East, the "undifferentiated aesthetic continuum" of which the philosopher Northrop writes.[24] From a naturalistic point of view there is considerable appeal in the image of man basic to many eastern religions in which—as Alan Watts sympathetically puts it—"the personal ego is hallucination" and the individual is seen "as some particular focal point at which the whole universe expresses itself," so that "the only real 'you' is the one that comes and goes, manifests and withdraws itself eternally in and as every conscious being."[25] The most profound mystical and theological expressions of the western religions tend to validate this image of man. But they also conceive of self-identity as having real meaning, as evidenced in such concepts as immortality, resurrection, and personal salvation. This belief in the reality of human identity includes and implies the concept of personal responsibility and conforms with our own subjective consciousness that we exist and are free.

Spinoza once wrote that "the more we understand individual things, the more we understand God,"[26] and the emphasis on individuality has always been a cornerstone of western thought. By contrast with the religions of the East, for Judaism, Christianity, and Islam the relationship of God and humanity has a directional aspect, since history has meaning. If individuals are real, there is a sequence of lives which can never be re-peated, and there is the possibility of development over time, as is uniquely symbolized for the Christian by the Incarnation, the direct participation of the fullness of divinity in both the physical universe and the historical process.

Thus there will always be an element of tension in the relationship between nature and values for western man, because both nature and man exist within the context of salvation history, in which nature is regarded as a means to the end of salvation, individual or collective. But it is possible for Christians at least to accept the maxim that "Grace perfects Nature" rather than wars against it. The medieval scholastic concept of natural law, however legalistic and overly deductive its methods and however simplistic and undeveloped its view of physical nature, was essentially sound in recognizing that creation was a manifestation of divinity and that, leaving aside questions of "eternal salvation," all men regardless of their theologies or lack thereof can, by looking into the visible words of God, including themselves, find the guidance they need for the creation of the good society and the good life on earth.

Value in Nature

A major difficulty for many people in regarding nature as a source of values is that by human standards nature seems to be profoundly immoral. To regard nature as in some sense divine or as a manifestation and channel of divine grace simply transfers to the natural realm the ancient theological problem of evil. Instead of struggling with the traditional problem of how an omnipotent and good God could permit suffering and death, we have to ask how nature defined as good could permit these things. To say with St. Augustine that all that is is good, and that evil lies only in negation, is to play a philosophical shell game which does not help anyone to deal with the real world in which the existence of pain and privation is a daily experience. How can nature, red in tooth and claw, be good? Is not "Mother Nature" in reality a "bitch," as one contemporary philosopher has suggested?[27] Is it not imperative that humans who rebel at her horrors be Promethean and defy her just as they would rebel against a Moloch who gloried in human blood and suffering?

That nature is a wanton and cruel mother there can be no doubt. Her creatures live in fear and pain, insofar as they can feel these. Most philosophies which postulate nature as the source of value place a supreme value on life. Nature created life, but nature is also careless of life. As the physicist Erwin Schroedinger has declared, "Nature treats life as though it were the most valueless thing in the world."[28] Her creatures all exist by eating one another. Some exist only by eating plants, but plants too are a

form of life, and, as we are learning, possibly a life-form of great sensitivity.[29] The integrity of nonliving elements of nature is destroyed in the
name of life, as bacteria and plants turn rocks and water into soil. Big fish
eat little fish, and the higher up one goes in the food chain the more likely
one is to see creatures feeding on one another. Nature seems to play no
favorites among her creatures. As the physiologist Sherrington writes, "Nature, though she has evolved life, makes no appraisal of it. She holds no
lives of higher worth or of lower because to her all lives are without
worth."[30] The balance of nature of which ecologists speak so reverently is
one in which there is not always cooperation but often the most violent
conflict among species. Aristotle was right, St. Francis wrong, Sherrington
asserts—conflict not cooperation is the norm.[31] Not all biologists and
naturalists would agree—symbiotic relationships of a cooperative character
play a vital role in natural processes. But nature is no Garden of Eden,
where the lion and the lamb always lie down together.

Not only did nature create life and yet is careless of it, but nature
created higher life forms and does not favor these as opposed to the more
lowly. Superior intelligence gives greater survival-potential, but is not an
unbreachable redoubt. When men of the Enlightenment such as Thomas
Jefferson, believing in the beneficence of nature, sought to find human
utility in the mosquito, they were straining credulity. What beneficence is
there in a natural "plan" in which, in order to serve the uses of the germs of
spinal meningitis and similar deadly diseases, hundred of thousands of
human beings must annually die in excruciating pain? "Naive thought,"
writes one commentator, "might suppose Nature would at least value transcendence in life, e.g., a man more than a protozoan speck, or than a
parasitic bacillus. But no."[32] It is profoundly amoral, and it is no wonder
that for millennia mankind has regarded nature as its enemy.

Yet as we have seen, nature has created morality. It took millions of
years of pain and conflict for man to develop. And man moralizes. Nature
itself is neither good nor bad on its own terms. One cannot add up the
pains and pleasures experienced by nonhuman creatures, however real they
may be, for these are not self-defining qualities morally; they require an
outside standard. Man is that standard for all of nature, including himself.
Nature—now that it has created man—is defined by man's values; just as
roses are red because there are women and men who see them, so nature is
moral insofar as it has created man, who is in and of nature.

Because natural forces create and destroy, bestow life and death, bring
forth beauty and horror, they cannot become simple touchstones of moral
action. On balance, nature is good, it has created life and man, and we
cannot repudiate it without repudiating our own existence. Our mother,
however we judge her actions, is still our mother. Nonetheless we have
been created to judge; this itself is a natural capacity. Nature, now that she
has created man, is defined by man.

For nature to acquire a moral dimension reflexively, through the creation of man, the process of man's creation must be assumed to have moral meaning. Natural history must have a direction, just as time has a direction in a cosmological sense, and it must be assumed that the antientropic forces represented by life and intelligence have value. This is a basic and undebatable assumption: Life is preferable to death, consciousness to lack of consciousness. Once this assumption is granted, the often cruel processes of nature take on a new meaning as the inevitable concomitants of the evolution of life and intelligence. Nature becomes normative once one assumes that the struggle for life is not an aimless process but an agony of birth, that it is not an eternal round but, through evolution, has a direction, and that that direction consists in the production of humanity, whose role it is not to conquer but to redeem and reconcile nature. The Garden of Eden was a myth of the past—an analogue of the myths of the golden ages of society which mankind traditionally situated in the past—but the true approach to Eden may be through the future.

What does morality within nature dictate? Nature builds into its creatures certain forms of behavior which are conducive to individual survival and above all the survival of the species. Survival of the fittest means, as Darwin's successors especially have emphasized, not individual capacity to survive but ability to procreate. But survival also means ability to develop and change. Species must be able to adapt to changing environmental conditions if their descendants are to survive. Single-celled animals are in a sense immortal; they propagate by dividing into two and will live after "reproducing" until consumed by enemies, but are immune from natural death. But as species they cannot evolve. Evolution demands a genetic mix, which requires more than one parent, so that variant types—possibly better endowed for survival than either parent—can come into existence. The purpose of sex is not simply reproduction but variety and progress.[33] With the invention by nature of sex another powerful force came into the world: death. The continued existence of parents was unnecessary and indeed got in the way of the presumably better-endowed offspring. Death is the price which nature exacts for evolutionary development.

But does progress or development really mean anything in the context of nature? Countless species have perished in the course of evolution, even before man, the supreme destroyer, came on the scene, but many have survived in restricted or humble ecological niches. There are creatures alive today which have not changed from what their ancestors were millions of years ago. How can one say that the newer species, including man, are any better than they? Many philosophers of science would deny that the processes which contribute to survival and evolutionary development have moral significance, because they deny either implicitly or explicitly that evolution has a direction.[34] For some all that evolution means is change, but directionless change, and anything that survives must be considered a success. Dandelions, for instance, have abandoned sex.[35] Who is to say that this, however regressive in an obvious "historical" sense, is bad, especially in an era when many advocate or foresee humanity abandoning sex, or at least sexual relations, as a means of reproduction?[36] To persons who deny direction in evolution, evolution is simply a random process, going nowhere, and all its results are of equal value. To impart value to the evolutionary process one must be willing to judge some of its results to be better than others. This kind of judgment is implicit in the comment of the naturalist Joseph Wood Krutch that, even aside from man, "recent animals are more intelligent, more intensively conscious, more capable of play, and joy, and love than the ancient ones were."[37]

Most commonly the argument against direction in evolution is that even if one can trace an historical evolutionary development leading up to humanity, it is impossible to assert that we are any better than "lower" animals, at least not on evolutionary grounds. There is some evidence, they would argue, with considerable cogency, that man may be dysfunctional and through war, pollution, or other means will destroy himself, thus leaving the field to other creatures, especially the members of the hardy and ubiquitous insect kingdom. Thus, measured in survival terms, man is not an evolutionary pinnacle. But implicit in this judgment is the premise that the intelligence we glory in is of no worth in itself, a premise as beyond argument as it is false. Explicit in this judgment also is the belief that man may not be able to use his intelligence in order to survive, a belief that can only be constantly refuted in every age not by argument but by action.

But if we assume that being human is worthwhile and that man can survive, we in effect affirm that evolution has had a moral dimension in creating man, and that the processes which have led to the emergence of man—the mechanisms of survival and development—have an ethical connotation. To deny that one can derive moral imperatives from nature is to deny the value of the humanity which it has spawned.[38]

Although we can say that nature and the evolutionary process provide moral norms, we must beware oversimplification. Not everything which is conducive to survival in the short run and in a particular environmental situation or ecological niche, is conducive to survival in the long run, much less to development. The record of nature is replete with examples of creatures which have come to evolutionary dead ends, usually through overspecialization. They have become overdependent on relationships with particular environments or in some cases with other living creatures and cannot survive when these are altered or destroyed or, like the fiddler crab with his cumbersome claw, they have stressed one particular useful element of their genetic heritage to the point that it has become dysfunctional to the whole.

Humanity has evolved over the past several millennia primarily in its cultural rather than its biological characteristics. But culture, which itself originated in the evolutionary process, can wander off into blind alleys as readily as genetic evolution, and sometimes does. Humanity's overdependence on particular rationalistic, mechanistic ways of perceiving the universe and on certain manipulative technological ways of dealing with it are leading us into an evolutionary blind alley of overspecialization comparable to that of the fiddler crab or the dinosaur. While this movement has been a response to some short-term norms of evolutionary ethics—such as our need to gain some measure of independence from certain of the vagaries of physical nature—it now runs counter to the larger imperatives of survival and development. Yet, because man has evolved into a creature capable of choosing, he can decide whether to remain human—to preserve his identity which depends upon a precarious balance between "mind" and "body," nature and technology—or whether to evolve toward a parasitical relationship of dependence upon forms of technology which can lead to the atrophy of many of his most fundamental characteristics and to the foreclosing of any future evolutionary options.

Whether the current direction of cultural evolution can be reversed is a moot question. Certain evolutionary theories, based on "Dollo's law," suggest that there is no turning back once a course of regressive development is entered upon.[39] Other evidence indicates that critical spurts of evolutionary change can come about through a kind of retracing, as it were, of the evolutionary path and starting over again, preserving elements of youthful development previously lost in adulthood, a phenomenon known as *neoteny*. (Many of the characteristics of human beings, for instance, are those of apes in the immature part of their individual life cycles, which in apes are later lost.)[40] Perhaps there is actually an evolutionary imperative behind the attention many contemporary observers have given to the possible role of youth in reversing current dominant cultural trends. Perhaps there is a profound natural basis for the biblical injunction that unless we become as little children we cannot enter into the kingdom of heaven.

Nature can be a source of moral guidance because it presents us with a process of evolution which upon analysis can give us clues as to how to deal with the crises of survival and development. In searching for a key to why human beings among all creatures have attained their present position of dominion over the rest of animate and inanimate nature, one finds it especially in human flexibility and refusal to overspecialize. Here, certainly, is an imperative for future conduct.

Nature also offers us more specific and proximate norms. Even those who argue that nature is amoral in itself must admit that it sets limits to what we can do. Values are "not resident in nature outside of man," writes a contemporary theologian.[41] But "nature does have built-in consequences."[42] "This is a universe of consequential validity. We simply do not get figs from thistles, as Jesus observed."[43]

Nature's norms are of surprising stability. Nature is highly conservative. Similar conditions produce similar responses, even in biological forms. Some Australian animals are marsupials, which have had a different evolutionary history from animals elsewhere. Yet, except for this one characteristic, they remain in other respects remarkably similar to comparable creatures found in the rest of the world, though somewhat weaker in direct contests for survival. There seems to be in nature a limited repertoire of possibilities, and it is no accident that human cultures show an analogous similarity in the limited range of responses available for meeting human needs.

Scientists often speak of built-in regularities as the consequence of "impotence principles," and some argue that the search for these limits to the possible should be the foremost task of science.[44] Impotence principles define what can happen and what cannot, and the central element of the current political crisis of mankind is that liberalism and its variants are coming up against the impotence principles inherent in the nature of the biosphere which supports human life, and that liberalism is incompatible with the cultural (including political) adaptations demanded by these limits.

One possible human response to the existence of such principles is familiar to virtually everyone: the rage manifested in the tantrum of a child who has been told, or found out the hard way, that touching a certain object will hurt him. A more reasonable response to limits, one which we associate with maturity, is acceptance and perhaps awe. Even as hardheaded a scientist as Garrett Hardin suggests that the current realization of the limits which nature places on human actions means that "in an operational sense, we are returning to a religious orientation toward the world."[45] However one conceives of the nature of ultimate reality, if one assumes that nature is a reflection of good, one's only possible response is to accept its dictates with respect and veneration.

We need not surrender blindly to all the consequences of the workings of nonhuman natural processes and meekly accept plagues and floods and drought. Nature, in creating man, created a morality by which her own actions can be judged and guided. But this moral judgment and guidance cannot run counter to nature itself taken as a whole; it must sustain rather than disrupt the total system of nature of which man and morality are components. Man's role in nature is not to reconstruct it arbitrarily, but to reconcile it to itself, to make of the jungle a garden.

The Human Value of Nature

Thus nature is more than a source of values for human beings. It is itself of value; that is, it is something to be respected, indeed treasured, in and for itself. "The great fault of all ethics hitherto has been that they believed themselves to have only to deal with the relation of man to man," Albert Schweitzer once commented.[46] Now we are called on to develop an ethic of man's relation to nature.

Nature is an object of value in two ways, often confused, especially in current discussions about the "preservation of nature." One way of regarding nature as of value is instrumental. If we act contrary to our natural limits and drink too much, we have a hangover the next morning. Therefore, in order to avoid the unpleasantness of a hangover, we should respect our nature and limit our drinking. This is the kind of respect for nature (defined as "what we are") advocated by Aristotle and the natural-law tradition. Since respect for nature is good for man, nature constitutes a source of values for us.

A second kind of instrumental approach toward nature—more prominent today—does not concern itself with the way in which the natures of particular things are instrumentally valuable but instead argues that nature as a whole—the entire natural universe—must be respected and preserved because humanity grew up in this natural world and will be spiritually and physically impoverished in a world of merely manmade things. Our stomachs cope better with "natural foods"; our bodies respond better to "natural" remedies. We need access to fresh air, sunlight, and open space; we need parks and beaches and wilderness to refresh ourselves physically and spiritually. We need the companionship of our fellow creatures to remain human and have an obligation to see that "no creature should pass from the earth through the instrumentality of man."[47] This point of view is well summarized in a recent statement that the "well-being of mankind is inescapably associated with a healthy, productive, and attractive environment. . . . With all his technological miracles, man is still basically an animal, with all the natural needs, reactions, and dependencies of an animal."[48] Or as Mexican President Luis Echeverria has remarked, "I am not being a romantic; man, for biological and spiritual reasons, needs his contacts with nature."[49]

This evaluation of nature as a vital instrument for human well-being is both valid in itself and necessary for the creation of a viable human future. As I will argue later in a discussion of the nature and ends of human society, contact with the natural, nonmanmade world is a basic human need, and the destruction of that world is a major element in our current social crisis. But to regard nature as being simply of instrumental value is not enough. In theory if not in practice, such an attitude is not wholly inconsistent with liberalism. It simply places fresh air and open space in the category of things which are useful and subservient to man—utilities to be apportioned and exploited through the liberal economic and political process. Nature is valued only insofar as it contributes to human social welfare. Indeed, this is the operational premise of most current discussions of environmental policy. It is akin to the attitude of those who view religion as a "good thing" for society—an instrument that is useful for promoting human betterment, a device for making people nicer and societies more stable.

Such an instrumental attitude toward nature—like the parallel attitude toward God—is necessarily largely self-defeating. Part of what human beings want from respect for and contact with nature is a sense of their own finiteness, of their kinship with other creatures and participation in a process larger than themselves—a touching of the deeper sources of their own being. Paradoxically, this instrumental value is realized only if people really believe that nature is something which exists and has value for its own sake, not merely for the sake of what contemplating or experiencing it does for them. Acceptance and awe are necessary ingredients of this self-regenerating experience of nature, and these require that nature be valued as something which exists for purposes other than man's.

Many who are devoted to nature as a value have learned to take pleasure in the simple fact of the existence of wilderness areas and living creatures they, and perhaps even other human beings, may never see. This is akin to wishing well people you will never meet, and in this context respect and love are the only proper terms. When ecologist Paul Shepard writes that "man is in the world and his ecology is the nature of that *inness*" and (following the philosopher Watts) that "we must affirm that the world is a being, a part of our own body," he is going beyond the level of science to that of religion.[50]

Foremost among those who have sought to develop an ethic of nature in modern America is the conservationist Aldo Leopold, of whom the poet and Trappist monk Thomas Merton could write, "Leopold brought into clear focus one of the most important moral discoveries of our time . . . the ecological conscience . . . centered in an awareness of man's true place as a dependent member of the biotic community."[51] And the language Leopold used was the language of veneration and religion: "Ability to see the cultural value of wilderness boils down, in the last analysis, to a question of intellectual humility."[52] His greatest ethical commandment is: "A thing is right when it tends to preserve the integrity, stability and beauty of the biotic community";[53] but, he warns, "We can be ethical only in relation to something we can see, feel, understand, love or otherwise have faith in."[54] The poet Gary Snyder, in discussing the need for a new society based on respect for nature, tells us that "The Goal of Revolution is Transformation."[55] Exactly. And such transformation means a collective, specieswide conversion to a new veneration for the total biotic community of which we are only a part.

It is vitally important to human identity and the restoration of human society that human beings learn to respect, preserve, and interact with nature without hostility or greed, and this in turn means that nature must become once more in some sense an object of piety. He who seeks his life shall lose it, and he who abandons it shall find it is one of the most profound paradoxes of Christian scripture. By the same token, only if we regard nature as valuable in itself can it come to be a source of value to humanity. A sense of oneness with nature and a realization that nature places limits on the possibilities of human history—both individual and social—can only become the basis for social structures and policies if we stop thinking of nature as something to be coerced, as in magic, but rather as something to be revered as in religion. To do so is not to downgrade our own humanity but to enlarge it, to become more fully human. To survive our current crisis, we must develop an "ecological psyche" that will make it possible for man to recognize that "he not only is *tied* to nature, but that he *is* nature."[56] Such a psyche would, for the first time in centuries, enable us to recognize what a wondrous thing it is to be human.

Nature as a Model of Society

If we wish to survive and preserve our identity as something apart from and superior to the technological mechanisms we have created, we must not only look to nature as a source of values and learn to venerate nature as an object of value, the field within which we exist; we must also remake our social structures and our social self-image in the image of nature. As the Marxists have long correctly held, theory and practice must be congruent, since one's means determine the ends one can reach. The natural environment cannot be preserved nor technology tamed by a society which is itself mechanistic.

What are the characteristics of nature which a natural society will emulate and reflect?

The universe is not a machine but a process. It is not composed of hard, discrete things or parts but of points, events which have existence or meaning not in themselves but only as they relate to one another in particular contexts. In this universe everything is related to everything else, every happening affects every other, yet the whole is always in some dynamic equilibrium. The ongoing process which is the life of the universe is not a deterministic one but contains basic elements of freedom. At every level of action there exists a freedom for lower levels which through their activity determine the properties of higher levels, yet do so within an overall pattern of order. Life, human life, and the human mind are emergent properties which have come into existence within the freely unfolding pattern of an evolutionary movement toward higher levels of order. All the elements of the universe are part of a process of mutual interaction which as a whole manifests certain regularities, and the character of the whole provides the impotence principles, the limits on what the component events can be. The continued existence of stars and species alike occurs within the framework of these limits on action which control not what will but what can take place. This universe—although every event in it is potentially free—does have a direction, movement in "time" toward an unknown goal.

The principles of order inherent in the universe are difficult for human beings to fathom, especially since all our ideas are conditioned by our own physical nature, and because we are involved in creating the "realities" we observe. Whatever integration exists at some level of order beyond human comprehension (a level which will always remain so because in order to understand the universe fully one would have to be the universe), we human beings are necessarily forced to think of the world as composed of several overlapping, and at times seemingly contradictory kinds of order, a "complementarity" in which everything is both "matter" and "spirit," both "determined" and "free."

If this is what the universe is like, one might suspect that this may be what human society, composed of beings who partake of the nature of the universe, is also like, despite the fact that we still think of society in mechanistic terms and therefore all too frequently try to make society conform to the outdated mechanical model we once had of the physical universe. Liberalism is outmoded not simply because it leads to a violation by society of the limits of physical nature, but because its image of society is analogous to the now-discredited mechanistic model of the universe. A postliberal society capable of dealing with the crises of growth, environment, and technological change must not simply reject the attitudes and policies of liberal society toward nature and the machine; it must in its social and political structure reflect the postliberal view of nature and the universe.

But before we can determine what such a society might be like, we must first look more carefully at the nature of man as a social animal, since any natural society must be in conformity with the nature of things not only at the cosmic level but at the level at which ordinary human beings live and die.

10

The Goodness of Man and
the Primacy of Politics

HUMANITY exists within and is part of the universe of nature. Human beings are composed of the same "world-stuff"—to use Julian Huxley's term—as are microbes and stars. But while the universe has overarching patterns of order, it is also a hierarchy of systems, each with its own internal laws. Human beings as individuals, and human society, which results from the interaction of those individuals, have their own natures which are the source both of regularities of behavior and ethical norms of conduct.

The Nature of Man

Man is a creature of the evolutionary process, whose body and brain evolved over time in accord with the forward movement of nature from homogeneity to complexity and from unconsciousness to self-awareness.

Biologically, *Homo sapiens* is one species, with a common genetic makeup and capable of interbreeding. Regardless of whether the progression from primate ancestors took place only once or at more than one time and place, all human beings today form a common interbred species. What are popularly called "races" do exist; there are different "genepools"—groups of human beings whose members interbreed with one another with greater frequency than with outsiders—with statistically relevant and empirically observable clusters of physical characteristics.[1] The extent to which particular mental, intellectual, and behavioral characteristics are genetically tied to particular physical differences is much less well established. The heredity-versus-environment controversy still rages today among scientists, and because of the complexity of the problem and the limited techniques available for its study, it is a dispute that will undoubtedly continue for generations to come. But one thing is incontestable. There is no pattern of culture that cannot be learned and practiced by any human individual or group of individuals (obvious physical abnormalities aside) of whatever genetic background.

Human beings are far better able to adapt to different environments than any other species, and this flexibility is a key element in humanity's ability to survive and prosper in a wide variety of ecological contexts. Human beings perceive their environment (and thus in a sense create it) through the mediation of culture, especially language and related forms of symbolic behavior, and not all human groups perceive reality in the same fashion. Anthropologists have compiled a vast store of information about the ways in which human cultures differ. But there are limits to human malleability. Every creature sees and creates the world differently. But, at some level, reality is more than a construct of individual perception. Neither human beings nor any other species could have survived if their views of the world—however partial or skewed—did not in some way conform to the world's underlying patterns.[2] Similarly, within the biosphere of earth in which humanity came into being and still lives, there are patterns of reality, including those which constitute the physical/mental makeup of humanity itself, to which differing human cultures must relate successfully if they are to endure.

Human nature consists of the common elements of our species's genetic inheritance and potentialities. It sets definite limits to how human individuals and groups can behave and still survive. Despite the differences among human cultures, there are isomorphic patterns of behavior common to every culture. Answers to common problems may differ from culture to culture, but the questions are always the same, and there are limits (impotence principles) to the kinds of answers that will work. Human nature is the same for all members of humanity, regardless of race, sex, or culture.[3]

The Essential Goodness of Man

Since human nature is a product of universal nature, it is by definition good. If we regard life as valuable, the question of whether man is good or evil is meaningless. But because so much attention has been devoted by political and other philosophers to this question, and because the assumption that man is in some sense evil has become a cornerstone of liberalism and a barrier to humanity's coming to grips with its problems as steward of nature and master of technology, it is necessary to say something more about it.

What does it mean to say that man is good or evil? A whole treatise could be devoted to exploring what we mean by the word *good*, to trying to find the common denominator of meaning implied when we speak of a good meal, a good day, a good novel, a good President, or a good woman or man. Some philosophers would argue that when we say something is good we are simply expressing an unreasoned personal preference, as in the case of the man who says he doesn't know anything about art but does know "what he likes." But usually the discussion of the goodness of man in a political or social context has revolved around the presence or absence of two characteristics—rationality and good will—which are capable of being defined and perceived without undue difficulty. We can usually tell the difference between a rational decision and lunacy, and between kindness and cruelty.[4]

Historically, the belief that human beings are neither rational nor benevolent has several sources. One is common sense. We are all aware that, as humans have observed since the dawn of recorded history, some people are stupid and allow themselves to be guided by ignorance and emotion. We are also equally aware that men and women are often selfish and cruel. In the West, Christian theology bulwarked this common-sense awareness with the doctrine of original sin. Since humanity was the direct creation of a God defined as good, humanity, like creation as a whole, was also good. Adam and Eve were not only considered to have been without sin but were assumed to have been perfectly rational, gaining knowledge without effort and so in control of their passions that medieval theologians speculated as to whether and how they could derive pleasure from sexual activity. But the doctrine also held that they had free will: the capacity, which other animals lacked, of being able to choose evil rather than good. And, according to the myth, they chose evil, defying God by eating the fruit of the Tree of Good and Evil, the symbolic representation of the human ability to choose. The result of this original sin was not only shame, pain and death, but a darkening of the intellect and a weakening of the will, and therefore a propensity to choose evil over good, a propensity which could only be overcome by the bestowal and acceptance of the gift of divine grace.

Common sense and this theological tradition gave powerful support to an attitude of skepticism toward human rationality and goodness. But a countertradition which stressed the equally observable fact that women and men are in fact capable of a high degree of rationality and especially of unselfish behavior also persisted throughout western history and played a particularly prominent role in the optimism of the Enlightenment. It was not the essential nature of man, many philosophers held, but the nature of human society which corrupted man's basic goodness and which was the cause of evil. If we would stop treating man as though he were evil, he would be good. This tradition, traceable as far back as the theologian Pelagius, was expanded by the Romantic movement, and provided the basis for anarchism and for the Marxian view of man.

In the nineteenth and twentieth centuries a powerful new impetus was given to the pessimistic view of human nature. Colonialism and industrialism offered ample evidence of man's inhumanity to man, and technology radically amplified the scale on which it could be exercised. The culminations of this trend were the Nazi extermination camps and nuclear warfare. In addition, strong intellectual currents made it easier to accept evil as more normal to man than goodness. The Darwinian doctrine of evolution (especially as propounded by its popularizers) made the struggle for survival of the fittest (viewed as necessarily carried on through conflict and cruelty) the very source of progress and of rationality. Freud undermined the confidence in reason that had inspired the Enlightenment by demonstrating the importance of subconscious motivation; in addition, his doctrines postulated that society existed in order to repress libidinal drives which if left unchecked would make civilized human life impossible, thus undermining the Romantic belief that humanity could flourish best in a context of freedom.

In more recent times, Darwinian and Freudian themes have been combined and expanded on by popular writers on ethology (the science of animal behavior) and on prehistory who have argued that the roots of human behavior are primarily genetic rather than cultural, and/or that our animal heritage is a legacy of force, domination, and murder. If man is in essence a tool-making animal, some would argue that his first tools were used to kill not only other animals but his fellow humans. Civilization began in human bloodshed. Where Freud in effect revived the doctrine of original sin in his theory of the killing of the primal father through which guilt came into the world, Robert Ardrey makes the creation of technology the original sin.[5]

The political consequences of the belief that man is primarily evil rather than good have been somewhat paradoxical. Christian thinkers such as Augustine and atheists such as Hobbes drew the conclusion that human depravity makes strong government essential. Man's propensity to act stupidly and selfishly can be curbed only by the wisdom and power of a strong ruler. Society rests on the hangman, argued the eighteenth-century French conservative theorist Joseph de Maistre. Terror is required to hold terror in check. One valuable intellectual achievement of post-Hobbesian liberalism was to turn this argument on its head. If men are as evil as generally believed, then how can they be trusted to rule their fellows, since rulers too are human? Since angels do not govern men, as James Madison observed in *Federalist* no. 10, the scope and power of government must be limited. The creators of the American Constitution, accepting man's nature as evil (as evidenced especially in the wicked designs the greedy poor have on the property of the rich), sought to create a governmental structure which would use men's egoism to check that very egoism. Even Calhoun, who rejected the social-contract theory of liberalism and embraced an Aristotelian doctrine of man's inherent sociability, conceded that though men might be inherently charitable, charity unfortunately not only began but usually ended at home, and that therefore men, however good in theory, are evil in practice. Accordingly, he advocated a system of constitutional checks and balances even more mechanistic and complicated than that of the men of Philadelphia. To this day, lack of trust in human nature has remained a cardinal operational premise of liberalism, especially in the United States.

All of the arguments about the relationship of strong government and human nature are, however, double-edged: Just as either an all-powerful state or a liberal constitutional one can be derived from the belief that man is evil, so the contrary belief that men are good can be used to justify either strong government—since good men, either individuals or majorities, can be trusted to act wisely and beneficently—or to justify anarchism—since rulers are unnecessary because people are naturally good.

Nor is it easy to decide, regardless of what humanity's inherent propensities may be, whether the experience of governing makes people better or worse. "All power tends to corrupt," wrote the famous nineteenth-century English liberal, Lord Acton, "and absolute power tends to corrupt absolutely." On the contrary, argue two contemporary American political scientists, power enobles men, making good rulers out of the mediocre and less than honorable.[6] Thus, in deciding whether one is for or against strong government, one can easily ignore the controversy over the alleged inherently good or evil nature of humankind and ask the more pragmatic question: What kind of government is best suited to deal with the pressing problems of a particular era?

However justifiable such an approach may be, intellectual consistency and the exigencies of ideological debate require that the issue be faced more squarely. In fact, there is ample ground for holding that it makes more sense to think of the human individual as good—even leaving aside the judgment that existence in and of itself is good—than to think of her as evil. Common-sense observation, for instance, tells us that humans are rational and kind as well as irrational and cruel. There are no statistics available to tell us whether people are more one than the other.[7] But we should keep in mind that we tend to be more conscious of evil deeds (especially when directed against ourselves) than of good deeds. Many people complain that newspapers are always full of bad news, but a recent attempt to publish a paper with only good news failed financially—not, one suspects, simply because it gave an incomplete picture of the world but because it must have been terribly dull. What is unusual, or perceived as unusual, attracts our attention: "Man bites dog" is still a better headline than the reverse.

The Roots of Aggression

There is currently much learned controversy over whether aggression is an innate characteristic of human beings, deriving from genetically controlled drives, or whether it is social in origin.[8] It seems highly probable that it is both—with innate potentialities socially triggered in particular situations. But when people argue that human beings are inherently aggressive, what they are usually saying is that people are normally and necessarily so. There is much merit in the argument of the philosopher Mortimer Adler that if aggression was a basic human trait, its expression would be as necessary to human life and hence as universal as rationality and sociability, and in fact this is not so; it can be repressed to an extent that other human characteristics cannot.[9] It cannot therefore be considered a basic human characteristic.

Similarly, the proponents of the doctrine of original sin cannot—and do not—deny that people can choose good rather than evil. The argument is about the prerequisites, supernatural or natural, for such a choice, and whether the ability to choose good can be conceived of as increasing over time for humanity as a whole. A powerful argument can be made that this ability is increasing. Rationality—despite the irrational characteristics of much of contemporary society—is probably more common in the developed world today than it has been in most primitive cultures or in earlier eras in history. Are we more beneficent now than in the past? Conservatives can point to Hitler, Stalin, or—if their partisan political orientation permits—current nuclear warriors as examples to the contrary. But we have eliminated slavery; the developed world generally no longer takes for granted the necessity of famine, child abuse, or even cruelty to animals; and a rising demand for equality all over the globe may be some indication that we are growing more conscious of human dignity, even if we have not yet solved all of the problems of human exploitation.

More recent challenges to the essential goodness of humanity can be faulted on their own ground.

When Engels wrote angrily that the Darwinian concept of evolution through competition was not a description of nature but simply a projection of bourgeois liberal social conditions upon the biological universe, he was close to the truth.[10] As a whole host of students of biology have argued since the time of the great Russian scientist and anarchist social philosopher, Peter Kropotkin, cooperation, both between and especially within species, has played as vital a role in ecological balance and evolutionary development as has competition. Humanity has achieved whatever success it has had in evolutionary terms not by competing with its own members but through cooperative activities, directed against other species but also against inanimate nature. That this competition against the non-human has, in our own time especially, gotten out of hand does not change the fact that human society is based primarily not on intraspecific but on extraspecific aggression.

Does the existence of the biologically conditioned drives discovered and expounded on by Freud mean that human society is necessarily repressive and must build aggression and domination into its very structure? Recent attempts to revise Freud—especially to marry the Freudian and Marxist perspectives—argue that this is not so.[11] But one can also go beyond this position and argue positively that other psychological theories give a clearer and more valid picture of human nature than that provided by Freud and his followers.

Is man merely a "naked ape" and society a "human zoo" in which only a thin artificial veneer of civilization overlies our essentially aggressive, domineering, and nasty animal nature?[12] The vast literature on this subject can be read in several ways, but there seems to be no clear warrant for agreeing that this is so. Human nature is based on genetically conditioned potentialities, but cultural evolution can and has taken place, and man's cultural characteristics are as real as his biological ones. Analogies between human and animal (especially primate) society are suggestive and useful, but there are severe limits on the extent to which they can be accepted as telling us much about human nature; and to argue that man is "nothing but" an ape wearing clothes is just as foolish as to think of him as "nothing but" a machine or a structure of protein molecules.

Actually the scientific validity of most attempts to derive human social behavior and structure from that of animals is highly dubious, based often on tendentiously selected data from animal societies or on observations of the behavior of animals in a captive rather than natural setting. One recent popular book, for instance, is based largely on comparing man with baboons rather than with his genetically and historically more closely related cousins among the primates, the apes, even though apes and baboons have been separate species for more than thirty million years, while humanoids developed from apes only two to four million years ago.[13] As a general principle, one must agree with Dobzhansky when he says, "It is as wrong to explain human affairs entirely by biology as to suppose that biology has no bearing on human affairs.[14] Men certainly can behave like apes or (as Hobbes argued long ago) like wolves to each other. But they need not do so. Indeed, most comparisons of men with apes or wolves are insults not only to man but to apes and wolves as well.

The argument that man is inherently evil ignores the fact that man can choose between good and evil. As long as that choice is possible it makes no sense to argue that he is necessarily evil. But is he then necessarily good? Of course not, but he is capable of being good.

Knowledge, Reason, and Virtue

How does a human being become good? Primarily through the use of reason. If it is possible to derive ethics from naturalistic premises, and if it is possible to define human life as good in itself, then whatever contributes to the maintenance of that life and the fulfillment of its potentialities must also be good. Discovering what in fact contributes to the survival and development of mankind is the task of human reason. Plato was basically right: Knowledge and virtue are the counterparts of one another. Good consists in the development of human potentiality, a potentiality which can be discovered by the use of reason, including what we know or can learn through the empirical sciences.

But, the classic objection to this essentially Aristotelian definition argues, do not human beings have potentialities for evil as well as for good? Cannot the skills and drives which make an individual a noble ruler also make him a master criminal? The answer to this objection is as old as the objection itself. All men, as Aristotle and his followers argued, seek the same thing: happiness. But some men seek it by means which are (in modern terminology) counterproductive. That there can be emotional satisfaction in being a tyrant, a murderer, or even, however momentarily, a suicide, cannot be denied. The drug addict feels a surge of relief as the needle enters his veins. But such satisfaction is temporary and illusory. There are criteria for happiness, which are objective and rooted in the universals of human nature. Health is a definable quality for both individuals and societies. Drug addicts sometimes want to be cured, the mentally disturbed often seek therapy, societies seek to reform themselves. Even when they do not do so voluntarily (and this is a crucial test), they are usually grateful for help after it has been forced upon them.

That individuals and societies often persist in their follies is an undeniable fact of life. But the first principle of any ethical system—to choose good and avoid evil—is an unarguable premise. It is possible to demonstrate that certain courses of action lead to death or illness, but unless a choice is first made in favor of life or health the argument is useless. All ethical argument assumes that rational living creatures will prefer life to death, health to illness. The evolutionary role of reason is to help the species survive and prosper. It may be necessary for an individual to choose death or illness for the benefit of others and so, indirectly, for the benefit of the species as a whole. Animals defend their young at their own peril. But that does not disprove—indeed it reinforces—the proposition that survival and the ability to function normally are in themselves desirable. The role of reason therefore is not to validate the choice of life over death. The choice is inherent in animal nature—including human nature—and an intrinsically undebatable premise of discussion. Reason can only help us determine what is conducive to life, functioning, and happiness.

But even if reason leads to virtue, this does not prove that virtue will inevitably triumph unless reason itself does so. Is human rationality greater than observable human irrationality, and is it increasing or decreasing? The pessimists will argue that irrational people have always been with us and always will be, and that they are a majority of mankind. There is no way of refuting the latter point directly. We cannot set up tests of rationality and administer them to all the billions of human beings generation after generation. But even if we assume that there probably has been no great (if any) increase in the innate mental capacity of women and men over the centuries, taken as individuals, one can argue that collective reason can be and has been increased through the growth and diffusion of scientific knowledge. If, for instance, we assume that medicine has progressed in the last two thousand years, we can now act more rationally to preserve life and health than in the past, and an individual who has access to that knowledge is, in an objective sense, regardless of his personal I.Q., more rational. And if, as we have argued, health is ordinarily a virtue, an ethical norm derived from nature, he is objectively more virtuous as well.

The greatest single objection to the possibility of human perfection is based on the observably high incidence—almost the universality—of selfishness. It is our egoism, our need to compete with others, which leads to domination, exploitation, and war. Yet what do we compete about? Human beings always have been, and perhaps always will be to some extent, concerned with status of various kinds and with the possession of certain unique goods—marrying a particular person, living in a certain house, being the world's acknowledged chess champion. But most competition is over scarce goods of a broadly defined nature: food, clothing, shelter, and the possessions that money can buy. Insofar as scarcity can be eliminated, either through increasing the availability of these goods in proportion to those competing for them or changing people's perceptions of how much is necessary, most human conflict can be eliminated or at least mitigated.

Money may not be the root of all evil, but it is of most. Even various forms of status are often valued because they lead to or reinforce or reflect economic power. To date technology has eliminated scarcity of the necessities of life for much of mankind and might do so for all of mankind if population growth could be halted. If this can be done and, concurrently, the drive to possess more than is necessary can be muted, human selfishness can be radically decreased. Egotism will remain, but it need not mean bitter conflict between nations, races, classes, or individuals over physical goods. As crime rates and common observation indicates, poverty and most virtues make poor companions, while contrary to many versions of Christian teaching, being rich is not necessarily the high road to moral turpitude (though becoming rich is of course a different matter.)

The increase in human rationality and the possible concomitant increase in human goodness is an illustration of a basic characteristic of human nature: It changes. Man is a progressive animal. The old saying, "You can't change human nature," is at best a half truth. Human nature grew out of primate nature, and is still developing, primarily under the influence of changes in human culture.[15]

There has apparently been little change in man's genetic makeup since the time of Cro-Magnon man at least. Physical evolution is still taking place in mankind, but almost imperceptibly.[16] Some biologists would hold that man is even degenerating genetically, as social welfare policies and modern medicine make possible the survival and reproduction of individuals who in ages of harsher competition would have perished without offspring. The evidence for this proposition is hardly overwhelming, however, and becomes difficult to evaluate when we consider that welfare policies and modern medicine are part of the environment against which fitness to survive and reproduce must be tested. But in any case if a child conceived by human parents of twenty-thousand years ago were to be born today, he would be indistinguishable from his fellow humans by his genetic heritage and would be fully as capable of absorbing any contemporary culture as the average child of today.

Some changes in human physical and physically conditioned mental characteristics have occurred. Because of better nutrition and medical care most citizens of developed nations are dramatically taller and heavier than their grandparents. On the whole they live longer (though that trend may be declining as a result of environmental pollution). One sees fewer ugly people around as obesity, malformations, and bad teeth are gradually eliminated. Because of better diet prior to birth and in childhood, we may even be becoming more intelligent than our ancestors. But physically we are still the same species.

Human evolution is, however, cultural as well as physical—in recent millennia, overwhelmingly so. Culture both reflects and alters human mental and psychological processes. Urbanization, electronic communications, and dependence on advanced technology are such relatively recent phenomena in human history that it is as yet impossible to assess their full impact on man's cultural nature; but that it is changing there can be no doubt. It is all too likely that current trends separating man from nature and assimilating him to the machine may soon lead to his physical and mental degeneration, taking our species up an evolutionary blind alley. This possibility is especially ironic in light of the conclusion that humanity for the past several thousand years has been gaining power over its world and increasing its opportunities for freedom, rationality, and goodness.

The Social Basis of Human Nature

Human beings are individuals in the simplest physical sense. Not only does each individual (save in the case of identical twins) have a unique genetic heritage, giving her or him a set of unique physical characteristics, but this pattern of uniqueness unites all the cells of each body.[17] One obstacle to successful organ transplants is that the body knows its own and rejects "alien" organic elements added to it by surgery. Individuality is rooted in biology.

We differ from one another in our perceptions as well, even at the physical level. Not only are some individuals color-blind, but human beings often react to different tastes and temperatures differently. We have differing tolerances for drugs and various kinds of external pressures on our physical bodies. Our psychological traits and the contents of our memories differ. Even identical twins reared together cannot go through life reading exactly the same books, seeing all the same things (certainly not from exactly the same angle) or having the same conversations with the same people. No one is exactly like anyone else.

But although we are different as individuals, we are not isolated closed systems in the universe of nature. We have been produced by and can only live within a larger whole, and everything which constitutes the unique pattern of our individuality has come from somewhere outside of ourselves. All of it ultimately derives from the larger web of nature in which we are enmeshed: the air we breathe, the food we eat, the sunlight we absorb, the people we interact with. We are all, as individuals, part of a larger ecological system. But most of what goes into making us the individuals we are is mediated through society. Our genetic constitutions do not come from nature as a whole but from our parents; and our physical and mental diet comes primarily through society. We have no more meaning as isolated units of our species than an individual electron has in isolation. Even Robinson Crusoe, the great Enlightenment figure of the isolated man, was the product of childhood nurture and had socially acquired skills without which he could not have survived on his desert island.

Not only our existence, but our very identity, is determined by our relationship with others. We can only come to a realization of our selfhood through interaction with others; and we can only define our individuality through an awareness of the ways in which we are like or different from others. We are points in a physical and social process. In the words of the physicist Erwin Schrodinger, " 'I' is the canvass upon which the collection of data are collected."[18] This is the principal lesson which modern science has for political and social philosophy.

Liberalism with its premise of the primary reality of the individual and the derivative nature of society was compatible with the worldview of

Newtonian physics. A worldview based on a concept of nature as energy in process demands a very different view of the relationship of the individual and society, one in which the individual is defined by society as much as society is defined by relationships among individuals. The knowledge that man is a social being by nature is not a modern discovery, of course, but part of the traditional wisdom of all civilizations. "A natural impulse is thus one reason why men desire to live a social life even when they stand in no need of mutual succour; but they are also drawn together by a common interest," wrote Aristotle over two thousand years ago.[19] The denial of man's intrinsically social nature in the heyday of liberalism was simply an historical aberration, and modern views of the physical universe now make that denial obviously intellectually untenable.

Man is essentially and necessarily a social animal. The rejection of this truth by liberal society is the root cause of most of our present difficulties. The problem of the human relation to nature and the environment cannot be solved by individuals as individuals. For example, only in the most limited sense do individuals create the air pollution which affects them as individuals. Society creates air pollution by the kinds of technologies and activities it fosters or permits. The individual cannot by himself do anything meaningful to clear the public air. He can try to escape from it (as many Americans do) through air-conditioning, at the cost of increasing urban heat levels, thereby making it worse for himself when he steps outside of his private shell as well as worse for others. But no one person can do anything significant to preserve wilderness, decrease traffic congestion, or keep dangerous substances out of the food the community eats or the air it breathes.

Nor can any individual as such control the nature of his relationship with the machine. The individual cannot live and function in contemporary society without using automobiles or telephones, without being aware of what is being transmitted by the mass media, without becoming a number in a computerized filing system, without working in or dealing with vast bureaucracies in business, government, education, or medical care. It is possible for individuals or minutely small groups to "drop out" of society and create their own desert islands like modern Crusoes, but they cannot as isolated individuals function within or affect the future of society. The image of man as an independent individual, dealing as an individual with problems created by himself or by other individuals acting as such, is completely foreign to reality. Clinging to this liberal notion makes the problems which beset societies impossible to solve.

Man's social nature is rooted in his biology. Human beings have the longest period of infant dependency of any animal. They must be fed, cared for, and protected for years before they even begin to be capable of fending for themselves, and during this period of dependency they require a high degree of personal contact with others. Historically this genetically determined aspect of human nature gave rise to the family, which provided sustenance, protection, and care to the young and those engaged in caring for them. Even if the functions historically associated with the family (a term which, of course, covers a wide variety of social forms) were to be filled by other means, whatever arrangements were made would still require intensive as well as comprehensive and long-lasting social interaction.[20]

But even adult human beings cannot operate effectively in isolation. Primitive man hunted in groups. Agriculture, especially where irrigation or the protection of fields against flooding is required, demands a high level of cooperative activity. Even the American frontier farmer—perhaps the most individualistic agriculturist the world has ever known—often enlisted the help of his neighbors to clear fields or to raise barns, as well as to worship God. From the onset of recorded human history people have lived in groups where the existence of specialized craftsmen necessarily involved some form of economic exchange. Whether struggling against or peacefully cooperating with nature, man has acted through organized groups from the dawn of history. Most animals are social to some degree, but man is especially bound to his fellows.

What was a matter of natural impulse and convenience in earlier eras has become an absolute necessity in our own. Primitive man might be able to survive as an individual or in a small family group tilling his own plot of land or hunting his own quarry with his self-made tools. Few can live that way today even if they wish to. If the billions of human beings alive today were to spread evenly over the earth's land-surface, each with his own little plot of land, without trade or industry, not only would their standard of living be reduced to the barest minimum capable of sustaining life, but most of the world's population would in fact starve to death, for there would not be enough land to go around which would be suitable for producing even at the subsistence level with the primitive tools that would be available.

Those of us who live in urban areas in developed nations are especially dependent on others for the production of what we consume and use. But we are hardly unique in this. Our species as a whole is locked into the web of mass technological society, increasingly so every day. The green revolution, touted as preventing or at least postponing famine in much of Asia, involves as a basic premise converting hundreds of millions of peasants into farmers dependent on world technology; and every advance in the battle against world hunger makes more and more humans dependent on the machine and on the social organization that produces and distributes its products.[21]

That individuals are dependent on others is such a truism that it hardly seems worth repeating, save for the fact that we often act as if it were not so. We still speak of self-made men—as if anyone ever existed without parents or teachers, to say nothing of fellow workers, employees, clients and customers. We still resent society taking "our" money through taxes—as though our incomes were part of our genetic endowment or the result of our physiological processes instead of being created by our role in a socially structured economy within which it is socially decided how much physicians earn in relation to garbage collectors or nursery school teachers.

What is less a truism is the equally important fact that just as we are dependent on others those others are dependent on us. We know our breakfast coffee comes from some distant tropical country, and we would be painfully aware of any conditions in the country of its origin which caused its price to rise sharply. But we rarely reflect upon the impact of changes in our consumption patterns on the economies of other nations. The increased use of instant coffee, for instance, has meant a shift of demand from Latin American toward African types, resulting in increased economic benefit to the remaining European colonial interests.

If most members of urban societies are only vaguely aware of the processes which sustain them, they are even less aware of the impact of their activities on other individuals and other societies. If we walk past a large building on a hot day we may be momentarily and unpleasantly aware of the blast of heat which its air-conditioning system is pouring into the city street, but we have little knowledge of what happens to our own garbage after it leaves our dwellings. The social nature of man, especially in modern technological society, means not only that his life is the summed-up effect of a myriad of causes external to him but that his own actions have a host of consequences for others near and far.[22]

But to be affected by someone else's actions is to be controlled by that person; and to affect someone else's actions is to exercise control over that person. No consideration is more central to understanding the nature of social relationships, especially in large-scale societies. We are accustomed to think of control as a one-to-one relationship—usually conscious, usually complete, usually involving person-to-person contact. An individual parent controls—or could control—the movies his children see, at least until they are a certain age. A driver controls the car she is driving. Society controls by explicit law the hours that bars can legally serve liquor. This kind of control we easily understand. But we find it difficult to recognize that motion-picture producers and distributors control the pictures our children will see if they are to see any at all. The new shopping center which fronts on the road between our home and our office controls how long it takes us to drive to work, perhaps forcing us to leave the house ten minutes earlier. Architects and bankers and construction unions control where we can live and what kind of home we can live in. Similarly we exercise control over the lives of others, even if we are not aware of their existence.

To be dependent on another is to be controlled to some degree by that other. Therefore to be mutually interdependent is to be engaged in a process of mutual control. I will return to this point later, when I define the meaning of freedom in a social and political context, but it is necessary to emphasize here that all social relationships are in a sense coercive. Talking about coercion, or force, is simply another way of talking about the application of energy in certain processes. Just as "objects" in nature are defined by their relationship to each other, so society consists of human beings— themselves complexes of energy—engaged in certain processes. What any particular social "objects," including individual persons, are is defined by their relationship to others and in turn defines these others.

Society and the Self

This concept of the social definition of the self is even harder for most people to grasp than the concept of the necessary interdependence of the individual and society. We think of ourselves as exactly that—as "selves." Yet as the American social psychologist George Herbert Mead pointed out over a generation ago, "The 'self,' as that which can be an object to itself, is essentially a social structure and it arises in social experience."[23] What we are—what constitutes our identity, our "self"—is the unique combination of the roles we play in relation to others.

One cannot wholly accept the position that there is no aspect of the individual self not defined by others, as Mead seems to assert and as the philosopher Alan Watts argues in ontological terms.[24] Our very consciousness of our existence is somehow something apart from the specific content of that existence (just as the ability to acquire language is something apart from and precedent to learning to communicate verbally). It is this consciousness of our actions and of society's reactions to them which constitutes what we call the self. But this does not mean that the content of our existence, including what is stored in our brain's memory cells, is not wholly derived from society, a product (even in our emotional response to events) of the interaction of self and society. Where else could the content of selfhood come from if not from society?

The content of our minds—our accumulated, recombined, and structured experience—can only be composed of socially derived and culturally mediated perceptions, or of direct individual perceptions of physical nature outside ourselves, or of psychic phenomena within ourselves. But even these latter are essentially socially conditioned. We look at a forest differently from the way a primitive hunter does. Even the visions that we create in dreams or under the influence of drugs or even perhaps as a result of mystical illumination can only be composed of images from our experience in a particular natural and social environment. We are not constituted in such a way as to "see" anything else. This is not to say that we may not create a pattern of our own from the material our environment—physical and social—gives us to work with; this is what individuality and unique-ness and even genius mean. We need not—indeed could not if we wished —be exactly like anyone else, much less a simple replica of some fore-ordained generalized social pattern. But our individuality exists only within and because of our social environment.

Why do we so often fail to realize this and think of our individuality as somehow more primary than our social nature and as self-generated rather than part of a larger social pattern? Here is the greatest irony of all. We think we are first and foremost individuals rather than social beings be-cause our society keeps telling us so. But *our belief in individualism is it-self a social product*, an ideology false to reality, which is the result of the very element of the human condition which it seeks to deny. By denying the extent of its control over us, society not only lies to us, it shields its own powers behind that lie.

Were we to realize the extent to which society shapes us, we could use our consciousness of this fact to make the process more fully reciprocal—less one-sided—and thus do more to shape our society and hence to shape ourselves. If I realize the extent to which watching trash on TV is brutaliz-ing or trivializing my mind, I can alter my TV diet, even if my aspirations for a better mind have a social origin. I can even ask that I be given better mental food to feed on, despite the fact that I can never be self-feeding. But failing to recognize that society shapes us as well as being created by us, we are like buyers who keep paying higher and higher prices without realizing that we are the sellers' only customer and could bring his prices down any time we chose, since he is as dependent on us as we are on him.

But whatever the character of our interrelationship with other men—whether it is consciously accepted and therefore capable of being made an instrument of reason and self-direction, or whether it is denied and we allow ourselves to be buffeted about by social forces outside our ken and therefore outside our control—one fact is inescapable: We are social crea-tures by nature. The free human individual living in a state of nature is a myth. The nearest approximation to the kind of individual postulated by a truly consistent liberal philosophy would be the "wolf-children" allegedly found from time to time, human children abandoned in infancy and nur-tured by other animals—snarling, autistic, dumb, covered with sores and lice, crawling on all fours. They are the only possible authors of a social contract entered into under the circumstances postulated by a consistently individualistic political theory. They may yet be its progeny.[25]

Human Nature and the Primacy of Politics

But man is more than just a social animal; he is also a political animal. Classical political philosophy in the West never made a basic distinction between the social and political orders, nor did ancient society.[26] Society consists of the interaction of human beings, an interaction based both on choice and on compulsion. Modern political theory tends to stress the identity between politics and compulsion, as in Max Weber's definition of the state as "a compulsory association with a territorial basis" having a "monopoly of the legitimate use of force,"[27] or in David Easton's definition of politics as the "authoritative allocation of values."[28] These definitions reflect the premises of liberalism, and reinforce our tendency to think of the political state as an organ of compulsion and of society as the area of non-compelled relationships, a distinction which is fundamental to liberalism.

But the difference between compulsion and consent is in fact arbitrary, a matter of degree rather than kind, for in nature all relationships are based on flows of energy, which is another name for force. Though he somewhat overstates the case, B. F. Skinner is essentially correct in arguing that all our actions are controlled by the world around us in some fashion, although we think of ourselves as free when the controls are not immediately evident or are internalized.[29] Which controls are internalized depends of course on our education in the broadest sense, including childhood socialization, formal education, and life experiences. Plato's *Republic* has often been discussed as a treatise on education (as indeed it is) because his ideal state was to be held together by socially promulgated myth and socially controlled childhood training rather than by force.[30] But all states are held together mainly by consensus (myth) rather than force; regimes based on naked force are short-lived. Aristotle spoke of politics as the "architectonic" science in part because the political order determined the educational order: The state shaped the myths which in turn made the state possible. For Aristotle, ethics was subordinate to politics because the polity already embodied and sought to reinforce ethical norms.

The early Christian period and the Middle Ages witnessed a divorce between society and the state because this era rejected the identity between ethics and politics. Since politics dealt only with man's earthly good and since man also possessed a supernatural destiny, there were aspects of human interrelationships to which politics was not central. The task of the ruler was to maintain law and order; God sent princes to punish the wicked. Augustine is the author of the prototype of this view in his separation of the City of God and the City of Man, and his model influences Christian (especially Protestant) thinkers even today.

In the late Middle Ages the Greek view of the world reentered the mainstream of Christian thought, and philosophers such as Thomas Aquinas could speak of the state as a "perfect" society in that it encompassed all others, a society with ends of its own to which even Christians owed allegiance. This attempt to promote the classical unity of society and the state, while maintaining the distinction between man's earthly and supernatural ends, entailed all sorts of difficulties both intellectual and practical, which need not concern us here, since the antipolitical bias of Christianity triumphed and became an important precondition of liberalism.

Liberal theorists have never been completely clear about whether the social and political orders were essentially the same. They could not be, since Locke and his followers wished to limit the power the king could exercise over the emerging bourgeois economic order. Yet these philosophers could not consistently defend a separation which was intrinsically absurd. Hobbes, a friend of royal absolutism, had no compunctions about postulating a single contract which created both society and state. Locke waffled on this point. His prepolitical state of nature in effect implied that society existed before the state came into being. His theory of revolution presumed that society could survive the state's destruction—and indeed be its judge—although common experience told him and his readers that even ordinary social relationships involve some degree of authority and compulsion.

But the fact that man is a political as well as a social animal—that indeed the distinction between the two is largely meaningless and often misleading—is of vital importance. If a good society is one in which man's needs are fulfilled, it is only through politics that this fulfillment is possible. Politics is not an end in itself, as some liberal thinkers would argue.[31] It exists to enable human beings to live and thrive. Politics is "all activity addressing itself toward the solution and alleviation of social problems."[32]

This does not mean that the specialized apparatus of political direction which we call a state, a regime, a government, or an administration, has the normal and necessary role of fulfilling all of its citizens' substantive needs. However, if the other systems of social interaction are not fulfilling these needs adequately, then the political system must, sometimes by compulsion, ensure that they are met or satisfy them itself directly. The naturalness and primacy of the political order is independent of the extent to which human needs, individual or collective, are met directly by the specialized political organs of society. But it does have the responsibility for ensuring that they are met and determining who shall meet them.

Politics is that form of social interaction which proceeds through conscious decision-making to determine how human beings shall interact with each other and, therefore, how they will fulfill their needs. All decisions in society are ultimately political decisions. Politics is what gives the social system its shape and when necessary maintains that shape by force; it orders the patterns of the system. To allow parents to beat their children, or to permit conditions which make it possible for even the children of loving parents to starve, is as much a political decision as the decision to prevent such conditions. This has always been the case, even though in societies in the past lack of knowledge and power may have restricted the ability of the political order to implement decisions which it made or would have made had the means to enforce the decisions been available. The growth of technology has vastly increased the power of the political order, and there are virtually no limits to what a modern state can do, except those based on lack of information or inadequate information-processing or on an inability to enlist the support of its citizenry.

Ideas shape political action and culture shapes ideas. But if ideas are to change society, they must become politically incarnate. A future society in which man and nature are reconciled and technology controlled depends on the spread of a new set of ideas about politics and the incarnation of this new political philosophy in political action.

11

Beyond Liberalism to Freedom

LIBERALISM is an outgrowth of the opposition of the rising bourgeoisie of the seventeenth century to the existing power structure in church and state. Liberalism's current hold over the political thought of the modern world derives from the conviction that liberalism and liberty are synonymous, and that the end of liberalism will mean the end of human freedom.

Freedom is the most widely used word in the lexicon of politics, especially in the western world. Yet no term in political discourse has been the subject of more confusion and disagreement nor is freighted with more emotional overtones.[1] Much of the difficulty we have in understanding what freedom really is is the result of fundamental errors in the liberal concept of freedom; and these errors are in large part responsible for the contemporary situation in which the human individual is increasingly enslaved within a society which claims to be based on the value of freedom.

Fundamental to the liberal misunderstanding of the nature of freedom is the insistence that freedom is an end in itself. Freedom is necessary for a fully human life—both at the individual level and that of society as a whole—just as breathing is necessary to life. It is a mode of activity, not something to be sought for its own sake. To make freedom as such the goal of life, individual or collective, is necessarily as unproductive as making happiness a goal in itself. "If life is experienced as concrete and meaningful, freedom will be part of it by being absent as a problem."[2]

Most people rightly think of freedom, or particular freedoms, especially political ones, as instrumental, as means to something else. We want to be free to go somewhere, to do something, to create a special kind of future; and we value freedoms, including political freedoms, because they help us get where we are trying to go. We are happy and free only in the course of fulfilling our other needs.

What we desire (the kinds of future states we aspire to create) depends, of course, on our identities: who we think we are and what we think we have the potentiality of becoming. Despite the implicit liberal assumption to the contrary, our identities are socially created. Yet we are no less free because of this. We cannot escape the fact that we are influenced by our desires and aversions, immediate and remembered, by our total environment and by our life histories. But these desires, aversions, environments, and life histories are what make up our identities; they are the base points from which we move into the future. They are the constituent elements of the self-conscious "I" which envisions and seeks to create the future. To be totally free from outside influences would be to cease to exist, for we are what we are: male or female, black or white, rich or poor, lovers of sport or of contemplation. Some of these conditions we can or may want to change; but a future in which they all disappeared would mean the end of ourselves as well. So freedom cannot be equated with an absence of limitations, for it is our limitations that give us our particular identities and our own special potentialities.

Freedom is perhaps best defined as *the ability to achieve a desired future state* or, in Kenneth Boulding's phrase, "the process whereby an image of the future is consciously realized."[3] Not only can freedom exist in an orderly universe and in an orderly society; it can *only* exist within an orderly universe and society. If we are to be able to create our own future, we must be able to choose from among alternative means those best suited for achieving our ends, and this is only possible in a context where the effects of our actions on people and other elements in our environment— and their reactions to our actions—can be predicted with some degree of certainty. In a chaotic world purposive action would be impossible, since we could never be in a position to assess what the outcome of any action might be. Supposed freedom of choice in such a situation would be nothing but an illusion.

An orderly universe is not thereby a predetermined universe. The laws of nature do not preclude freedom. The structure of each level of reality, however orderly it may appear as viewed from another level, is created by the free interaction of its constituent elements. The free movement of electrons is what makes up atoms; the activity of living creatures creates ecosystems. Each subsystem of the universe interacts freely with other subsystems, in this way constituting still more inclusive ordered levels of reality. Orders of reality such as the living human body and the human mind are the emergent properties of their constituent elements.

Within this physical universe, certain human characteristics have emerged—consciousness together with its concomitants, language and culture—which enable people to act in an especially and consciously nondeterministic fashion. Nature in creating man has created him capable of acting freely. People cannot do absolutely everything they wish at any given moment; their freedom of choice is necessarily exercised among the alternatives structured by the physical context of the universe and its laws: We cannot fly without making ourselves wings, we cannot drink poison without getting sick or dying. But we can make choices.

What is true of members of the human species within the order of the nonhuman universe is even more true of human beings within the social universe. We live in societies which have patterns that are the result of the free interaction of human beings over time. We ourselves as individuals have identities which are both aspects and outcomes of these patterns, and we act within the constraints imposed by our identities and by the social contexts in which we live. But we are free to choose among alternative courses of action within society, just as we are free to choose alternatives within the larger universe of nature. And in exercising our ability to choose we change the pattern of society, we create new futures not only for ourselves but for other individuals within society and for society as a whole.

Freedom and Power

In modern society our ability to exercise freedom—to participate in creating chosen futures—is weakened by the second great liberal error about the nature of freedom. Beginning with Hobbes, liberalism has identified freedom with absence of restraints imposed by other human beings. If a man is too sick to rise from his bed, he is still free in the Hobbesian view as long as no one is holding him down.[4] Social freedom, liberals would still argue today, is simply the absence of the control of one human being by another.[5] But freedom requires more than an absence of restraint. It requires the presence of power if it is to be meaningful.

Power is as difficult to define as freedom, but it can perhaps usefully be thought of as the ability to alter the state of other elements in the universe, living or nonliving, in some desired direction. Power, like freedom, is future-oriented. It is, as Bertrand Russell once defined it, "the production of intended effect."[6] Freedom from external coercion, the usual definition of political freedom, is meaningless if one is without the power to do whatever it is that one would do if coercion was removed. If the law permits me to build a million-dollar home, this has no meaning for me if I am penniless and fated to remain so. Freedom necessarily implies power. This conforms to the average person's common-sense notion of what freedom means, as embodied in our everyday language when we say we can't do something, regardless of whether we mean it is forbidden by the power of others or our own lack of power to do it.

Freedom and Interdependence

Just as freedom, contrary to the usual liberal conception, necessarily implies the power to effectuate goals, it also necessarily implies the control of human beings by one another and by their total environment. Freedom cannot mean absence of control. Whenever we act in response to an anticipated reaction of forces outside ourselves, we are in a sense controlled by them. This is an obvious fact which seems to scandalize some otherwise intelligent people when it is pointed out to them by behaviorists such as B. F. Skinner. But, after all, we are none of us isolated omnipotent beings, and we all live within a social context.

This "control" which society (and nature as well) imposes on us is not something alien forced upon us by outside forces: We are its product and its creator. It is an aspect of the total system of society and nature of which we are a part. "The system *is* the social control, it does not 'impose' a control. . . . The controls are nothing but the relations of mutual dependence."[7] In politics, for example, my vote for a particular candidate is a free action even if the structure of the political parties and the electoral system (created by the free actions of human beings in the past) sets limits to how my vote will be cast and to whom it can be given. Even in nature, systems are not fixed entities whose actions are strictly predetermined. Especially insofar as they are composed of living beings, systems evolve as the result of changing patterns of interaction among their constituent elements and between themselves and their environments. Human beings are constantly engaged in a process of controlling others and being controlled by them, controlling nature and being controlled by it.

Human beings are controlled not only by nature and one another but by the extent and character of the technological development of their societies. A society's technology is the product of human choices, but it also determines what choices are open to the members of the society. There is no question but that the inhabitants of rich, technologically developed societies are in many respects freer than those of poor and backward societies because, having more power at their disposal, they have more options open to them. Even a rich American industrialist's son in 1865 could not have chosen to be an airline pilot. At the same time such societies, both because of their social arrangements and the way they deal with their physical environment, may also effectively foreclose certain choices. If I live in Southern California, even if I am rich I am virtually immobile unless I am willing to use an automobile.

The problem of freedom in technological society is one of creating the optimum balance between gains and losses of freedom, making it possible to choose between being a farmer or an airline pilot and at the same time between using an automobile or other means of transport. The problem is one of adding to the sum total of human choice as much as possible rather than simply foreclosing some options in favor of others. Our freedom of choice is conditioned by the alternatives which society makes available to us. It may be impossible, for instance, for a person to work at a particular occupation, live in a certain kind of house, and also walk to work. Insofar as society, by commission or omission, is responsible for restricting our choices, to that extent it limits our freedom.

Maximizing freedom therefore requires both (1) increasing the number of choices open to us and (2) increasing our ability to determine what these choices shall be, that is, giving us more opportunity to choose how our choices will be structured. To the extent that the social and physical arrangements which establish what choices are available to us are under the control of others, to this extent our freedom is diminished—to this extent we are under the control of others just as much as if they could tell us which of the alternatives they have left us we must choose. As long as society by default permits technological and economic forces to have the power to structure our choices, we have lost our freedom to these forces. Political action may and in many cases will be necessary to return to society as a whole the power to control the structuring of choice, a power which now rests largely with technological or economic forces or those specially placed individuals able to influence them. The liberal philosophy which sanctions the legitimacy of the power of the forces that restrict our choices and which blocks political action designed to enlarge the scope of meaningful choice is an enemy not a friend of freedom.

Freedom and Coercion

Historically, liberalism was the product of a society in which there was a great deal of direct coercion of individuals by the state and its agents. Economic activities, religious practices, and the expression of political beliefs were subject to regulation, prohibition, and censorship, and people were highly conscious of these restrictions and of their source. Then as now, the knowledge that one's future was in thrall to the power of other human beings was especially galling, since it was clear that these others were simply men like themselves, no more rational, and subject to the same selfish passions. As a reaction to such control, the definition of freedom focused narrowly on the prevention of the coercion of individuals by other individuals.[8]

Insofar as freedom is limited by the ability of others to control our actions directly, by intervention before or punishment after the fact, any diminution of such coercion means an increase in freedom. But it would be a terrible mistake to confuse the reduction or elimination of personal coercion with freedom itself, as liberal political philosophy does. For, if freedom is the ability to realize chosen futures, it may in any given situation be menaced more by lack of power and knowledge and appropriate social structures than by direct coercion. It can be argued that in modern liberal society we have lost much more freedom to the forces of economic and technological determinism and to subtle social pressures than we have to the coercive powers of other individuals or of the state. Automobile manufacturers determine what kinds of cars we can buy. The food industry forces us to eat chickens injected with hormones. Corporate planners move our jobs to the West Coast. We can no longer purchase our favorite newspaper or magazine when it is forced out of business. Our favorite trout stream is turned into a noxious sewer by industry; a quiet residential street becomes a funnel for traffic to a new shopping center; our hard-won skills become obsolescent; smog kills the tree outside our window. As Paul Goodman has written, "Social and technological considerations determine behavior in every detail; the way they lay out the streets is the way we must walk."[9]

Moreover, and here lies the key to the role of coercion in technological society, interpersonal coercion may be the most effective means—in some cases the only effective means—of restoring and enlarging our freedom in general. One contemporary political theorist, in seeking to defend human freedom, argues that the only time coercion is justified is to prevent coercion.[10] But this is nonsense. Even as doctrinaire a nineteenth-century liberal as John Stuart Mill recognized that laws prohibiting the employment of children in factories or requiring that they be educated, although they involved the coercion of factory owners and potential workers, of children and their parents, did in fact enlarge the degree of human freedom in society by making self-realization (the creation of desired futures arising out of one's nature) possible for more human beings in the long run.[11]

But the premise on which Mill's liberalism was based was his assumption that actions affecting only oneself could be separated from those affecting others, that the first should be free of social control and only the latter should be the subject of public regulation. Such a distinction is unrealistic in any human society, but above all in an interdependent technological society such as our own. Every act has consequences, either directly or indirectly.[12] And as long as our actions have consequences for others—and they all do—they cannot be subject solely to our own volition without impinging upon the freedom of others. If I burn my garbage in a backyard incinerator, it infringes upon my neighbor's freedom to breathe clean air; if he guns his motorcycle, it infringes on my freedom to enjoy a quiet morning. If I am allowed to park my car at work it will contribute to making bus service economically unprofitable and deprive my neighbor of the freedom to take a bus to work if he wishes. If he is allowed to run all his electrical appliances at maximum power it can contribute to a brownout which deprives me of the freedom to keep my food unspoiled in my refrigerator. And so on, ad infinitum.

If we understand that all social interaction—including interaction with technological and economic forces created and influenced by other persons —is a process of mutual control, we can recognize that interpersonal coercion is only a special form of control exercised by human beings over other human beings, only one form of limitation by some on the possible futures of others. It is often unpleasant—both to those who are subjected to and those who administer it—and it is often inefficient as a means of social control.[13] But it is hardly unique as a menace to human freedom. Coercion which on balance allows more freedom—more choice of futures—to more people by coercing some, or by coercing all to some degree, can be an instrument for maximizing freedom rather than diminishing it. Lack of coercion—certainly lack of social rules which have coercion as their ultimate sanction—is no proper measure of a society's freedom.

Our present inability to recognize this fact is the result of long intellectual conditioning and is the root problem of freedom in liberal society. We identify freedom too narrowly—as the absence of overt, direct interpersonal coercion such as that embodied in the governmental apparatus of laws, courts, policemen, and jails. Thus liberal political leaders can convince people that less government—at least in the economic sphere—means more freedom when what it really means is giving the future of the average person not into his own keeping but into the hands of corporate executives, labor leaders, or the blind forces of economic and technological determinism.

For reasons which are at base psychological we are willing to let impersonal forces control our lives to a degree we would consider intolerable if the control were being exercised by identifiable persons or social institutions. If an individual or the flesh-and-blood employees of a government agency tell us we must accept unemployment or else move we resent it; if we are told by leading citizens and the press that "economic changes" make this necessary we docilely accept it. If automobile traffic makes it impossible for us to cross a nearby street with safety we accept it; but if an individual policeman forces us to use a crosswalk we resent it. Obedience to abstract economic processes and technological forces disturbs us less than obedience to our fellow human beings. In part this curious reaction stems from sheer pride. We are more willing to subordinate ourselves to things bigger than we are or alien to us, things with which we are not in competition for psychological status, than we are to other men and women.

There are some indications that this way of reacting to the various sources of limitations on our freedom may be changing. Today many persons—the young especially—resent the degree to which their lives are controlled by faceless bureaucracies and impersonal computers and are increasingly coming to realize that, although human beings may have passions and personal interests, they also have a capacity for understanding the feelings and responding to the needs of others in a way machines do not.

Freedom and Hope

We can influence other human beings more readily than we can influence the technological and economic forces which today set most of our choices. Only by acting in concert with our fellow human beings can we hope to affect these forces and gain a greater measure of control over our own futures. The first step toward maximizing freedom therefore necessarily involves an increased willingness to substitute mutual, conscious control by human beings of one another—including, at the margins, the use of coercion—for control by abstract forces. Such mutual control is the essence of politics. And the restoration of politics is the first step toward individual liberation.

If we are to act effectively to control the forces that currently determine our choices, we must avoid succumbing to the self-validating belief that the opposition is unbeatable. If we assume that our actions to recapture our freedom are doomed to defeat, they are as good as defeated. Therefore a vital step toward maximizing goal-seeking behavior is a belief that the future is not rigidly predetermined and that our actions can contribute, in however small measure, to creating the future. Such a conviction is not empty optimism but simple logic. If we accept the premise that any state of a system, physical or human, is simply the result of the behavior of the elements which constitute it, then every action has consequences. Our individual behavior can affect the total state of a system only to a degree, but it does change it, and it also changes the system's effects on ourselves.

Recognition of our ability to influence other actors also implies recognition of the necessity of being influenced by others. Most liberal models of decision-making processes are inherently static in that they fail to take account of the fact that the process of social interaction inevitably changes the desires—the images of the future—of the actors;[14] for this reason outcomes are not the simple resultant of the forces initially present, because in the course of the process these forces have changed their character. *Compromise and adjustment do not necessarily involve any diminution of freedom because they alter the selves which engage in these processes, so that the new selves thus created can freely accept a future which the old selves might have rejected.* The process of action changes the identity of the actors, but it is a growth process; it enhances rather than diminishes the self-realization of the participants.

Freedom to act—to enter into the process of action and change which is creation—can be inhibited not only by a sense of objective powerlessness but also by low self-esteem, a sense of personal worthlessness. The roots of passivity are many and complex, but it is clear that passivity (like activity) builds on itself. Nothing fails like failure. One requisite for creating the ego strength necessary for the exercise of freedom is to persuade the young, by precept, example, and experience, that they themselves are capable of taking effective action, of influencing their environments. They also need to know that in the process they too will change—that to be free they must be open to new experiences, ideas, and insights. For all of us are on a journey to Tarsus and, like St. Paul, must be prepared for the moment of illumination, the radicalizing and restructuring experience.

We can change the future. We can regain control over our own choices, but we must first reject the liberal assumption that our freedom is menaced only by coercion on the part of other human beings; we must come to realize that whatever affects our future qualifies our freedom. Freedom lies neither in the absence of coercion by other human beings nor in not being influenced by other human beings or by nonhuman forces. Freedom resides in our conscious interaction with our total environment so as to achieve desired goals. Human society is not a deterministic system but a collective learning process. *We become free by consciously choosing how we will relate to our physical and social surroundings so as to affect the total future state of the systems of which we are an integral part.*

"A person is free to the extent that he has the *capacity*, the *opportunity*, and the *incentive* to give expression to what is in him and to develop his potentialities."[15] Insofar as these potentialities are multivalent, freedom offers the opportunity for the actualization of many possible self-generated future selves. In the words of the Russian Orthodox philosopher Berdayev, freedom is "the opportunity for creative activity."[16]

What is true of individual freedom is also true of the freedom of human societies and the freedom of the human species as a whole. Humanity exists within a larger ecosystem composed of human beings, the nonhuman world of nature, and the technology man has created. There are limits to the choices available to humanity given the world into which it was born and the knowledge and tools it has itself created. But within this larger universe, humanity can choose how it will interact with the human and nonhuman world—how it will change the world and itself in the process, learning by observation the results of its actions and freely creating its own collective future.

12

Human Needs
in Social Perspective

WHAT are the basic human needs and desires, which must be fulfilled in a social context through political means? How do we set about discovering what human beings need in order to be human?

If life is good, then whatever is conducive to survival and functioning is also good. A naturalistic social ethic must proceed by examining human nature and determining what human beings need to survive and function. These requirements for survival and development are the basis for moral imperatives, all of which can be summed up in the basic principle, Do what is good (life-giving) and avoid what is evil (life-diminishing). "Good is all that serves life; evil is all that serves death."[1] Social institutions and policies which are life-enhancing are good, those which are life-diminishing are evil. The task of social and political philosophy is one of determining which policies and institutions are best able to fulfill the needs of man's nature, insofar as these can be determined through reason and empirical observation.

This would be a relatively simple task if it were not for a basic problem: Human nature is so constituted that its most significant single characteristic is its adaptability, an essential aspect of its progressive nature. Other animals flourish best in states of essentially static equilibrium with their environments. Human beings flourish in a dynamic, moving equilibrium in which humanity changes and changes its environment, physical and social, at the same time.[2] The balance humanity lives in is not that of the jeweler's scale one sees in the traditional portrayals of "Justice" but the balance of the tightrope-walker who is continually moving from side to side as he moves forward. As a result, the needs of human beings cannot be thought of simply as a sequence, the fulfillment of which automatically creates satisfaction. Many human needs are contraries which must be balanced against each other. Ideally these needs are taken care of virtually simultaneously, since too violent oscillation between behaviors designed to fulfill opposing needs could cause a fall from the wire into psychological or social disaster. We all feel a need for both work and play, both security and adventure, both satisfaction of desire and the kindling of desire. Attempts to deal with human beings as if all they wanted was to satisfy a certain limited set of desires—to be in balance like a scale—ignore the fact that humanity seeks to remain in balance while moving forward.

Earlier I argued that the individual is in some sense a myth, since all humans are creatures of nature and society, points in a natural/social matrix. But the individual we speak of in ordinary language does exist for certain purposes. Just as the cell is an identifiable subsystem within the body, a process of internal interaction which has boundaries at or through which it interacts with other cells, so human individuals are subsystems within nature and society, bounded organizations of energy. But, like the cell, they are open systems which in order to continue to survive and function must exchange elements (physical and mental) with the world outside themselves, and exchange these elements in certain ways.

This is not to say that human society is an organism analogous to the body as political theorists since Plato have often held. Organic theories of human society have always involved concepts of fixed specializations and hierarchy. They mislead because of their dualistic concept of mind and body, ruler and ruled; the body is much more a democracy and the roles of its parts much less specialized than organic theorists postulate.[3] The individual, like the human cell, is not solely a "part" of society but possesses an identity and a degree of freedom, and this gives rise to individual human needs.

These individual needs fall into two categories, physical and psychosocial.[4] Unless our physical needs are satisfied we perish as systems and our constituent elements return to the totality of nature, as when a decaying corpse becomes food for maggots. Unless our psychosocial needs are satisfied, we are crippled in our ability to function as complete human beings, sometimes even to the extent of becoming the kind of human vegetable one can see in mental institutions.

The physiological needs of human beings are relatively simple to state, although the margins of these needs vary from individual to individual and from situation to situation. The individual requires a certain amount of food with certain nutritional qualities. He needs water to drink. He needs opportunities to eliminate waste products. He must be able to breathe and get rid of used air. He has to be able to maintain a certain body temperature and therefore, as circumstances dictate, requires clothing, shelter, fuel for heating, and so on. He must be protected against bodily injury or disease of certain kinds if he is to survive. A certain amount of rest and sleep is necessary as well. Our physical desires—the subjectively felt aspects of our needs—ordinarily do not war against each other. We may have to choose which ones to give temporary priority to, but the problem is minor. Largely as a result, there is little dispute about the nature of human physiological needs. Any intelligent child could draw up a list, and any competent physician could put the items on the list into quantitative terms for the average human, even specifying how long the need could remain unfulfilled before death took place.

Some Definitions of Mental Health

Disputes about what human beings need arise when we turn from physiological to psychological and social needs. We all recognize that these latter exist, but we differ as to what they are and how one should measure their nonfulfillment. Different schools of psychology define human nature differently, and as a result we lack agreement even on a common terminology to describe, much less prescribe, the social-psychological needs of humankind.

For the Freudians, man has two basic drives, ultimately biologically determined: *eros* and *thanatos*, love and death. He perceives the lack of satisfaction of his needs and reflexively seeks to obtain what satisfies the desires his drives generate. When and as long as he is satisfied, he can and does lapse into passivity. This is of course an oversimplification. But, essentially, desire and satisfaction combine to create, as it were, a closed system, a balance which can only be upset from the outside or by the decay over time of the ability of an object to provide satisfaction. We need food and sex, we are hungry or lustful, we find food or sexual objects, and we are in balance until with time we become hungry and lustful again.

For the behaviorists, James W. Watson and his followers such as B. F. Skinner, there is no such thing as mind, only behavior. Man is even more simply a passive creature than he is in the Freudian system. There is a stimulus from the outside environment and a response from "within," and equilibrium is restored.

Both theories, the psychoanalytic and the behaviorist, regard human personality as "a homeostatic system addressed to tension reduction."[5] Despite such refinements of the stimulus-response (S-R) psychology as operant conditioning it remains a closed system, postulating a scalelike balance.[6]

Recent developments in biology and psychology have undermined this simplistic view of human nature. "Present day biological theories emphasize the 'spontaneity' of the organism's activity which is due to built-in energy."[7] We do not come out of the womb a passive object, waiting for stimuli from the outside to respond to. We come out kicking and screaming, like a bull out of a chute into a rodeo arena, preprogrammed with desires and aspirations. We seek not to return to any womblike stasis, any quiet balance, but have an inchoate forward thrust toward goals we cannot imagine. Living man, as Dubos argues, does not merely "react with" but "responds to" his environment, and uses the environment for "self-actualization."[8] We are physically and mentally antientropic, not closed but open systems seeking to seize energy from outside of ourselves to create new patterns of order—to grow. We do not know what is at the end of the tightrope, but we seek to move along it, not to balance in a stationary position in one spot. Our needs can never be satisfied.

The thrusting nature of life, especially human life, makes the task of defining mental health and hence social ethics exceptionally difficult. Actually, even physical health is not as easy to define for humans as it might at first seem, partly because it is intimately interconnected with mental health. Unlike other animals, human beings can choose to seek ends other than health and make health not an overriding but simply a subsidiary, instrumental value. For human beings health is not the purpose of existence but a prerequisite of activity. We all know "health nuts" who spend so much time and energy, physical and mental, staying in good shape that they have no time for work, true play, friendship, or love—any of the things one presumably wants to be in shape for. Busy men take risks with their health because what they are doing with their lives seems more important to them, at a given moment, and we all understand. Being in good shape is rarely an end in itself for the people we respect. Some minimum of physical health is essential for maintaining existence and hence activity, but it must be balanced against other ends of life. For the sociologist Ernest Becker, health is "the striving for more and more complete facilitation of human choices . . . a synonym for human liberation in a wide sense."[9] For Rene Dubos, health "demands that the personality be able to express itself creatively."[10]

What is true of physical health is also true of mental health. Aristotle rightly defined happiness as the end of man. But our definitions of happiness are often puerile notions of inactivity, unconscious bliss, typified by the "happy moron." If seeking physical health for its own sake is a marginal concern of truly alive human beings, seeking happiness as such is to follow a will-o'-the-wisp. We do not want to be happy; we want to be alive, active, and creative—only this can make us happy. The profound scriptural admonition that he who seeks his life shall lose it and he who loses his life shall find it, quite clearly applies to happiness. Happiness has no generalizable substantive content of its own; it is our symbol for activity that we find rewarding.

Thus while various psychologists have attempted to define the components of mental health in a manner analogous to the components of physical health, most such lists are of limited use. They can help us know what is necessary to maintain our balance, but they do not tell us much about the direction in which we want to move. Freudian formulations are especially useless, since they are essentially negative. Even when valid, they focus on barriers (such as neurotic behavior) to healthy functioning but tell us very little about what we should seek once we rid ourselves of the chains of illusion and compulsion. As a result, neo-Freudians such as Fromm have had to move far beyond Freud in order to find a more progressive definition of man.

For behavioral psychology what is good, what constitutes mental health, is defined ultimately in terms of individual and species survival. Skinner defines the good as that which on balance gives us positive reinforcement, basing this prescription on the assumption that over the eons of human evolution, we have learned to be reinforced by what is conducive to species-survival. This is fine as far as it goes. Here Skinner is simply reiterating, in however disguised a fashion, the traditional position of the natural-law theorists and indeed of common sense.[11]

The Nature of Human Needs

But psychological health means more than maintaining mental equilibrium. It necessarily involves positive goal-seeking—what a third group of psychologists, the humanists, speak of as the search for being, meaning, self-actualization or even transcendence.[12] Abraham Maslow, the leading theorist of contemporary humanist psychology, speaks of love, self-esteem and self-actualization. He also attempts to rank human needs in an order of priority, an order in which the most essential are in a sense the least noble.[13] We need to fulfill physical needs in order to have a psychological life, and within our psychological needs there is a progression from lower to higher.[14] "Erst kommt der essen, und den kommt die moral," is the way a Bertolt Brecht character puts it. Thus human needs can be thought of as a hierarchy, or they can be thought of as a set of requirements all of which must be met if an individual human being is to maintain his psychological balance while at the same time moving forward toward meeting his needs in a more complex and developed fashion.

For the child, security means his mother's arms; for the adult a sense of competence in meeting the challenges of social living. Adventure for the child is crossing a street or climbing to the top of a jungle gym; for the adult it can be travel to a strange land, or taking up a new profession, or confronting new ideas. For the child love is possessive; for the adult it is identifying with others. Many of our desires, like the craving for both security and variety, seem contradictory and must be met in different ways at different times.

What are the psychological needs of all human beings which the good society must make it possible for the individual to fulfill? The following is an attempt to glean and summarize from the work of others the kind of information about human nature which has political relevance and will enable us to better understand the basic requirements of the inhabitants of a fully human society.

Identity and activity are the two fundamental needs of human beings, and these two needs are not really separate but two sides of the same coin. *The human person is a purposive animal who seeks goals through his activity.* The goals which human beings seek can reasonably be classified under the broad categories of *security, self-esteem,* and *variety* (or adventure).

Security as a goal involves the need for subsistence or what is necessary for physical survival; a considerable degree of orderliness, regularity, and consistency in our environment; and an ability to count on the good will and reliability of other persons. *Self-esteem* requires that we have an opportunity for sustained loving relations with others—in our families, in our friendships, and in our communities—and that we have opportunities for meaningful activity—activity that has value for ourselves and for our society. The desire for *variety* is satisfied by opportunities for adventure (actual or vicarious), for play, and for engaging in new activities and trying out new roles in a nonthreatening context; variety is not only valuable for what it contributes to the balance and growth of the individual, it also has value for society as a whole, since it encourages the flexibility and adaptability which have played a primary role in enabling the human race to survive and develop.

Maslow and his followers tend to treat self-esteem as the single overarching need of all human beings. The need for self-esteem is obviously much more basic than any supposedly inherent competitive drive, and this approach to the problem of human needs gives us some insight into the reason why the need to compete—to win, to be on top—is so important in contemporary liberal society as well as why competition need not necessarily be as important in a society that is differently structured. But whether it is justifiable or useful to reduce the desire for security and variety to the desire for self-esteem seems at least doubtful.

Like all such classifications, the one proposed here is necessarily arbitrary; but the substance of the needs is real. These are individual needs fulfilled by individuals as such. It is the obligation of society to provide a context within which these needs can be met as fully as possible given the conditions of a particular time and place. The obligation to do this is one aspect of the obligation of the political order to provide for the common good.

Identity and Activity

Just like the electron, we are what we do. What we do determines who we are, and who we are, or think we are, determines what we do. Our sense of self requires both identity and activity. *We create our identity through our activity* and *our identity generates our activity*—these two are parts of a continuously interacting whole. In much the same way activities oriented toward obtaining security, self-esteem, and variety are not only goal-determined actions in themselves but can be (and usually are) the necessary preconditions for envisioning and achieving still more complex or advanced forms of these goals.

SECURITY: PHYSICAL AND PSYCHIC

Security is a prerequisite both for identity and for activity. A child gains a sense of self from her interaction with her environment and especially from the way in which the important persons within her environment respond to her. It is only by means of her environment—the things and persons surrounding her—that she is able to formulate any coherent picture of the world, of her place in it, and of what the results of any actions she may take will be. The infant cannot survive unless her physical needs are met, and how these needs are met will condition to a large degree the child's ability to reach her full potential as a human being. The experience of warm loving care and an ability to trust in the essential orderliness and stability of her environment are vital if a child is to feel capable of venturing forth in search of the new experiences she will need for continued development.[15]

Obviously not only children but adults as well require a sense of security if they are to fulfill their human potentialities. Without this sense of security, adults may focus all or most of their activity toward trying to create a more secure environment for themselves (often by means wholly unsuited to reaching the desired goal). If we are to be truly human, we should not have to spend all or most of our time striving for security. Security like health should be primarily an instrumental means to other more strictly human goals.

All human persons need to be certain that their physiological needs for air, food, and shelter will be met. It has been possible in past historical periods, and is still possible in some parts of the world today, for individuals to provide—on a limited and impoverished basis—for their own physical needs, provided they have access to land or the products of the land or sea. But specialization and exchange is normally a characteristic feature of human society, and this has never been more true than it is today in the developed world.

Exchange is a social process which proceeds according to rules: regularized forms of behavior. Liberal economic theory assumed that these rules could be arrived at and maintained on a purely voluntary, indeed automatic, basis, and that in this way all would have their needs met. (Individuals incapable of entering directly into the process of production and exchange, such as small children and the infirm, were considered to be special cases to be taken care of by supplementary arrangements.)

The issue has never been whether the provision of what is necessary for individual subsistence is or is not a social process. It obviously is. The issue is the extent to which this process, if it is to achieve its goals, must be politicized, that is, made subject to the self-conscious rule-making processes of the political order, which can when necessary ensure compliance with these rules by direct overt compulsion. Even classical liberals considered it proper for government to intervene to prevent fraud, breach of contract, and restraint of trade.[16] Today throughout the world the paramount role of politics in governing and managing the economy is acknowledged by most economists.[17] Even in the United States, the last stronghold of at least rhetorical allegiance to classical liberal "free enterprise" economics, the controls placed on wages and prices by the Nixon Administration in 1971 were accepted with hardly a murmur except from a few diehard ideologues.

This general abandonment of previous liberal ideals has a variety of causes. In some instances it represents a recognition by even liberal economists that the structural rigidities within large-scale economies are such that the automatic adjustments which were supposed to make the classical economic model work cannot take place—major automobile manufacturers, for instance, cannot go in and out of business the way corner grocery-stores do without disrupting the economy to an intolerable degree. In addition, government spending—the result of centralized, large-scale, long-range and therefore relatively inflexible decision-making—plays a major role in all modern economies, a trend accelerated by military spending, and thus skews the working of classical market-theory.

Less fully appreciated is the fact that the principal role of the political state in "capitalist" economies has always been to promote the power and profits of "private" enterprise. Governmental regulations and subsidies are the principal means for creating and maintaining monopolies and other kinds of privilege. The issue therefore is not whether we ought to politicize economic activity—it is already (and inevitably) politically based—but what the ends of this politicization should be: whether it will be used to make the social process of exchange serve the subsistence needs of all individuals within society or just those most advantaged; whether it will be used to shore up a mismanaged and corrupt Lockheed Corporation, or to put food into the mouths of the children of migrant laborers. Whatever is done—or not done—the fact that political action is required to enable or force the economy to fulfill the subsistence needs of society can no longer be in any doubt.

Human beings are creatures of flesh and blood. Prick them and they bleed; fracture their skulls and they may die. Security against physical injury is thus a necessity for survival. Because human beings also require physical objects—food, clothing, shelter—in order to meet their physical survival needs, to maintain their sense of identity, and even to act effectively, they must be protected against theft as well as assault.

Liberal theorists such as Hobbes and Locke made physical security a preeminent reason for abandoning the state of nature and entering into civil society. While the state of nature is a fiction, there is no question that security is a paramount human concern which can only be fulfilled in society, and that social controls are necessary in order to provide this security.

For the most part the need for physical security is taken care of by the strong bonds that exist among human beings. If security rested upon the individual's limited ability to protect himself or on the coercion of large numbers of potential criminals by the state, it would be unattainable (as it is in certain neighborhoods in some American cities today). The problems of security are magnified in large-scale societies where impersonal, non-face-to-face relations constitute a high proportion of all social interactions. The United States today is a prime example of such a society. In 1972, one third of the Americans living in large cities reported to poll takers that they had been the victims of serious crimes within the previous year, and one fifth of those living in suburban or rural areas had had the same experience. The coercive apparatus of the political state was judged (with reason) to be so unable to protect their security that fully half the victims never even reported the crimes to the police.[18] Human beings can of course survive physically even where high levels of crime exist, but only at the expense of sacrificing higher psychological goods and aspirations.[19] Crossing the street to avoid groups of strangers at night does not add to psychic well-being, nor does living behind locked doors or in controlled-access communities. A basic precondition of a human society therefore is the creation of physical security for its members. There is no more basic human civil right than to be able to walk the streets unmolested.

While crime is an especially serious problem in urban America, where it has strong racial overtones, it is a rising problem in all industrialized societies, socialist as well as liberal.[20] It can perhaps be mitigated by compulsive measures at a high cost in terms of values other than security, but it can only be reduced to a humanly tolerable level by the reinstitution of real community, the breakdown of which is its single most important source.

Physical security can also be menaced by conflict among political entities. In previous eras, how much security one had against such dangers was determined by the power of the political unit to which one belonged, with members of weaker political systems enjoying less security than members of stronger ones. In an age of nuclear weaponry, which may in time become universally available, the ability of the political state system to provide physical security for its members has been gravely weakened, if not destroyed. Physical security therefore demands the creation of an international political system in which large-scale violence is abolished. The one place where the state of nature has in fact existed historically, as philosophers such as Locke have correctly pointed out, has been in relations among nations; and the creation of a world civil society to replace the present international "state of nature" is a prerequisite for individual human physical security.

But individual security is threatened not only by crime and war and the lack of fulfillment of physiological needs, but also at the psychological level, by overly rapid change. Human beings are flexible and adaptive animals, but too great a demand for adaptation can be destructive of identity rather than a source of growth. A child needs stimulation (variety) as well as security (a pair of contraries which build on each other) if he is to develop. Though the lack of challenge stultifies, too great a challenge can destroy a child's confidence and self-esteem. The same thing is true, of course, of adults. A considerable degree of stability is essential to human survival and growth. If there were no sequences of events which could be depended upon to take place, memory would be useless; the world would be like a code in which no word ever meant the same thing twice, a code we could never hope to break. If nothing takes place regularly we cannot even develop any sense of our own bounded identities; we melt into a chaos of perceptions. The political order must make it possible for individuals to meet their needs for a reasonably stable environment.

PHYSICAL IDENTITY

The child's first sense of identity is physical, but he soon learns to downgrade this aspect of his identity as he begins to cope with the social world around him. All societies necessarily divert the child to conscious mental activity directed toward what is outside his body. But modern industrial society especially teaches us to downgrade the sense of our own physical existence and thus to lose esteem for part of ourselves, to lose part of our identity. The current mushrooming of classes in Yoga, Tai Chi, massage, and similar activities, evidencing a new interest in our physical bodies and their capabilities, suggests that a major revolt against the old dualistic attitude that our bodies are somehow shameful or else merely machines in which our minds are only sojourners may currently be underway.[21]

SEX AND IDENTITY

Careful readers will have noticed that I omitted any discussion of sex in the section on physical needs. Strictly speaking, sexual activity is not a physiological requirement for the survival of individual human beings— though it has so far been a requirement for the survival of the species. But a secure sexual identity and sexually satisfying activities are important both to our sense of self-esteem and to our desire for variety, for the expansion and intensification of experience. Our sexuality is central to our physical and social identity. Sex has always played a role in human and animal life which transcends and in some cases is only marginally related to procreation. Sexual activity is naturally polyvalent.[22]

In contemporary society the need for a sexual component in our lives suffers from a twofold distortion. On the one hand, it is the last vestige of primary, private bodily life admitted, however grudgingly, in an increasingly rationalized and mechanized society, our "last green thing."[23] As a result it has had to bear the whole burden of our longings for a sense of bodily awareness and gratuitous physical activity. In addition, in North America, as the philosopher George Grant has observed, sex must support a further burden. North Americans are all relatively recent interlopers on their continent. The European or Asian can seek his own primitiveness in his ancestral forests or caves, but the sacred places of the soil in America belong, spiritually at least, to the dispossessed Indians. North Americans, therefore, find it especially difficult to fulfill the desire for spiritual communion with nature through contact with the land. We remain somehow alien. Sex thus becomes our one "chthonic" thing, our one close contact with earthiness; and the cultural revolt against loss of harmony with the natural world and with our bodies formed from the earth concentrates on sexual expression.[24]

At the same time, we have come to think of sexual expression as primarily and exclusively meaning genital intercourse, that socially necessary expression which even industrial society permits while it continues to reduce the ability of its members to respond erotically to a whole range of other physical and social experience. A more fully human society would place relatively less emphasis on the sex act itself and more emphasis on a reeroticization of the whole culture, an essential precondition for a heightened awareness of our identity as bodies, single systems of spirit and flesh.[25]

COMMUNITY

Identity requires not only that we know ourselves as individuals but as members of society as well. Self-esteem is the product of our interaction with our social environment.[26] The nature of that interaction is crucial in determining how we evaluate our own identities. Our valuations of ourselves are being constantly confirmed or denied by the behavior of others toward us. The need to reaffirm our positive valuations of our identities—the need for self-esteem—is a basic motivation of all human beings. Both history and common observation bear witness to the extent to which people will choose to affirm their self-images even at the cost of physical life itself.

For a sense of identity and self-esteem, both children and adults apparently need the security of close and prolonged ties with other persons—in families, in small groups, in a community. Some social commentators regard a high degree of social mobility and rapid movement in and out of close relationships as unavoidable or even desirable,[27] but a secure identity requires stronger ties and larger memberships than such an arrangement can ordinarily offer. The social environment ought to provide many opportunities for interacting with others in a broader context than one based on abstract ties or the enactment of highly specific, limited roles. As Becker has observed, "It is fundamentally degrading to man to be reduced to his roles."[28] We need relationships with others for whom we are whole beings. Love and mutual respect are the coin of such encounters, and larger interactive groups than families or pairs of friends are necessary for their fullest exchange. Community membership is a requisite for identity, and the cement of community cannot be calculated expedience as in liberalism but only some form of love.[29]

Human beings are attracted to each other, as are all "objects" in the universe; love is as universal a force as gravity, as well as being its analogue in other ways. Community is the norm.[30] But how wide can community be? If our affections normally draw us together, our interests pull us apart. Communities in modern society tend to be based on special interests or attributes. The special bonds which pull some of us together set us apart from others: the black community against "Whitey," the local community against "outsiders," the college fraternity against "barbarians." Some would argue that community can only exist when there is an external adversary—"us" against "them"; our most basic contemporary community, the family, is often viewed as a fortress against the outside world;[31] and we see the manifestation of political community above all in war, a bloody expression of communion in which mutual support and devotion reach heights usually unattainable in peacetime.

The universal desire for community—for strong ties with others—takes many forms in contemporary America. It was one of the ideological underpinnings of the counterculture of the sixties. The encounter groups and the communes which are part of the legacy left by the counterculture are witness to the felt need of many for closer bonds with their fellows. The restoration of community plays a major theoretical role in the rejection of liberalism by the New Left. But this same desire for community is found in contemporary conservatism as well, not only among conservative political theorists but among ordinary Middle Americans, and it is manifest in their political actions. Those who are politically liberal (in the everyday usage of that term) often fail to realize that much opposition to school integration—on the part of both whites and blacks—to public housing projects, and to the extension of federal powers into areas which were traditionally local governmental functions, represents not only racism or unthinking opposition to change but a genuine desire to preserve community and identity as embodied in neighborhoods and local communities and common life styles. Ethnic politics—not only among blacks and Jews but among the so-called "new" ethnics as well—is closely related to this desire for a rooted identity and for relatedness to others.

It is through participation in small groups that one becomes capable of relating to other individuals, so small subgroupings are the essential base for participation in the larger political community. Yet such groups are always potentially divisive, possible threats to the broader political community. How, short of uniting against a common national enemy, can community be fostered on the level of the national political system as a whole?

EQUALITY

One important answer lies in Aristotle's comment that friendship is possible only between equals. In discussing the nature of political community, he pointed to the dangers inherent in the division resulting from great discrepancies in wealth, which cause the rich to confront the poor in necessary mutual hostility. Rousseau too warned that a large measure of equality is a prerequisite for social unity.[32]

Even if income disparities are greatly reduced, blacks, Chinese, Irish, and others may still feel a special bond with those who share a similar genetic and cultural heritage. The same thing is also likely to be true of Ivy League professors and Chicago steelworkers. But much of the hostility inherent in the interactions among different groups would be removed if material envy was not present to stoke intergroup conflict.

If the need of the individual for community—if his desire for the security and self-esteem which stable human relationships engender—is to be met, the political order has two apparently contradictory tasks to perform. The state must, insofar as this is possible, avoid actions which disrupt existing communities and which cast individuals adrift in a sea of strangers. At the same time the state must use its power to minimize the economic and social inequalities endemic in present-day liberal society.

Like most of the words we use every day in talking about the nature of society, the meaning of the term *equality* is slippery and ambivalent. Things can only be equal in terms of a common measure. A pound of iron weighs the same as a pound of feathers because we have scales attuned ultimately to the earth's gravity. But iron is not feathers, and in other respects than weight they are not much alike. Equality is usually spoken of in a social or political context in terms of justice. To treat people justly is to treat them "fairly." But this substitution of terms simply rephrases rather than solves the problem. What is justice?[33]

In his *Ethics* Aristotle distinguished between two kinds of justice, radically different in their implications, which human beings throughout history have in practice had difficulty reconciling. One he called *commutative justice*, strict mathematical equality.[34] In this sense justice is served if every human being is given the same ration of calories a day—say 2,700—or is paid the same amount for his labor, say five dollars an hour. But some people are of course bigger than others and normally need more food, so to give all human beings equal rations would leave some hungry and others surfeited. Some people work harder or more effectively than others, and to pay all equally might reward sloth and incompetence and could conceivably discourage skill and diligence.

The normal human sense of equity (at least in western culture) rebels against both these actions as unfair. So Aristotle posits a second kind of justice, *distributive justice*, which (following the Platonic definition of justice as giving each person his due) would treat people in terms of a different measure of equality. To feed people equally might mean giving a larger person a greater ration than a smaller person, just as a parent gives teenaged children larger portions than infants. To reward workers equally might mean paying one more than another in terms of accomplishment, measured either in quantitative or qualitative terms, as our market system often does.

The difference between these two concepts of justice is no mere theorist's quibble but the substance of bitter everyday social conflict. In the United States in recent years this clash between standards of equality has been acerbated by the fact that certain kinds of inequalities are racially linked. Currently argument rages over whether students should be admitted to universities in terms of their grades—treating equal achievement equally—or in terms of racial quotas—treating racial groups as such as equal on a basis of proportional equality, or simply as equal undifferentiated persons, selected by lot if more apply than can be accommodated. Housing is normally allocated according to what people can afford to pay for it, which means the rich get one kind and the poor another. Controversial attempts have been made to "equalize" housing opportunities by subsidizing the rent payments of the poor in order to provide a more equal opportunity for them as persons—rather than as earners—for better housing. Every such debate turns on the question of what standard of equality ought to apply.

As a practical matter, any attempt to treat people equally in distributive terms involves constant dispute over which differences are really relevant, and which therefore ought to be the basis for reward—or punishment. Any attempt to treat people on a basis of strict formal equality will usually either involve a violation of the customary sense of what is just or will necessarily entail efforts to eliminate those differences which are due to social experience or even heredity.[35]

If there is so much uncertainty as to what equality means, if (as Aristotle in effect held) the nature of justice is not absolute but relative, then what is just will differ from society to society, since every society has—indeed is based on—a particular concept of justice. How then can we reasonably say that equality is a fundamental human need which society must fulfill and which politics must ensure that it does in fact fulfill?

In the last analysis, only an almost innate, subrational, or suprarational sense that some ways of ordering human relations are more just than others can inspire our constant attempts to do justice in human affairs. Like other elements in our sense of what is just—such as our concepts of "due process" or "cruel and unusual punishment"—humanity's ideas about what inequalities are supportable change—they evolve. Recognition of what forms of inequality are destructive of individual self-esteem can develop and grow within a society, and it is possible that, as our awareness of the effects of particular inequalities deepens and becomes more subtle, society through the political process will progressively eliminate these barriers to personal self-esteem and to the achievement of community.

Gross inequality—especially in economic terms—is an original and necessary element of liberal society as envisioned by Locke and as realized in our own history. The glaring inequalities in income distribution which exist in the contemporary United States are rarely the subject of political debate, and evidence exists that even members of the nonélite, disadvantaged groups have a principled or simply visceral distaste for the concept of equality.[36] Yet throughout history the desire for equality breaks through. One political scientist, James C. Davies, has even argued that this "species-wide desire" has a biological basis.[37] But this universally felt need, he postulates, is not a commitment to a particular concept of equality but is a generalized desire for a "sense of equal intrinsic worth, equal dignity as a person—and a public and private recognition of that desire."[38] There is no reason why the form this recognition takes cannot grow and evolve insofar as society continues to evolve in the direction of permitting and encouraging the fullest possible development of the human potentialities of its members.

While it may be difficult to discover any specieswide consensus on what equality consists of, we can identify those inequalities which are insupportable in the same way that, although we cannot define justice directly, we do have a real sense of injustice.[39] We can know the limits of wealth and poverty which are tolerable in a given human society just as we can define certain forms of legal action, such as arbitrary arrest or bribery of judges, as palpably unjust. Insofar as the actions of social systems lead to degrees or forms of inequality which violate our sense of what is necessary for self-esteem—our own or others'—we must take action within the political order to redress these inequalities or at least to prevent the political order from serving as instigator and bulwark of inequality, as is all too often the case at present.

The problem of equality is of crucial importance today, not only because the escalating demand for equality on the part of an increasing number of groups threatens existing structures of political and economic power, but because if this demand is met in the wrong way the long-term viability of human society will be endangered. The principal means employed by liberal society to meliorate the problem of inequality, especially in the economic sphere, has been a growth pattern which enables individuals to feel that they are making progress, that their personal worth is receiving greater social recognition, even while others may be vastly increasing their share of the society's wealth and power. Instead of slicing the economic pie differently so as to give those with the least a more equal share, its total size has been increased. Crises of distribution within society are met by increasing the demands society as a whole makes upon the life-support system of the planet. Growth is substituted for social justice. If the assumptions underlying this growth pattern are fundamentally unsound (which is the premise of this whole critique of liberal society), then the human desire for a degree of equity which coincides with the individual's need for a sense of self-worth will have to be fulfilled by means other than growth, and the political order may have to take action to radically revise relative differences in wealth and status within society in order to make this possible.[40]

PRIVACY

To know who we are and to feel secure in our identities we not only need close and sustained relations with others; we also need opportunities for privacy. Privacy will be an especially difficult human need to fulfill in the crowded world we are creating. If, as one biologist half-seriously suggests, we are destined to evolve genetically so as to be at home in a regimented and teeming world,[41] we will no longer be truly human. Societies differ with regard to how they define and how much they value privacy, but some degree of privacy is necessary for the maintenance of the human self.[42]

CONTACT WITH NATURE

Just as we need to know who we are as individuals and as members of a particular social community, we need to know ourselves as members of a species. Part of our identity is simply our being human. We are distinct from other creatures and distinct from the rest of physical nature. In the past this sense of difference could be taken for granted. Humans were conscious of being among other animals—the beasts of the fields and the birds of the air—and they believed in the existence of a spirit world distinct from that of man and physical nature alike. We were one distinct form of life among many. Today it requires more of an effort to relate to nature, to recall that we are creatures among creatures and not one kind of machine among other machines. As one psychoanalyst notes, "The non-human environment, far from being of little or no account to human personality development, constitutes one of the most basically important ingredients of human psychological existence."[43] To maintain our sense of human identity we must be able to interact with nature; indeed, as Dubos argues, "The humanness of life depends above all on the quality of man's relationship to the rest of creation."[44]

If we allow all other species to perish, or limit ourselves to the preservation of domestic animals, we will not be able to recognize ourselves. If we cover the earth with our artifacts, we will forget these are our creations and not ourselves. Encounter with nature is a necessity for human identity; if we lose it we will lose ourselves in a maze of distorting mirrors, as in an amusement park, and forget the lineaments of our own visage. To know who we are, we must maintain our distinction from the machines we have created.

Some primitive tribes fear the camera of the explorer, afraid that whoever captures them on film has stolen their souls. Their reaction is not wholly baseless. Our image can be captured in our artifacts and our integrity destroyed. In an age of humanoid robots we might not be able to tell ourselves from our creations, and vice versa.[45] Robots which are humanoid in form are unlikely to become common for the same reason that automobiles are no longer built to look like carriages: inefficiency of design. But we can change ourselves to be like machines. If we spend much of our time talking to or through computers, their language will become the only one we can speak seriously. And it is a poor language, intolerant of redundancy, incapable of ambiguity or irony. To maintain our own identity, we need to keep our own languages and emphasize the disjunctions between our nature and that of machines. By definition they are deterministic systems; by nature we are not.[46]

WORK AND PLAY

Identity gives us our consciousness of who we are. But it is only one side of a coin, the other side of which is activity, through which we reveal who we are. But activity does more than reveal our identity; it is the major factor in creating it. From the moment we are born we respond to the actions of others and to others' responses to us. We try from the outset of our lives to test the world about us, looking for novelty and adventure as well as security and approval. When we stop seeking opportunities for expansion of identity through new experiences, we have already given assent to our death.

Human activity is essentially of two kinds, work and play. Work is outside-oriented; we seek to do things which will be of use, for ourselves or others. Play is oriented toward altering (however temporarily) our own state of consciousness to some condition we prefer. How important work as such is to a healthy personality is a matter of dispute. For Freud it was fundamental, and some neo-Freudians have found it difficult to accept the idea of a society in which work did not exist, since work in the Freudian system is one of the repressive factors which holds personality and society together. Much of the difficulty vanishes once we recognize that we require work for the sake of self-esteem; we need to feel of use—of acknowledged use—to others. If others appreciated the value of our play, this basic desire for esteem could be fulfilled just as well.

Inactivity is intolerable to normal human beings. The much-lamented decline of the "work ethic" in modern industrial society is not the result of any change in human nature but a recognition that much modern work is unfulfilling—useless both to society and to the individual (though this is not the same thing as saying that most work always was so).[47]

Because regular work (or purposeful play) is essential to the security and self-esteem of the human individual, society has an obligation to ensure that opportunities for meaningful activity are available to all. Over-rapid change can mean not only a diminishment of individual identity, but also the destruction of the ability of the individual to engage in purposeful activity as well. A certain amount of stability—of regularity and dependability within one's environment—is essential if one is to engage in activity that is purposeful and satisfying. "Stability," in Geoffrey Vicker's words, "is not the enemy of change. It is the condition of any change which can hope to be welcome and enduring."[48] It is impossible to engage in purposeful activity if all is flux. If one cannot predict with reasonable certainty the outcome of one's actions, then action necessarily becomes random and meaningless, wholly devoid of purpose—and thus no longer really human. It is a vital political function of society to ensure that opportunities for participation in work of socially acknowledged worth are available to all its members and that change within society is not so rapid or of such a nature that it deprives the individual of such work without providing him with some at least equally acceptable alternative.

It is worse than meaningless to train people for jobs which will not exist in the near future, to impart skills which cannot be exercised, to urge people to save money which will become worthless or to have children with whom they will not be able to identify. Even the young, the bearers of change, suffer from disorientation and despair when they are unable to plan their futures as a result of such uncertainties as the Vietnam war draft, or the possibility of nuclear war or environmental catastrophe. Social change is more discontinuous than we usually think, but a future which is radically discontinuous with the past can deprive human beings of both identity and purpose.

We have always recognized that one man's work can be another man's play. Artists and sportsmen and amateurs of all kinds do for pleasure what others do for a living. Youths who avoid regular jobs like the plague will spend weeks of arduous labor customizing their automobiles. Beyond the seeking of bare physiological necessities, what human beings do is determined by a variety of motives, and throughout history we have carried on activities—including what might be regarded as work—for playful motives, even if we pretend to others—or to ourselves—that our motives are purely utilitarian.[49] Because modern western, and especially American, society regards pleasure as bad in itself or else only justifiable as a means for helping us to work more efficiently, play is downgraded and disguised.

Of course, not all play is personally fulfilling or socially desirable. Children often maim and kill small animals as they play with them. Grown men play at being business executives or soldiers with sometimes disastrous consequences for humanity. Scientists play with the universe. The potentially negative character of some play cannot be overestimated. The assaults on our natural environment and the natural fabric of human life perpetrated in the name of economic growth and technological progress must be recognized as stemming not only from utilitarian motives but from playful ones as well. Supersonic transports, genetic manipulation, trans-Alaskan pipelines, and electronic battlefields in Vietnam exist in large measure because to those who design and operate them they are fun. Attempts to subject economic and technological activity to control on behalf of humanity as a whole cannot succeed if such activity is criticized as though it stemmed solely from a desire for private or public material profit. Much of such activity is in fact play—work engaged in for its own sake—including, as in the Soviet and especially the American space programs, collective play. Play is as fundamental a human need as some form of work is, but all play is not necessarily conducive to the long-run satisfaction of human needs any more than all work is.

Play is one way in which the human desire for variety can be satisfied. It has social as well as individual value because it permits individuals to tackle new tasks, to assume new roles, or to think new thoughts within a context that is minimally (if at all) threatening to their already established self-images and to their ability to function in whatever roles currently give them their sense of social worth. Variety, as the oft-quoted proverb asserts, is the spice of life. Indeed, it is an essential ingredient of human life. Variety can mean refreshing change, release of tensions, or opportunity to discover new kinds of purposeful activity—opportunity to "go outside" oneself to seek relaxation or adventure. It provides a chance for change in a situation in which "coming back" is still possible and even expected. But it also has a creative potential for being a source of permanent change in the individual or in his environment. In this way it helps preserve the flexibility and adaptability of the human individual and of human society that have always been our key evolutionary advantage.

The political order needs to make certain that all members of society have opportunities to change as well as to maintain their identities and activities. Provision of opportunities for reeducation at all ages—whether vocational, avocational, or cultural—is one obvious way in which the state can help satisfy the desire for variety and help promote the flexibility essential to individual and social survival. Encouragement of "sabbatical" months or years is another possibility. Various kinds of antidiscrimination laws could play a role in opening up new options to individuals. The existence of a variety of recreational facilities, from city parks to wilderness areas, can offer opportunities for diversity in the activities of individuals and for the acquiring of new skills and competences. The possibilities are almost endless, but society through its political process must take thought about how best to make provision for variety as well as for stability in the life of its members.

Both play and work insofar as they are meaningful to those who undertake them can be subsumed under the more general concept of creativity. Human beings are entropy-reducing organisms. They seek to create new order in the physical and intellectual universe, to leave it somehow different from the way they found it. Sometimes this impulse leads to behavior as objectively undesirable as someone carving his initials in subway cushions. Sometimes it is as humble and mundane as refurnishing the living room ("property is the art of the poor," G. K. Chesterton once wrote). Sometimes it creates the rose window of Chartres or the general theory of relativity. But humans seek, as individuals and as groups, to move the world forward toward some goal of greater order or at least greater "made-ness." Human creativity in the past has often wounded the natural or social world in which it took place (Hitler's "new order" was the epitome of social breakdown). New forms of creativity must be found that do not destroy man's habitat, like a giant fireworks display burning down the wooden grandstand which surrounds it. But man's creative impulse is as ineradicable as it is essential to his continued evolution.

All creativity is based on a fundamental human characteristic: goal-orientation or purposefulness. Just as the human species as a whole is moving forward in an evolutionary process, so also individual human beings are forward-thrusting creatures. All human behavior is motivated by anticipation of the future.[50]

Man is a time-binding animal; he looks forward. He cannot escape being haunted by a linear image of the world and his life in it. Paradoxically, the existence of death gives meaning and shape to life. Few human beings can live simply for the sensations of the day, flies of a summer. To be able to savor each moment of existence as a gift to be gloried in is a necessary grace, but there is in all human beings an innate hunger for what comes next as well.

Human purposes vary and clash. They are individually chosen and individually achieved. Yet they take place within—indeed they constitute—society. And the ability of individuals to fulfill their needs for purposeful activity depends on their social environment, which in turn depends on the political order.

13

The Common Good:
Present and Future

BECAUSE they share the same human nature, all human beings have the same basic needs. Satisfaction of these needs provides norms for social ethics; the good society is one which maximizes the possibility of the satisfaction of individual needs for identity and meaningful activity, for security, self-esteem, and variety, within a context of freedom. Politics is a means toward satisfying these needs; politics is the conscious mobilization of the community to achieve satisfaction of human needs. But these needs are the needs of individuals as such, satisfied on an individual basis and to some extent in competition with the satisfaction of their needs by other individuals.

Besides the requirement that the good society provide a context within which individual needs can be met to the fullest extent possible, there are also *common needs* which must be met by society through the political process if it is to survive and develop, and if the needs of individual members of society are to be met. These common needs can and must be satisfied in a fashion which does not require and indeed precludes competition among the members of the system. Taken together, these societal needs constitute what has been called by political philosophers the common good, or, in the more usual modern usage, the public interest.

The Common Good and Its Enemies

The concept of the common good is as old as political philosophy itself, and in the western tradition it can be traced back to Plato (though it is not a term he stressed) and to Aristotle. For the latter the common good was the measure of justice and the goal of politics: "The good in the sphere of politics is justice; and justice consists in what tends to promote the common interest."[1] Belief in such a common good persisted throughout the Middle Ages, and was a staple of the political thought of scholastic philosophers such as Thomas Aquinas. For Aquinas, "Every law is ordained to the Common Good."[2] With the advent of liberalism, however, the concept of the common good was abruptly, though never completely, abandoned.

Liberalism postulates the primacy of the individual, who is necessarily engaged in a competitive struggle with his fellows over particular goods. It holds that people have such radically different desires that, as the influential nineteenth-century utilitarian philosopher Jeremy Bentham put it, "The only common measure the nature of things affords is money."[3] Therefore, the only possible common good for mankind is the liberal panacea of general economic growth.

A few idealist philosophers, such as T. H. Green (who was strongly influenced by Aristotle as well as Hegel), continued to take the concept of the common good seriously.[4] The school of economics known as "welfare economics" has tried to define a common good within the context of economics, but it has been a small minority among economists and has never enjoyed much acceptance among its determinedly liberal colleagues.[5] Some liberal theorists, such as L. T. Hobhouse, held views on the common good which included some substantive elements,[6] but by the mid-twentieth century the only content left for the common good was the preservation of those institutions of civil society which controlled the conflict of individual interests sufficiently so that the struggle for power and wealth could proceed.[7]

Liberal political theory was never a strong enough intellectual force to overcome the dictates of common sense, and ordinary political discourse—including legal and administrative decisions—continued to have reference to the public interest as a meaningful norm for public policy. But formulations of policy based on appeals to the public interest are more and more suspect, and their credibility has been weakened by a growing skepticism about the intellectual validity of the very notion of a public interest.

Modern Anglo-American political science, which is increasingly dominant throughout the western world, has carried this skepticism to an extreme, consistent with its origins in liberal society and its role as the intellectual bulwark of that society. Save for those theorists whose intellectual roots are in classic political thought,[8] only the followers of Rousseau have been able to take the concept of the common good seriously. "The general will alone," Rousseau wrote, "can direct the state to the object for which it was instituted, i.e., the common good . . ."[9] But the concept of the general will has been dismissed along with and for the same reasons as the concept of the common good, and Rousseau's teachings have had little effect save in socialist countries and in developing nations which share the Jacobin, Continental theories of democracy.[10]

Attempts to create a scientific theory of politics in the early twentieth century provided apparent empirical evidence of the fallacy of thinking about the public interest as something real. So-called group theories of politics and their more sophisticated variant—pluralism—postulate, as in Hobbes's state of nature, that politics is a dog-eat-dog conflict among interests, except here the contenders for power are no longer individual human beings but groups of individuals with shared interests.[11]

Groups can be based on ascriptive characteristics, such as sex or race, or on common life-circumstances, such as occupation or income. Their joint interests may be purely material, as in the desire of labor unions and industrialists for high tariffs or, in later formulations, as psychologically subtle as the "interest" of the aesthetically sensitive in the creation of less ugly public buildings. But the assumption or contention is always that there are no interests which are common to all members of society. Therefore, any time the term "public interest" is used in political discussion, it is regarded either as empty rhetoric, betraying the naïveté of the person using it, or is assumed to be a deliberate cloak for the interests of a portion of the population in conflict with the rest.[12]

Above all, there is no such thing as a substantive public interest—better roads, cleaner air, less inflation, for instance—though some theorists would admit that there is general public interest in maintaining the process of politics through which this clash of special interests is resolved. The rules of the game itself constitute a truncated kind of common good, but the game is one in which no final score is better than any other from the point of view of the community of players as a whole. Indeed, to postulate otherwise, to argue that certain substantive results of the public policy-making process are intrinsically preferable to others, is considered anti-democratic or hostile to freedom, since it is assumed to imply (as of course it does) that in certain cases the rules of the game might have to yield to the requirements for producing a certain final outcome.

This modern denial of the existence of a common good or a public interest is fallacious for the same reason that the liberal philosophy of life which it reflects and supports is fallacious. If, as I have argued, human beings have a nature, examination of this nature enables us to make ethical judgments (however fallible, like all human enterprises) about the substantive outcomes of actions. Such judgments can be made not only about individual but about collective actions. Since individuals are members—indeed creations—of biological and social systems, there must be public—systemwide—interests as well as individual interests, public needs which must be met if humanity is to fulfill its inherent collective potentialities.

The common good then consists in the fulfillment of the collective needs of human societies, including global society (humanity as a whole).

It is made up of interests which all human beings share and share equally, the fulfillment of which is not a competitive but a communal process, since the fulfillment of the needs of one individual in this context does not conflict with the fulfillment of the needs of another, and may indeed depend on it. The common good consists of those properties of the total social system which must be maintained if the system is to function, survive, and develop, and if the individual interests as such of any member of the system are to be served.

The Nature of Common Goods

Some skepticism about the existence of common goods in human society stems from the fact that while all members of society have a common interest in the existence of such goods, some individual members may benefit more from them than others. The progress of medicine, let us assume, has benefited the human species as a whole, but not all human beings have equal access to expert medical care. Yet as long as the provision of medical care to some does not absolutely worsen the health of others, the race as a whole has benefited. All members of society have a common interest in some level of public order, but some members of society are thieves. However, even the thief would be chagrined to return home with his spoils to find his house burned because arson is unchecked. The existence of differences in the degree to which individuals or groups share in the common good does not mean, therefore, that there are no goods common to all. Skepticism about the existence of common goods sometimes derives from the fact that different means for achieving common goods affect various individuals differently. All may benefit from police protection, but the taxes which support it may be levied in an inequitable fashion. The inequity of taxes, though, does not affect the desirability of the police protection and can be adjusted without diminishing the good achieved.

A more compelling objection is that in some cases there seems to be a direct connection between helping the community as a whole and hurting some of its members. An efficient system of traffic control may result in greater convenience for people living in most areas, but in locally increased congestion for others. Economic progress toward lower costs of producing or marketing basic necessities for the population at large may injure the immediate particular interests of those who produce or sell goods at necessarily higher prices. If the individuals disadvantaged in such cases were simply and solely residents of particular streets or producers of certain goods, it could plausibly be argued that the concept of the existence of a common good in efficient traffic movement or cheaper prices was erroneous. But all members of society are engaged in multiple relationships within society, even if this does not always involve multiple formal affiliations with interest groups, and it can reasonably be assumed that the relative losses suffered by individuals in the course of particular activities which promote the common good are balanced by their gains in their other capacities. The new traffic light which causes congestion in my neighborhood may be a functionally necessary part of a total pattern which enables me to get to work earlier; even if this is not so, the new system may lead to a decrease in accidents and a reduction in my insurance premiums, decrease the transportation costs of goods I buy, or reduce the amount of smog. In a society organized so that gains in collective economic productivity are equitably distributed, my loss of a job as a result of new methods of production which make my skills obsolete can be compensated for socially. If it is cheaper in total economic terms to keep me on the job than to retrain me or give me a pension, the change which resulted in my unemployment was not a contribution to the common good of the total society in the first place. Injustice in the distribution of common goods does not vitiate the claim that they exist and can be identified.

Group theory postulates that men belong to different groups because they have different interests stemming from different functions, different roles in the social system. Our automobile driver and displaced worker can be thought of as members of groups if any others share their fate. But as Mary Parker Follett noted half a century ago, "Man has many functions or rather he is the interplay of many functions."[13] The conflict of groups which supposedly confounds the existence of a public interest is actually a conflict within individuals. Usually "it is more appropriate to talk about the mental conflict between the interests of an individual as an individual and the interests of that same individual as a group member, that is, in his social relationship."[14]

What is true of the relationship between the individual as an individual and the group of which he is a member is also true of the individual as a member of a subgroup within society—a farmer or a devotee of skiing— and as a member of the total society. Social policy must serve the needs of individuals in all their group identities. The good of individuals or groups is subordinate to the common good because man exists, in John Dewey's phrase, as a member of a public, a public which is a product of the total pattern of interaction among human beings in all their particular functions and roles.[15] Thus the more extensively we participate in the life of the total society, the more we become aware of the existence of the common good and of the breadth and legitimacy of the claims it makes upon us. Political democracy, Dewey correctly argues, "forces a recognition that there are common interests, even though the recognition of *what* they are is confused."[16] Walter Lippmann has defined the public interest as "what men would choose if they saw clearly, thought rationally, acted disinterestedly and benevolently."[17] He is right, except in his insistence on disinterestedness. Men and women do not have to be saints in order to seek the common good, since the public interest is what people would seek to effectuate if they always saw clearly and sought rationally their own best interests as human beings. A principal purpose of politics—and a major justification for democratic, participatory politics—is to enable people by interacting *consciously* with others (interact they always will, whether they are aware of it or not) to enlarge their recognition of their common interests. Thus, "politics," as Robert Maynard Hutchins rightly asserts, "is the science of the Common Good."[18]

The existence of common goods becomes more readily apparent once one recognizes that not all interests are strictly material, in the commonplace usage of the term. If we accept the broad definition of interests which even group theorists have been forced to adopt,[19] it is obvious that material differentials which may exist among individuals as a result of the collective search for the common good through politics or other means can be and are balanced by psychological factors; indeed, material and psychological goods are broadly interchangeable. In fact, psychological goods can take precedence; once certain basic survival needs are met, people usually do give priority to psychological satisfactions in the broadest sense, as humanist psychology implies.

Even the willingness to give priority to collective common ends with which one identifies—call it public spirit, altruism, or whatever other term we choose—is a more common behavioral characteristic than theorists of competition who use the clash of economic interests as a model for politics can admit. I may find psychological satisfaction in the progress of medicine, the maintenance of public order, efficient transportation, or rising general living standards, even if I do not share directly in their benefits equally with everyone else or am in some aspects of my existence inconvenienced by society's pursuit of these goals. Pride in national accomplishments is often strongest among the least favored members of society, as political leaders who call attention to national shortcomings find to their sorrow. Human beings have long accepted differentials and sacrifices in the name of national glory, especially in wartime, or even accepted them to some extent in the name of general, specieswide accomplishment. However dysfunctional war and some of the activities embraced as progress may be, people do in fact conceive of themselves as being benefited by activities which bring them no direct material good, and the satisfaction they derive from such goods is as real to them as any others.

In order to be able to deny that there are elements of systems which involve goods common to all members, one would have to find individuals who are absolutely and totally worse off—physically and mentally—as a result of the existence of such goods. Since many elements of the common good are interrelated—economic prosperity requires political order, for instance—one would have to find individuals who were totally worse off as the result of the conjoint maximization of the elements of the common good within a society. Such individuals and groups of individuals may indeed exist within (or in association with) societies and the species as a whole, but if they do they are persons who have been denied full membership in the society—nonintegrated elements who have been so cut off from participation in the life of the whole community that they are the obvious victims of the most flagrant exploitation, oppression, and injustice—and the existence of such persons is a remediable condition which can and must be changed.

The Common Good in Ecological Perspective

The existence of common goods is a commonplace throughout living nature, which consists of systems of interdependence of interacting elements. Unicellular organisms unite to perform actions necessary for survival which no individual could perform alone; primitive slime molds consist of individual organisms which, when collectively joined, can move in search of food, something which the individual members cannot do. Various species of living creatures have symbiotic relationships with other species in which survival depends on mutual assistance. But nature does not always know best. Despite the fact that balance is a common good for all the species which are members of an ecosystem, in some living systems short-run competition among different interests can menace survival. Normally species in competition with each other within an ecosystem are in balance—predators keep down the numbers of grazing animals with the result that there is enough grass to feed them. But if what biologists call "competitive exclusion" occurs and all the grazing animals are killed, the predators themselves will go hungry, and within the particular local ecosystem both species will cease to exist.[20]

What constitutes the common good for a human society? Despite the rejection of the concept of the common good by most political philosophers today, common sense will not be denied, and various social scientists have attempted to set forth the elements of the common good or public interest under other names. The rise of "functionalism" in anthropological and sociological theory has led to a number of such formulations.[21] Two weaknesses of these efforts have been their tendency to concentrate on the obviously social aspects of social systems with a consequent neglect of the material bases of society, and their emphasis on the maintenance of existing structures, processes, and norms of particular societies, rather than on those mechanisms for learning and change within societies which help ensure their survival and development.

What follows is a generalized theoretical statement of the basic needs of society as a whole which, taken together, constitute the common good.

All that is said here applies, unless otherwise noted, both to individual national political societies and to the all-encompassing society consisting of humanity as a whole. For our purposes the common needs of human society can be most clearly set forth under a threefold classification of subsistence, order, and purpose, somewhat parallel to the individual needs for physical survival, personal identity, and purposeful activity. Though for analytical purposes we must deal with these various components of the common good separately, they are interrelated and interdependent; none can be achieved without the others. The common good consists of their necessarily joint existence.

Subsistence

Human societies are groups of human beings whose physical survival depends on utilizing the resources available within their natural environment by means of their technology. The level of their physical existence depends on the interrelationship between the resources available and the technology involved in their utilization. Physical resources are natural endowments; technologies are created by societies in accordance with the laws of nature.

The natural environment consists of a fixed sum of matter and energy which makes possible the existence of the biosphere of which humanity is a part and of an essentially fixed sum of energy which enters the biosphere on a regular basis as a result of the sun's radiation. Air, water, soil, and (in developed societies especially) minerals are necessary for human survival and subsistence. But the major resource on which human subsistence depends is energy,[22] the factor which enables people to use all other resources.

Throughout human history, most of the available energy has come from the sun. Primitive and ancient peoples recognized this fact and some accordingly worshipped the sun as the source of life. Human beings feed on the energy of the sun by eating plants which derive their energy from the sun and by eating animals which derive their energy from plants. Humanity has also been able to command the energy of animals directly by making animals work for them. Humans burn wood, releasing the stored energy of the sun. Modern industrial society is based on the release of the energy of the sun stored in former plants and animals in fossil fuels such as coal and oil. To a limited extent industrial society has, through hydroelectric power, been able to use the solar energy which has gone into the meteorological cycle of evaporation, rain, and consequent water storage at certain altitudes. Marginal amounts of energy have come from sources other than the sun: from the gravitational pull of the moon on the oceans— tidal power—and from the continued heat of the physical constituents of a still partially molten earth—geothermal power. In contemporary times a vast store of non-sun-derived energy is beginning to be tapped through nuclear power plants which utilize the energy "trapped" in the constitution of matter itself, matter after all being only energy in certain relationships.

Human beings, and all living animals, also require another resource: oxygen. Their lives as organisms are based on using oxygen to help convert food into energy to maintain internal life processes. The supply of oxygen in the biosphere is of course vast, but it is also definitely finite, and the supply available can be and is affected by man's activities.[23]

The subsistence level of any human society depends on the kind and quantity of resources available to it. Maintenance of levels of subsistence over time depends on the maintenance of levels of resource utilization over time. Some resources are relatively renewable, even finite supplies of minerals can be recycled and reused. But all resources can be dissipated to an extent where they are for practical purposes unrecoverable, and even the soil upon which vegetation depends can be destroyed. Some sources of energy are naturally renewed—like the direct energy of the sun—but are difficult to use in comparison with their availability. Other more accessible sources, like those on which contemporary industrial society rests, are finite; once used, coal and oil are gone forever. Subsistence systems based on their use are therefore highly unstable, since they cannot be maintained indefinitely. The energy available from atomic fission is limited also to the small amount of highly fissionable forms of uranium available.

If nuclear fusion becomes practical as a power source (and research toward this end is now under way both in the Soviet Union and the United States), other forms of uranium could be used, and it could provide an inexhaustible fund of energy for the foreseeable future, though its use might have insupportable side-effects. However, any society basing itself on any form of nuclear energy would be making a tremendous gamble on its own future stability. To maintain an economy based on nuclear energy would require a high degree of order for the proper use and safeguarding of radioactive materials, including radioactive wastes. In a nuclear-based society, civil war could mean total disaster. Just as a society based on agriculture is committing itself to remaining in being until the next harvest, and one based on large-scale irrigation to a high degree of social peace over time, a society based on nuclear energy necessarily makes a commitment to and on behalf of future centuries.[24]

Space is a prerequisite for the physical maintenance of the human species.[25] Space in its broadest sense includes the total physical context in which human life exists, in which all of the events and processes relevant to human existence on earth take place. For instance, maintenance of oxygen levels in the earth's atmosphere depends on the existence of space given over to the plant life which converts carbon dioxide created by the breathing of animals and man back into oxygen. No organism can live on its own waste products as such. They can be recycled or made reusable by treatment involving the use of energy. But the process of recycling also requires space. Moreover, all the energy that we use produces heat as a byproduct, and life-support systems in order to maintain themselves must somehow get rid of this excess heat.

One theoretical limit on the number of human beings the planet can support under any assumption of the availability of resources is the amount of heat that can be radiated into outer space. Another theoretical limit to the number of human beings who could exist on earth even if they were piled one on top of the other is based on the ratio of the weight of human flesh to the mass of the earth. Attempts to calculate the absolute limits established by physics concerning human population size and utilization are of course ultimately silly, useful only as an intellectual exercise,[26] or as a reminder of the inherent stupidity of the argument that there are no limits to the possible population the earth can support. Millennia before any such limits were approached, human society would break down and human life in any meaningful sense would be at an end.

The continued subsistence of any human society is dependent upon the relationship between (1) the amount of energy available and (2) the space necessary to permit its use and to absorb the byproducts of its use. These are absolute limits for the human species as a whole, granted the assumption of the planet Earth as humanity's habitat. Even if elements of the species emigrated elsewhere there would still be limits to the subsistence capability of the home planet, and any possible input of energy or other resources from extraterrestrial sources (there is already serious talk about mining the moon) would not eliminate the space-linked limits the planet puts on their utilization on or within it. Actually, we know relatively little about the precise limits which the biosphere places on the growth of human population and activity. What we do know is that there are such limits and that some limits can be overstepped in short order when the pressures on them are increasing at an exponential rate. Just as an additional degree of heat converts hot water into steam, so an additional small increment of pressure on the ability of natural systems to regenerate themselves can totally destroy that capacity. Unlike the conversion of water into steam, this could be an irreversible process. There *are* straws that break camels' backs.

In addition to the need for energy and space, the ability of a social system to attain certain levels of subsistence for a given number of members is determined by a third factor, technology. Technology is "the institutionalization of utilitarian norms," the "application of rational principles to the control and reordering of space, matter and human beings."[27] Technology is a natural outcome of man's genetic endowment of intelligence and his adaptive, goal-seeking behavior. Primitive man had his technologies. Even other living creatures have technologies. Birds, animals, and insects build nests and use tools. The difference between man and other animals in respect to technology-creation is like that in respect to consciousness and other mental capacities, a matter of degree rather than of kind.

Technology exists to serve human purposes. Some of these purposes are other than mere subsistence. Primitive and modern men alike created technological means to serve psychological needs, as in art and religion, but technology has above all expanded man's ability to subsist. It has enabled human beings throughout history to increase the amount of energy they can utilize for their own physical needs and desires: to hunt animals and make them food or beasts of burden, to better utilize the sun's derived energy stored in plants through agriculture, to work wood and iron into useful tools and artifacts and dwellings, to liberate the energy of fossil or atomic fuels. Technology has also, but so far to a tragically limited degree, helped humanity to meet the problems of the finiteness of space and of other resources by enabling us to reuse materials or dispose of wastes. Technology is as necessary to the existence of the species as a whole, or of any social system within it, as energy and space. Without technology humanity would be a precariously surviving species, subsisting on the food it could gather from untended plants and the animals it could catch with its bare hands, ravaged by disease, naked and shivering, sheltering in caves or under trees.

The common good of subsistence therefore depends on the energy, space and technology available to any given human society. In the case of societies smaller than the race as a whole, it depends on the resources (especially the energy resources), the space, and the technology available or generated within their territorial boundaries. These can be and are supplemented by factors of subsistence obtained from other sources. Western Europe and North America get oil from the Middle East and other underdeveloped regions. They export wastes in the form of pollutants which go into the atmosphere of the total planet. They export technology to less developed countries. Imbalances within societies can be compensated for by outside factors; nation-states are normally open rather than closed systems. But for the planet as a whole, balance is necessary. Subsistence and survival on the level of the species depends on a balance between available energy and space (constituents of nature) and the technological utilization of these resources. It depends on a balance between technology and nature. This is a point on which all informed students of the human present and future—both pessimists such as those associated with the Club of Rome and optimists such as Buckminster Fuller—necessarily agree.

The balance between technology and nature is subject over time to certain instabilities which can impair subsistence levels and even threaten species survival. The consumption of finite resources leads eventually to their exhaustion. The ability of technology to compensate for this by making new resources available to fill the same subsistence needs is not infinite and is dependent on time factors which may not coincide with the demands put on the system. The Mayans could have survived as a society, given radical cultural changes, if they had had new technologies, but these were not available, and Mayan civilization perished.[28] Many of the technologies which might solve our contemporary difficulties may—even if intrinsically feasible—not be available when they are needed. Ironically, the very creation of these technologies for the future may be menaced by the high consumption demands of humanity on presently available resources, a situation similar to that of poor individuals who cannot accumulate capital for future needs because they must use all of their assets for immediate survival.

The amount of space—including above all the area necessary so that plant life, both cultivated and uncultivated, can recycle waste products—is limited. Physical space suitable for isolating certain nonbiodegradable wastes is also finite. And technologies which increase the amount of available energy at any given time may radically acerbate the problem of finding space for waste recycling and removal.

Basically, the subsistence needs of a social system are the joint product of its level of population and its cultural patterns. Increasing levels of population and culturally conditioned forms of resource consumption create imbalances which cannot be remedied even by technological advances, since finite amounts of utilizable energy resources and above all space cannot support infinitely increasing population levels or infinitely increasing energy consumption.

The common good of subsistence is the prerequisite for any other common goods, just as physical life is a prerequisite for psychological processes at the individual level. Subsistence demands an equilibrium among four factors: energy, space, technology, and population. This equilibrium need not be conceived of as static and fixed, as in the homeostasis of the human body. It can be a moving equilibrium within fluid limits. But at any given point in the history of a particular society or of the race as a whole balance is a necessity. Whatever the precise physical limits of the capacity of the planet Earth to provide for human subsistence may be, such destabilizing factors as increasing consumption of space and an increasing population cannot continue indefinitely without the destruction of the common good of subsistence and survival.

Order

Societies require order for survival and development. Social life is a process of interaction among human beings. Social structures consist of regularities in the patterns of interaction. Since individual existence and identity is a function of social interaction, then both the individuals and the societies which their interaction constitute can dissolve unless relatively stable patterns of interaction persist over time. Since human societies are composed of conscious, perceiving individuals, *perceivable patterns of order are necessary* if order is to exist at all. If ordered relationships of production and exchange did not exist, for instance, the physical subsistence needs of society could not be met. When civil wars break out, those people farthest from farms are in danger of starvation, and must use force, often with only limited success, to seize food from agriculturalists as normal economic relationships disintegrate.[29] Insofar as patterns of order require individuals to act in ways which they do not perceive as being useful at any given moment, individuals must have physical power to control each other's actions, and social relationships therefore take on the characteristics which we call political. The political order of society is not a separate set of social relationships, but a particular characteristic of social relationships.

Common perceptions of the physical and social world constitute cultures. For a society to have the order necessary for survival and growth, *a common culture is required.* This common culture, by structuring the perceptions of its component members, leads them to interrelate in certain kinds of patterns. Not all members of a functioning society need share all cultural characteristics with all other members; a certain amount of cultural pluralism is possible.[30] But normally the members of a particular society must share a common perception of the possible and desirable outcomes of interrelationships if the society is to function effectively.[31] If I think an outstretched hand is a sign of friendship and you think it denotes hostility, we can never be friends.

Not all differences are socially dysfunctional. Some differences in perception can be complementary or compatible or lead to outcomes which are functional for maintaining societies. Many agreements are arrived at because of a difference of perception—especially of judgments about the future—between buyer and seller, husband and wife, or contending political or social groups. Sometimes different preferences are necessary for social efficiency. Because Jack Sprat and his wife had differing tastes in meat, they were able to lick the platter clean and cut their food bills. But culturally conditioned agreement on the ends of actions—especially collective actions—is generally the best guarantee of stable patterns of interaction.[32]

In certain situations sheer political power—with its implicit threat of coercive sanctions—can maintain patterns of social interaction necessary to social stability (ships at sea are traditionally steered not by universal consensus but by their captains' orders); but coercion is a limited and highly inefficient method of social control. If societies require order to be able to survive, they require a certain level of cultural consensus as well.

But societies exist for more than survival: They are vehicles for the development of their individual members and of the human species as a whole. And even survival requires that societies be able to meet new challenges which arise from changes in their environments, including the activities of other societies.[33] This being so, societies cannot depend simply upon the regular patterns of automatic interaction enshrined in particular cultural systems, the way we depend on our lungs to breathe normally without our conscious attention. *Societies must possess a collective intelligence and will.*

The terms *intelligence* and *will* do not here imply any physical or metaphysical entities apart from the ordinary workings of society. They simply refer to collective self-consciousness which exists on an observable basis in what we call leadership, government, and public opinion. Societies do in fact perceive collectively: Leaders and members of a hunting tribe recognize that game is becoming scarce; leaders and many members of contemporary national societies become aware that their currency is overvalued on world exchanges and their economies are suffering. Societies also have collective wills: Hunting tribes decide to move to new areas in search of food; governments devalue currencies so that their nations can compete in foreign markets. Collective will and intelligence are operational, observable facts of life.

But man is a future-oriented, purposive animal. Therefore social will and intelligence must be future-oriented. Such future-orientation is what is called *planning*.[34] Just as individuals exercise their freedom to create desired futures by channeling their current actions in such a way as to achieve future goals, so do societies. Without planning, freedom cannot exist at either an individual or a collective level.

Shared perceptions (culture) and shared will (planning) are necessary elements of social order. Together they make possible that collective action which constitutes society.

A final element of social order is *participation*—what is sometimes called democracy. Coercion is not only unpleasant, inefficient, and unreliable, it negates the very concept of social self-control. Where coercion is a normal rather than a marginal means of maintaining patterns of interaction, society does not really exist. Instead, what is present are two societies, those of the controllers and of the controlled. Differentials in power among individuals and social groups may permit the persistence of such patterns over varying periods of time, but they are inherently unstable. Such divided "societies" are highly vulnerable to radical change under the pressure of outside forces and have low survival value; and they are incapable of development, since any new cultural or economic forces threaten their inherently brittle structures. Not only do such social arrangements inhibit the development of the coerced partial societies within them, but the expenditure of energy required to maintain social relationships based on physical force rather than shared perceptions and goals also limits the development of the dominant individuals and groups as well. Living systems grow and bend; nonliving systems remain inert and break. Tyranny— the classic Greek definition of which is government in the interest not of the community but of the tyrant—is the negation of social order, conducive neither to survival nor development.

Purpose

The concept of social purpose as such plays a limited role in liberal political philosophy. The fascist doctrine that man exists for the state met with scorn in the western democracies, which, together with their collectivist Soviet allies, destroyed the state which proclaimed it. Yet it is normal and natural for human beings to seek ends beyond their own personal lives. When the fascist doctrine was rephrased by a personally attractive American President as "Ask not what your country can do for you but what you can do for your country," it was eagerly embraced by a democratic people. The tragedy of Vietnam is now leading to a reevaluation of the facile grandeur of the Kennedy years,[35] but the yearning for social purpose is as basic as are the related desires which cluster around the institutions of religion, and just as persistent. Social systems cannot long endure without a sense of purpose any more than can individuals, even if that purpose is simply the perpetuation of the systems themselves and of the cultural values they incarnate. Even static societies such as Confucian China and caste-ridden Hindu India have had a purpose in that sense.

The social purpose of liberal society has been to increase individual autonomy and economic growth. But even liberal societies can have special national purposes. The concept of Empire and Pax Britannica made life meaningful for millions of ordinary Englishmen as well as for their aristocratic rulers. The French "mission civilatrice" has survived even the vulgarities of the Third Empire and the Third Republic. And "making the world safe for democracy" has constituted an American purpose broader than its particular use as a wartime slogan. "Building socialism" today holds together the social systems of many nations; and "national liberation" still has meaning in less developed countries, even those which are technically legally independent. Group liberation or "power" gives meaning to the lives of members of many racial and ethnic minorities, while keeping alive the People of the Covenant does the same for most Jews. Small nations with little hope of influencing the future of the world, such as Denmark or Trinidad, can focus on simple survival or a happy hedonism for their inhabitants, but most nations seek a more future-oriented role in human affairs.

Purpose is, of course, the future-oriented expression of identity and inseparable from it. Common bonds of language, culture, and history usually hold nations together, but in some cases they alone are not enough. Britain today suffers from a crisis of identity brought on by its rapid decline as a world power. France under Gaullist régimes has sought, somewhat fatuously but understandably, to cling to the remnants of "grandeur." The United States today suffers from a dual identity-crisis resulting from uncertainty about its world role and uncertainty about its national identity, as blacks, Chicanos, and other ethnic groups reject the accepted meaning of American history, while women's liberationists and many young people contemptuous of the work ethic reject its traditional cultural patterns, and an administration elected in large part as an embodiment of traditional moral and cultural attitudes is exposed on national television as a cesspool of lawlessness and corruption.

Purpose is necessary for societies because, while society consists of the patterns of interaction among its members, the equilibrium achieved is not purely homeostatic but dynamic, kept in balance by the movement of the system as a whole. Other animal groups can remain in equilibrium by trying to survive as individuals and to keep the species alive. Human societies are self-conscious and must have a guiding vision of the future.[36]

The vision which underlies purpose, being future-oriented, must center on posterity. Most human beings take this for granted. We worry about the future of our children and our country. Few of us would care to live lives in which, as one political theorist approvingly suggests, "we would all write in disappearing ink."[37] But why *should* we care about the future?

We care about the future because society is not a mere partnership for the production and sale of goods, but a union of women and men—and not only those alive today but those dead and not yet born. So argued Edmund Burke, and he was right. Just as we can amount to nothing as individuals in isolation from our contemporary fellow humans, we are nothing if we forget our ancestors and neglect our posterity. We are all part of the living stream of history and of the evolutionary process, kin to our grandfathers and grandchildren, to the amoeba and whatever type of human being may follow us.

This assumption of the moral value of continuity must be central to any philosophy of life appropriate for dealing with the problems of post-liberal society. For the essence of the unhappy legacy of liberalism is that we are in the process of despoiling our patrimony and impoverishing and perhaps even genetically perverting our progeny.

It is possible to argue that we do not have any moral debt to posterity.[38] What has posterity ever done for us? Posterity, save in children already born, does not exist. We would not be able to recognize our distant descendants any more than we would be able to recognize our distant ancestors, or they us. We might well not approve of their values or they ours. Even if we admit that we have some tenuous obligation to the hypothetical men and women of future generations, is it not more remote and less compelling than our obligations to our billions of contemporaries? To what extent can we justify policies which might diminish the lives of the poor in less favored lands today so as to provide food and space and options for our great-grandchildren or theirs?

Some might argue that whether or not we have any obligations to posterity we cannot predict and therefore cannot rationally influence the future, that prediction in the social sciences is impossible and futurology is bunk, worse than science fiction. Others might argue that we have no right to try to influence the future: Our descendants will have different values, and we should not impose our ideas on those yet unborn. As one writer puts it, we can't predict the utopias of our grandchildren "no longer . . . sitting inside the hothouse of one of the historic high cultures."[39] If such arguments are valid, attempts to influence the future lead only to injustice in the present and tyranny in the future. Though the concerns and premises differ, the results of accepting either the argument in favor of the present against the future, or the argument based on our presumed lack of right or ability to influence the future, are the same—we have no operationalizable obligations to those who come after us. Posterity can damn well take care of itself.

Can we know the future? As I have argued earlier, we can, or at least we can foresee what the future will be like in the absence of action on our part to alter it. Should we try to do something about the future?[40] Are we morally obligated to worry about whether our descendants will have water to drink or air to breathe or whether they will come off an assembly line in a human hatchery rather than being born? Do we have an obligation to pass on to those who come after us the opportunity to live human lives?

The argument in favor of posterity can be stated rationally. If, as I have contended, the universe has meaning, one of its imperatives is the integrity and progress of the human race. We are trustees for God or nature or the historical process. Nor is acknowledging this obligation a matter of pure altruism. Present societies depend on the future as much as it depends on them. Even if the inhabitants of the future might not always appreciate our actions, we have a selfish motive for being concerned about posterity, one which they in their turn will share. Individuals and societies that give no thought to the future are as selfish as Plato's city of pigs and are likely to wallow in filth even in their own lifetimes, creating nothing of value. The desire for fame and a good name in history, as the Greeks knew, is one of the things that makes men behave decently in the present and keeps social life tolerable. If we give no thought to what those who will come after us will say or feel—to the verdict of posterity—life in the present could become a jungle.

But, it could be argued, many high-minded men and whole schools of thought have throughout history assumed that the world was coming to an end and that one should give no thought to the morrow, yet their followers behaved decently. The early Christians expected the Second Coming momentarily and only after the hope of Parousia receded into the distance did the Constantinian captivity of the Church, with its attendant evils, begin. But these early Christians were able to act indifferently toward their earthly future because they believed in a future beyond history. To commit oneself to belief in the inherent value of life on earth is to commit oneself to concern over its future.

Many religions in both East and West hold explicitly or tacitly that the future is in the hands of Providence and that we should not worry about it, certainly not to the extent of changing the hallowed ways of the past or the cherished habits of the present. "Inshallah," says the Muslim peasant who disdains better seeds and fertilizers because the will of God which it is impious to flout can alone determine what crop his land will yield. Christians often echo this fatalism. It is possible to feel love for posterity and yet not take thought for the future. But not if one believes that we are destined to be the collaborators of Providence, its agents in building and maintaining the earth.

Some people would implicitly argue that the best way to take care of the future is not to worry about it as such, but simply to perpetuate the past. Following tradition is often a good way to take care of tomorrow; its survival value has been tested. Societies and individuals necessarily do seek to retain existing norms as well as to achieve new futures.[41] But in an era of technologically induced change amounting to a planetary crisis, the most valuable traditions of the past can only be preserved and extended into the future by conscious future-oriented actions designed to preserve them.

Ultimately though, the decision to care about posterity and to do something about the future must be made on nonrational grounds. The great imperative of animal life is not individual self-preservation but preservation of the species; the struggle for existence is, paradoxically, a cosmic school of love. We humans today remain blood brothers to the humblest creature dumbly giving her body to save her brood from a predator, to the most brutal protohuman risking danger in the hunt so all might be fed. We share in their desperation and pain. Even if we believed we had no stewardship to answer for to a higher power, that the universe was as meaningless as some philosophers contend, or that our descendants might reject our sacrifices on their behalf, we would, as societies, still have to follow the example of our cousins and ancestors or we would simply not be human; for we would not be animals serving life. Caring about posterity is as much part of human nature as breathing and needs no more justification. To fail to care is as much an example of pathological deviance as would be the refusal to draw breath.

But, just as we are kin to all those fellow creatures who serve the evolutionary process by struggling to keep their species alive, we have a special burden. Our intelligence is our prime instrument of survival. Its existence is the result of the upward struggle of others. Part of what sets our species apart from those of our cousins is our conscious orientation to the future. Of course we cannot know exactly what the world in the year 2074 will look like despite all our social indicators and computers, or how our progeny will regard the future world we shape today. Neither could our most primitive ancestors know what would happen if the tribe moved in search of water or game or better soil. But they read the signs of the earth and the stars as best they could, took counsel with each other, wagered the lives of their fellows and their own, and made their choice. We too must choose, for not to decide is to decide. We must not only care about the future but act as if we could do something about it. To do otherwise is to betray not only generations of children unborn but also that primitive band huddled in the flickering council firelight and to ally ourselves with the alien and hostile eyes watching from the surrounding darkness.

The Global Common Good

The concern with posterity which is necessary to give purpose to society must become in the next few decades a concern not merely for the future of national societies but for the human species as a whole, threatened as it is by war and the joint crises of pollution, overpopulation, and resource depletion which, together with the ever-present shadow of nuclear holocaust, raise the specter of ecocide. It must be a concern not merely for the physical survival of humanity but for the survival of human cultural and biological continuity.

National purpose takes on new meaning in this perspective. The leading nations of the world, including the United States, are the ones who have the power to destroy the race by nuclear warfare, to destroy the biosphere by unchecked use of scarce resources or space. They are the nations in whose laboratories and hospitals the genetic future of the race will be decided. They are also the nations whose overutilization of scarce resources mocks the poverty of less-developed nations and makes attempts at population control in the latter appear meaningless and futile—the nations whose overgrowth at the expense of the total planet undermines the possibility of any democratic world-order. But smaller and less-developed nations too can find new purpose in concern with the future of the whole race. They can become pilot plants for social experiments in relationships among persons and between humanity, nature, and technology, and they can lead by the power of example and ideas. The overriding national purpose of all nations in the coming generation must be the restructuring of their own societies and of their roles in the world so as to enable human society as a whole to fulfill its collective needs for subsistence, order, and purpose—to achieve the common good of mankind.

Our consideration of the needs of society thus brings us back to the point at which we began our discussion of human needs and their fulfill-ment: the primacy of politics. For the subsistence, order, and purpose of human society are interrelated needs which can only be met within a political context. The economic order on which subsistence depends can-not, especially on a global basis, function automatically but must depend for its regularities on decisions implemented by means of the collective intelligence and will of society. The world economy is necessarily a politi-cal economy. It can only be restructured in such a way as to provide the necessary redistribution between rich and poor and the necessary checks on unsustainable runaway growth through political decision-making within and among nations. The creation of a viable world social order will depend on the breaking down of barriers and patterns of dominance among nation-states—a matter of political decision.

The sense of purpose involved in creating a human future for all man-kind presumes a movement toward a world political order. The world common good demands the existence of a world community, a planetary public consisting of all those affected by the actions of others on earth. Such a community, a total society binding together all the lesser societies within it, need not find expression in a single world "state" in traditional legal terms, but it must be an ordered community with authority of a political nature—that is, the ability to shape patterns of human interaction in accordance with a collective will.

The creation of a world community to serve the world common good is a political act. But its prerequisite is a cultural transformation. Such a world community cannot flow from or even make sense in terms of liberal political philosophy or the scientific and metaphysical assumptions underly-ing liberalism. It can only be justified and worked for in terms of a new philosophy of nature, man, and society.

14

Ecological Humanism:
A Political Philosophy
for Technological Man

THE vandal ideology of liberalism cannot provide guidance for the solution of the problems of human beings living in a technological society. Nor can such alternative political philosophies as conservatism, socialism, the New Left, or anarchism. The rejection of politics espoused by the counterculture or by religiously inspired political nihilism is no answer either; instead, it is a surrender to the forces of dehumanization and destruction. Technological man—man living in a technologically based society, which he must master if he is to remain human—needs a new political philosophy. This new philosophy is ecological humanism.

This philosophy is ecological in that it is based on the conviction that man, if he is to continue to exist and if he is to fulfill himself as a human being, must live in a conscious ecological relationship with nature and with other men, and that the ecological perspective on the natural order provides a necessary analogue for the social order. This philosophy is humanistic because it holds that men are self-conscious, that they relate to nature and to each other through culture, and that the ecological balances within society and between society and the biosphere cannot be left to automatic, unconscious mechanisms as in other natural systems, but must be the subject of human knowledge, judgment, and will—and ultimately the subject of conscious political action.

In the preceding chapters I discussed the natural and social bases of the philosophy of ecological humanism. In the succeeding chapters I attempt to apply this philosophy to the current crises facing America and mankind by suggesting some of the public policies it implies and the kinds of political and social action that will be needed to implement its tenets. It would seem desirable at this point, therefore, to sum up the major themes of ecological humanism.

Man and the Universe

(1) The universe is a process moving forward in time. Matter and energy, the material and the nonmaterial, are aspects of that process. All natural entities are defined by and exist as a result of natural processes.

(2) Life and all its attributes consist of processes. Living entities are defined in terms of processes, that is, in terms of their relationships with other entities.

(3) Life is an emergent property of matter; consciousness is an emergent property of life. Both have evolved in the course of the ongoing development of the universe.

(4) Processes which maintain themselves over time are structures, that is, patterns of order which persist because of relative equilibria among their constituent elements. The solar system is such a pattern. The biosphere of the planet Earth and the various ecosystems within it are such patterns. Human individuals and human societies are also processes in equilibrium.

(5) Change in nature is the result of the interrelated changes among the elements of natural systems. These changes of relationship over time alter the characteristics of natural systems, living and nonliving. Thus the equilibria are not static. Over eons they break down. But changes in their elements and in the relationships among these elements must remain in balance as the total system changes its characteristics, or it will quickly disintegrate.

(6) The world and humanity are one entity, one system in equilibrium. Earth is humanity's only home; humanity is one people in relationship to the earth. We are all passengers on the same spaceship. The capacity of the earth to sustain human life is finite and the relationship between human numbers, human technology and the earth's resources is the most basic factor in maintaining the total system of the world and humanity in equilibrium.

Human Beings and Society

(1) Societies consist of processes of interaction among human beings which persist over time. Collective decisions and social institutions are

emergent properties of the interactions of individual human beings.

(2) Human beings are the product of societies. They attain self-consciousness and identity and have their existence only within societies. Their rights, freedoms, and duties are statements about their relationships to other constituent elements of their societies. They necessarily both control their societies and are controlled by them simultaneously. Individual potentialities can only be realized within the framework of the needs of the total system of man and nature.

(3) Human societies and humanity as a whole exist within a natural system. The world, including mankind's physical habitat and other living species, serves human needs, but the natural world has rights of its own. It is not a servant but a brother and a collaborator.

(4) The existence of human societies depends on the maintenance of balances among natural and social processes.

(5) Human beings have a discernible common nature which is the source of human values. The human body and the human mind are aspects of the same reality, and both have needs the fulfillment of which constitutes the content of human values. Each individual and each human group has a unique destiny, but a destiny which can only be fulfilled within the balance of the system of society and nature as a whole.

(6) Human beings are self-conscious and future-oriented. Freedom is the ability to actualize desired futures.

(7) Culture is the self-consciousness of human societies as wholes, their common perceptions and attributions of value, and their common responses to the world they perceive.

(8) Technology is that aspect of culture through which social systems relate to nature. It controls the balance between man and nature. Technology has been created by humanity as its servant, to aid it in creating desired futures.

(9) Social change results from changing relationships among human beings which over time alter the characteristics of social systems. Changes in social systems both result from and create changes in cultural systems.

(10) Freedom—for individuals or societies—consists of the ability to create desired futures. Freedom therefore presupposes power, and can only be realized in a process of mutual interaction and mutual control among human beings.

Politics and Humanity

(1) Because of the shared nature of humanity, there are universal goods valued by human beings as individuals, including subsistence, security, self-esteem, and variety, which are achieved in and through social

processes. Individuals and groups have needs which they ought to have the power to fulfill insofar as their actions respect the needs of others and the balance of the whole.

(2) Politics is that aspect of culture by means of which social systems are kept in equilibrium. Politics is ultimately responsible for defining the structure of relationships among human beings.

(3) No political system can be based on rational calculation of individual interest alone. A political system, to be viable, must be based on love for others and piety toward the community.

(4) Because of the shared nature of humanity, there is a social common good which consists of the subsistence, order, and purpose of particular societies and of human society as a whole. All individual human goods are necessarily subordinate to the achievement of these common goods, since individuals can only survive and develop if the social system as a whole survives and develops. A common world requires a common order to seek the common good; a planetary political order is needed to deal with matters of planetary concern.

III

A Planet Fit for Man

15

Necessary Utopia I:
The Politics of Ecological
Humanism

THERE is only one alternative to the subversion of human civilization by alien forces, and that is the creation of utopia.

Inspired in large measure by the growing control over nature which science and technology were bestowing upon mankind, utopianism flourished during the nineteenth century. World War I was a severe shock to this optimism, and the expression of fears about the effects of science and technology on human life became more frequent. The antiutopian tradition was embodied in works as diverse in inspiration as Zamiatin's *We*, Huxley's *Brave New World*, and Orwell's *1984*. In the post–World War II era and especially in the United States since the early sixties, a new utopianism has sprung up, motivated largely by attempts to find alternatives to a technological society increasingly rejected by intellectuals and by elements of the young; and this new utopianism has found expression not only in literary works but in living social experiments.[1]

It is possible to object to utopianism on a variety of grounds. Many utopias posit the creation of social conditions which are at odds with what we know about human nature; most, while describing some ideal society, fail to tell us how to get there from where we are now. A third objection, often leveled by idealists who reject the first two, is that utopias, even if they are intrinsically possible and can be created, are undesirable because they limit human freedom in a total and final sense, by creating rigid, static societies.

The next several chapters—indeed this book as a whole—are an exercise in utopian thinking. How then do they meet the objections raised against utopias in general? The earlier parts of this book offer a perspective on human nature against which the future, sketched in Part III, can and must be measured; if this perspective is valid, then our future society is intrinsically possible. Chapter 18 deals directly with the problem of how the new society is to be created. But what of the charge that any attempt to create a utopia necessarily restricts human freedom and in a sense puts an end to human development and thus to history itself?

Most utopias are overly detailed and precise about the shape of the ideal society. It is this precision which makes them rigid and static. Systems consist of the interrelationships of the entities which constitute them. If one postulates in detail all the relationships, one creates a system which will function because it is in equilibrium (assuming that one actually understands how the different components affect each other). But it will be a system so tightly organized that it cannot change, like the social and political systems postulated by most "functionalist" social science theory. Quite often this thrust of utopians to create a possible future which is static because it is so highly detailed is the result of an overly elaborate agenda for reform,[2] since utopias are ordinarily alternative societies, created in reaction to the adjudged ills of the present and since utopians usually have a myriad of grievances against things as they are. The aim here is to create a relevant utopia—not to bring to pass our dreams but to give the lie to our nightmares; to create what perhaps might be called a provisional or demi-utopia. I am not concerned with trying to sketch out an ideal world in which to live but one in which future generations can live, a society which will be in equilibrium to the extent required and for the time required to prevent certain current tendencies from rendering the planet uninhabitable for truly human beings.

We live in the midst of an historical crisis in which our choices are between what some would call utopia and the end of civilization, in which we are presented with the existential paradox of a necessary utopia. In this necessary utopia, social conflict will occur and human beings will still have the option of being unhappy. But at least there will still be human beings. Not everything that might ideally be desirable is thereby necessary, and in our utopia-building we are not called upon to perform the impossible task of predicting or prescribing in detail the total future of humanity, nor enlisting all humanity in a crusade to create our own private version of the good life.

Of course, the relationships among cultural and social traits and institutions are such that what is necessary for the survival of human society will require changes in a wide range of habits and patterns of behavior.

Certain claims of nationalism, certain institutions of sovereignty must yield to the need to avert nuclear war and to bridge the economic gap between rich and poor nations. Large families cannot be an ideal in a world faced with a desperate population problem. The private profit motive and infatuation with technological change for its own sake cannot continue to determine acceptable economic behavior in an economy of limited growth. But within broad limits there are many divergent ways in which a world society, based on the principles of ecological humanism, can be created and structured. Just as one can arrive at the sum of 6 by adding 5 and 1, 4 and 2, 3 and 3, or 6 and 0, or build an automobile which performs in a certain way by using various structural materials, energy sources, engine and body designs, it is possible to create a future human society in a variety of ways. But not in an infinite number of ways. And not without knowing what results we wish to achieve. What is important is not to sketch in the precise details of the future society, but to stipulate what is necessary if society is to meet basic individual human needs and provide the common goods of subsistence, order, and purpose.

Respect for the freedom of choice of human beings in the present and the future and acknowledgment of our own limited (individual and collective) ability to predict the outcome of our actions are not the only reasons for a certain tentativeness in describing and prescribing the future. Flexibility is a key element in ecological humanism, since this philosophy postulates that balance is achieved through the interaction of free, self-conscious human beings, and that the future is an emergent property of this interaction. The good society is, above all, an active, learning society, which is constantly creating its own new patterns of equilibrium among society, technology, and nature as it moves through time.[3]

Therefore, what follows is only meant to be suggestive in its specifics, however necessary its basic outline. It is an attempt to describe some of the necessary characteristics of a social order which could replace liberalism and make possible the restoration of humanity. It is also an outline of a demiutopia which has the most immediate relevance for the major world powers and, in its details, for the United States. The ability of a national society to determine its own future, especially in a technological world, is dependent on a certain population size and especially on the size and variety of its resource base. The nations which are "great" or potentially great powers (the United States, the Soviet Union, and China) and a few other nations (Japan, India, Brazil, and perhaps a hypothetical United Europe) can create their own futures. The rest of the nations will be more dependent on what happens outside their borders.[4] Smaller nations can act as valuable laboratories for all of humanity: Tanzania can perhaps provide a model of a socialism of austerity and fraternity valuable to other developing nations; Canada, a model of a society which combines an advanced technological system with civility and cultural pluralism. But Ghanaians are not asked to decide whether future human beings will be conceived in laboratories or nurtured in artificial wombs. Paraguayans will not be the first to deal with the problems which the computer poses for the preservation of individual privacy. Cubans will continue to grow sugar for world markets. Kuwait will continue to produce oil. Even if by selling most of their national assets the people of Barbados or Nepal might conceivably gain possession of a few atomic weapons and cause great destruction with them, such nations will not decide whether or not human civilization is destroyed in a nuclear holocaust.[5] The challenge to create a truly human future society which can be a model for the planet as a whole is one which addresses itself primarily to the larger, richer, more powerful nations.

Toward a Politics for Tomorrow

Human societies are living systems. Like all natural organisms they must carry on vital functions. Just as the individual body has physiological processes which must be in balance, society must maintain a balanced economic life. Just as the individual human being must be able to perceive, analyze, and decide in order to act and develop, so must society. Because human societies are self-conscious and purposive, the political function of choice must control the economic function of subsistence. Unlike human beings, who can allow their metabolic processes to go on without supervision except in emergencies such as illness, human societies must shape their physical processes to chosen goals.

Political and economic systems have a protean character. Structures are simply processes looked at in slow motion, and they change over time. Different societies will have different political and economic institutions at different times and places as a result of human interaction within differing technological and environmental contexts. It is therefore impossible to do more than tentatively suggest how any society, including the contemporary United States, should reorder its political and economic institutions so as to be able to survive and to promote the fullest possible development of its members. On the other hand nature sets limits to possible human actions—impotence principles—and these apply in social as well as physical nature. Some social policies and institutions are mutually exclusive, and some features are basic to any properly operating and stable society.

There is a complex relationship between forms of government—processes viewed as structures—and the substantive social policies which affect human individual or collective welfare. In utopia-building it is necessary to be careful neither to overestimate or underestimate the effect of particular governmental structures on the creation of a human future. To become infatuated with form for its own sake is a waste of time and energy and can have dangerous consequences; to neglect the impact of differing governmental structures on potential policies can be equally dangerous.

"O'er forms of government let fools contest, what's best administered is best," wrote the eighteenth-century English poet Alexander Pope. Pope's aphorism emphasizes that the function of politics is to serve human needs. But it is somewhat misleading, since the processes of government also affect the quality of human life directly; people find self-actualization in part through participation in the political process. To deny them this participation impoverishes their lives. Nor can the process and the outcome really be separated; different kinds of governmental structures will tend to produce different policies reflecting different social interests.[6]

Though the political process is everywhere basically the same, structures of government vary from nation to nation. Particular governmental structures are the outgrowth of the underlying cultural and social forces in different countries. The British parliamentary system had its origin in the long struggle of the aristocracy against the crown, and the vestigial House of Lords is a reminder of the role social class still plays in British society. The American system of government with its separation of powers reflects the fear of liberal theorists of effective majority rule, while the federal system has deep roots in American history and social ecology. The European parliamentary tradition owes its origins largely to the French Revolution and the Jacobin concept of democracy and is found even in the showpiece legislatures of socialist states. The one-man rule (usually rule by a narrow elite of intellectuals and technicians) of most so-called developing nations is an expression of fragile nationhood and deep social cleavages.

Most utopian thinkers tend to ignore or reject the deep roots and tremendous inertia of political structures and to waste both intellectual and political energy trying to replace existing structures with totally new, historically rootless, rationalistically ideal governmental machinery.[7] Some kinds of political structures are intrinsically more effective in meeting the challenge of creating and implementing the policies necessary for the common good than others, either in general or in particular respects, and the systems of many societies, especially that of the United States, may have to be drastically changed in order to meet the challenges of the coming decades. But political systems, not least of all the American, are flexible and adaptable. Given the necessary changes in basic cultural attitudes and social values and the political mobilization needed to translate these changes into policies, any structure of government can accomplish what is necessary, and its form will evolve to deal with its new functions, as happens in the evolution of all living systems. No structure of government, on the other hand, can be an effective agency of the common good if the will to achieve the common good is absent among leaders or citizens.

This being the case, a primary task of political reform is the creation of an understanding and acceptance of the functions which government must perform in the world of tomorrow. This will not be an easy task, especially in the United States. The American system of government is based on the worldview of the Newtonian physics dominant when liberalism was born. In this view the world is composed of hard objects, distinct from each other, which can act on each other only directly and through the use of physical force. Basic American constitutional concepts such as separation of powers, checks and balances, the bill of rights, and federalism are directly derived from this Newtonian worldview.

The worldview of contemporary science, in which entities are dependent for their existence and characteristics upon their relationship with others and in which life is a continuing process of mutual interaction and influence—a fluid, ever-changing relationship within the boundaries of certain relatively stable patterns of form and order—makes the American Constitution intellectually archaic, as usually interpreted. Yet we all find it difficult to talk about politics, or most other things for that matter, in language which is suitable to our new vision of reality. Our inherited structure of language and conceptual thought is still essentially Newtonian. Only gradually—as evidenced by the increasing popular use of such terms as "feedback" or "input"—are we beginning to develop a vocabulary more appropriate to a more flexible world. Therefore, what we are forced to do today is to try to describe the essential functions and activities of government, and the economy as well, in language which adequately reflects reality, and then translate it back into the conventional, traditional language of politics, economics, and law.

Given this perspective it is possible to assert that the political order, by which we mean society interacting self-consciously in a certain way, must have certain characteristics. It must have *political intelligence*, the ability to perceive reality. It must have *political will*, the ability to decide how to relate to perceived reality. And it must have *political power*, the ability to act upon the world in such a manner as to implement the decisions made. The reformation of the American government, or that of any other nation, requires the creation of these characteristics, their localization in one or more structures of government, and the development of appropriate interrelationships among these structures.

Political Intelligence

A serious weakness of American government at this time of world crisis is its lack of political intelligence.[8] The President and the Congress read newspapers and books—or should. They receive reports from government agencies. They are subject to a flow of information from the electorate about popular perceptions and desires. Yet they find it difficult to understand the world in which they must operate. The reason for this is a simple phenomenon common to modern technological society: information overload. They do not lack information; they have too much of it, and they receive it in a form which makes it impossible to organize and assimilate. Their world is that of the newborn child suddenly faced with an overwhelming cacophony of new sensory stimuli, a world which is a booming, buzzing confusion.

In order for government to operate, it must have information which is ordered to action; this means information structured in terms of the problems to be dealt with. The information must be future-oriented, since decisions always take place in the future relative to the information they are based on and it is always in the future that decisions are implemented and that their consequences are felt. Given the fact that the interrelated nature of life in technological society makes an ecological, systems perspective of reality mandatory, the information available to government must be so organized that it presents an adequate picture of the interrelatedness of problems and of data. As various futurists have suggested, we are desperately in need of social institutions designed to collect information about current trends and future possibilities—social "lookout" stations of various kinds.[9] Such institutions can be public or private, national or global, but they must concentrate above all else on the implications for society of present and potential developments in science and technology.

In the United States the Nixon Administration took a giant step backward by destroying the scientific advisory system created by previous administrations to keep the President abreast of developments in what is increasingly the most important frontier of social change. The American Congress authorized an Office of Technology Assessment for the purpose of keeping Congress up to date on new scientific and technological possibilities and possible ways of controlling the probable social consequences of technological innovations, but the actual establishment and staffing of the OTA was long delayed by fears that it would come under undue influence from particular political factions, despite the fact that similar bodies such as the Legislative Reference Service and the General Accounting Office have long been objective and effective servants of Congress as a whole. Creation of political intelligence for the American government, no matter how it is structured, is an absolute necessity if American society is going to be able to control its collective future, which is only another way of saying, if it is to remain free.

Modern technology provides valuable tools for compiling, storing, and evaluating information for the purpose of increasing social freedom. The computer makes possible calculations about the relationships of elements of problems which can lead to new understanding of possible policy alternatives and of the consequences of adopting particular alternatives.[10] But regardless of the technical facilities and expertise available, no specific institutions or structures designed to raise the level of political intelligence can be effective unless their personnel, the members of the executive and legislative branches of government and, in the long run, the general public, better understand the systems approach to the world and better appreciate the possibilities and limitations of attempts to foresee and control the future.[11]

Creating Political Will

Equally important, though logically dependent on the existence of adequate intelligence, is the creation of political will. To be free means to be able to have a role in creating the future. To be able to create the future means being able to decide. A society which cannot decide is not free. Yet while there is little objection, in principle at least, to creating more effective political intelligence within the American system, there is real and deep-seated objection to the creation of political will. One of the essential tenets of liberalism is the denial that effective government is either possible or desirable.

The political will that the founding fathers most feared was the will of a majority united to attack the privileged positions of minorities. In creating a system designed to frustrate such a will, they made effective government on behalf of any concept of the public interest, however derived, virtually impossible. Subsequent political evolution has accentuated this feature of American government.[12] As James Reston once commented, "Much has been written since the Constitutional Convention about the 'tyranny of the majority,' but wherever we look today willful minorities seem to be dominating public affairs and even prevailing by force over the majority."[13]

Congress and the President represent different constituencies—in a sense different ways of ascertaining the majority will[14]—and no talk of responsible party government can bridge this gap. Responsible party government, once the darling of many American political scientists, is essentially incompatible with the separation of powers, and no tinkering with electoral or party structures will alter this fact.[15]

Congressional committees are normally dominated by those groups whose interests are the most directly affected by the legislation they handle, while the interests of the public at large go begging.[16] Administrative agencies in the executive branch are ordinarily examples of the same phenomenon. The ability of special-interest groups to exercise effective control over particular segments of the government has made it almost impossible for even a supposedly strong President to reorient their activities. The American President as commander-in-chief of the armed forces has de facto power to order the Strategic Air Command to strike any target in the world, initiating a third world war, but he cannot effectively control the Army Corps of Engineers, who go about reshaping the American earth in accordance with the desires of contractors, construction unions, and local politicians. As one sociologist has written, "The basic fact about the American system is that while we have a modern technology and economy, ours is an antiquated Tudor polity. . . . What we have in this country is not decentralization but disarray."[17]

How can American government be restructured with a minimum of effort within the historically foreseeable future, so as to be a more effective vehicle for mobilizing a popular will toward the goals of ecological humanism? The primary burden of focusing popular consciousness on social problems and presenting alternatives for political decision must rest with the Presidency. The debacle of Watergate has caused many Americans to lose respect not only for Richard Nixon but for the Presidency itself. Speculation is rampant that the period of executive ascendancy which began in recent times with Franklin Roosevelt and the first New Deal has come to an end—and none too soon, some would add—and various constitutional changes have been proposed which are designed on balance to increase the power and prestige of Congress and downgrade that of the President.

But this would be a mistake. The Presidency must bear the major leadership role in reorienting American society. To replace the President with a prime minister on the model of other nations in order to reduce some of the monarchical aura of the office would involve a political revolution few Americans would support. Such a change would necessarily destroy the equal status the Senate and the House have vis-à-vis each other, or else would make for political chaos if a prime minister was acceptable to one house of Congress and not the other. Even if a ministerial system were adopted, there would still have to be a ceremonial head of state. No constitutional monarch is available, and to create a new Presidency on the current French or German model would simply reintroduce the problem of deadlock in a different form, as President and prime minister vied for power.

Yet as long as a President, directly elected by the whole American people, exists, no leader of a congressional majority can substitute for him as national spokesman. Even if the House should once again create an almost omnipotent Speaker on the model of "Uncle Joe" Cannon, the Senate would not follow his leadership, nor could it provide an alternative leader. A collegial leadership of either or both houses of Congress, no matter how strong the majority they represented, could not fulfill either the necessary tasks of day-to-day decision-making which the President now performs or his symbolic function.

For reasons of history, popular psychology, and political dynamics, the best opportunity the American people have of expressing their beliefs as to what general direction public policy should take in the future lies in their choice of a President. This may be an unclear mandate, but it is far clearer

than the results of the confused, localized, and often obscure process by
which individual members of Congress are elected. The President—
because of his fame, his office, and the simple fact that he is one man
whose actions can be observed and whose existence can be grasped—is also
in a specially advantageous position to dramatize the choices open to the
nation by his advocacy or opposition, action or inaction, and to solicit or
simply absorb public reaction to his policy positions. To destroy the special
role of the Presidency would be to cripple the political will of the nation.

But, contrary to the implications of the Newtonian view of nature and
government, the process of political struggle need not be a zero-sum game
—in which if one advances others fall behind—any more than the struggle
of a party of mountain-climbers to reach their goal is. The continued and
even increased ability of the President to elicit and represent the popular
will need not lead to any diminution of the role of Congress, any more than
better political intelligence in the Executive need make the Congress less
well informed. Congress in a reformed government can continue to affirm
or reject alternatives set before the nation by the President and can also
propose alternatives of their own which the President can accept or oppose
through his legal and political powers.

If Congress is to play an increasingly useful role in forming the political will of the nation, several things are necessary. In addition to increasing
the amount of political intelligence available to them, members of Congress must be willing to take clearer stands on public policies so that their
election or rejection by the voters can have more meaning as an expression
of popular will. They will have to create an internal organization which
will make it more possible for their leadership to represent the center of
gravity of congressional—and presumably national—opinion than is often
the case now when seniority, in part the product of representing safe and
sociologically unrepresentative districts, plays such a large role in determining legislative power.[18] Above all, Congress must drastically reform the
committee system, replacing the present standing committees with broader
committees oriented toward dealing with the real problems of man's future
in technological society, or possibly abolishing the standing committee system entirely in favor of ad hoc bodies. While longer terms for individual
congressmen may be desirable, some members of both houses should come
up for election at frequent intervals, perhaps every year on a staggered
basis, so as to provide more adequate feedback of information—and will—
from the public to its government.

Planning and Political Will

A major necessity for the creation of political will is the existence of
adequate national planning. Plans are simply will extended over time.

Without plans to implement them, decisions are meaningless. Not only must goals be chosen, but their integrity must be maintained. Societies, like individuals, often choose goals without accepting the means toward them, and goals can be whittled away to meaninglessness by qualifications, exceptions, and crosspressures.

Perhaps no term in the lexicon of politics creates greater confusion and apprehension, at least among Americans, than does planning. We almost all plan ahead—for a weekend at the shore, for our children's college education, for retirement. Businesses plan next year's products and sales campaigns. Large businesses plan a decade or more ahead. But the concept of social, economic or political planning on a societywide level fills us with visceral horror. Yet the essential characteristic of a free society in control of its future is planning; the test of the utility and indeed the continued viability of political institutions in the years to come will be their ability to plan. When we think of planning we think of regimentation and coercion, the nightmare world of Stalin's Russia or Orwell's *1984*. But this is absurd.[19]

A political system committed to policies of restraining economic growth, controlling technology and population, and redistributing the social product more equitably—the basic policy goals for planning which logically follow from the premises of ecological humanism—need not operate like an old-fashioned army where everyone wears a uniform, eats identical meals at the same time, works in an assigned job under close supervision, and has every minute of his time controlled by a central authority. Regimentation, uniformity, and compulsion are the essence of incompetent and inefficient planning. These characteristics are not found in the ecological balances created by nature, yet planning is only consciously created balance.

What is efficient planning? How is balance achieved in a society whose purpose is the greatest possible self-actualization of its members?

In order to understand the nature of planning for balance, we first need to reject two widespread and dangerously misleading concepts of planning. One false concept stems from our familiarity with rigid physical planning. An example is the architect's blueprint. One individual or a small group draws up the plan for a building, in consultation with the person financing it and possibly, though not always, with the people who will eventually occupy it. Every specification is set—spatial relationships, materials and, to the extent possible, the timing of construction. Ideally the builders follow these specifications to the letter without modification. When the building is finished, it is a complete actualization of the original blueprint. Its occupants must adjust to it. They may be able to make a few minor structural modifications, but their options are negligible. Plumbing, wiring, windows, doors, stairways, heating plant are all there, and the user's patterns of life and activity must adapt to them. Freedom is at a minimum; only the architect and the financier exercise freedom. Not only buildings, of course, but whole cities and social institutions can be planned in this way, but most human beings find life in such constructions constraining, and the idea of a society planned in such a way fills them with dismay. The building or social institution may or may not work efficiently, but whether it does or not, rigid planning minimizes the self-actualization of those who must conform to its results.

A polar opposite of rigid physical planning is what is sometimes called indicative planning. Rather than try to make the future conform to his will, as the finished house conforms to the architect's blueprint, the planner seeks to anticipate the future and adjust to it so as to maximize certain real or assumed desires of those on whose behalf he plans. A manufacturer seeks profit; he tries to anticipate future demand and plans to meet it by producing certain items to sell at a certain cost at certain times. A school system expects more students, so it plans to build more schools and to hire more teachers. Much social planning, including what passes for city planning, is of this nature. It is primarily reactive to events outside the control of the planners. Freedom here is also at a minimum, though this is usually not realized within liberal society. Real freedom—the ability to control the future—is essentially in the hands of those forces to which the planner reacts. Such planning is more flexible than rigid physical planning. It can make many minor and even major adjustments as anticipated events fail to materialize and unanticipated ones do. This is why it is sometimes called, or blends into, incremental planning.[20] But it usually results in a future radically different from the one desired by the planner or by those on whose behalf he acts. It is really just keeping, if possible, one step ahead of events, like a squirrel storing nuts for winter. Such planning fits in with the ethos of liberal society in that it involves a minimum of direct control of human beings by other human beings, but it seldom involves any real control by anyone of the future and hardly deserves the name "planning" at all. It has no goals but sheer day-to-day survival.

Planning is a human activity and assumes that human beings are different from other animals. It assumes that we are self-conscious and can evolve in a chosen direction rather than simply react to events in our environment or seek growth blindly. The problem of any adaptive organism is to get from an existing state to a desired state.[21] Planning for a human society is therefore necessarily normative; it is based on human desires. It is an act of the social will which creates a desired future in accordance with politically chosen goals. Like both the architect and the indicative or incremental planner, the normative planner operates within environmental constraints: outside forces such as the laws of nature which he cannot alter, his own respect for the integrity of nature, and the initial state of the system in which the planning takes place. In addition, unlike the physical planner, the social planner—whether an individual, a group, or the total society politically organized—should respect the insights and desires of those who will build the future and live in it. In Herbert Simons's words, a good plan is not a blueprint but a recipe, necessarily respecting the nature of its human and physical ingredients and how they act.[22] The desires and reactions of the human beings who are affected by planning can and must be just as much a part of the input (initially and thereafter) into the planning process as are the nature and reactions of the physical environment in which the plan operates.[23] Planning is itself a learning process.

But while an institution which plans must learn from both the human and physical environment in which a plan operates, it must not forget that it itself is a willing, purposive entity. Rigid planning must be discarded, but "disjointed incrementalism"—simple reaction to reactions—must be avoided as well. What is needed is "jointed incrementalism"[24] in which reactions and adjustments take place, but never in such a way that the ultimate goal of the planning process is forgotten. Planning designed to curb growth which itself leads to unchecked growth is pointless. Planning designed to control technology for human ends which makes the machine's values paramount is social treason.

Learning societies will change their plans in the course of implementing them, and in the process will change their intermediate goals. Planning and social learning are immanent processes which lead to the emergence of new social conditions. But change in basic goals must always be a self-conscious process in which society collectively reflects upon what its activities are teaching it; no system less than the whole should have the power to change the basic goals of society. The varied desires and activities of individuals and groups within society cannot be allowed to so distort the plan that the actions taken cancel each other out and there is no movement toward a socially chosen goal, nor can society allow the various incremental decisions which are made in response to particular pressures to result in a change in goals.

Planning is as much a political process as is government itself, of which it is an essential part. Just as politics is not merely a clash of interests but a process in which interests are redefined in terms of common purposes, planning assimilates particular needs and views into a common pattern which has its own definable meaning. The goals of planning can be changed, just as we scrap our indvidual plans for a vacation abroad and decide to stay at home instead. But goals must be changed self-consciously as an act of collective will—because a goal appears impossible, undesirable or too costly in terms of other goals—not because we have been unconsciously and cumulatively buffeted by events to the extent that we find ourselves on a plane headed for an unchosen or even unknown destination.

While we must respect the freedom of future members of human society, we must recognize that no matter what we do—or don't do—we are establishing the foundation upon which the future will be built and are thereby limiting its possibilities. Or, as one political scientist has phrased it, "The 'openness' of a society. . . . [is] only meaningful within the context of its present development."[25] To ensure that humanity will have a future and one in which some freedom of choice and action will exist, it will be necessary to try to prevent the actualization of some futures where disaster waits.

The United States is virtually the only nation in the world today without a national planning process.[26] This is especially ironic, in that we refuse to give aid to developing nations which do not have national plans, and our intelligence agencies, using raw data, available options, etc., have done extensive work in mapping the possible future national plans of Communist nations. It is doubly ironic in that planning plays so large a part in the decisions of the large corporations which so greatly influence our economic and political life. Many have called for effective national planning; several years ago a congressional committee produced a forthright recommendation for a comprehensive National Resources Planning Commission.[27] Yet such appeals continue to be ignored.

The reason they have been ignored is that social planning under government auspices is an attempt to implement political will. But political will must be created. We have argued that it can be incarnated in both the Presidency and the Congress in the American context, but it must first exist within the larger political and social process.

How is political will created? As I argue later, the creation of the political will needed to reorder society will not occur through capturing existing political parties, starting new ones, or tinkering with existing party and electoral systems. It will be an immanent process of change in public opinion, based on two factors: increasing acceptance of the substantive arguments of ecological humanists in favor of a balanced society, and increasing realization that the rigid political balance exalted by liberalism leads to deadlock and inaction, that pluralism "is socially valuable, and yet counter-productive when elevated to a principle of government,"[28] and that "democracy depends on the blending, not the balancing of interests and thoughts and wills."[29]

The Role of Leadership

But even though a new political will must come into being through an immanent process, this does not mean there is no need for leadership in politics and society. Choice of social goals is not a purely rationalistic process. Such choice is highly conditioned by history and tradition, and involves symbolization and responses to symbols on the part of the public. Above all, it requires leadership. The role of leadership is not to substitute the leader's will for that of those who choose to follow him, but to articulate their unarticulated desires, to translate the ends which he has helped them perceive into means, and to rally their support against forces which would block the accomplishment of their goals. Real leadership, whether exercised by an individual or a group, must anticipate problems and frame solutions for them and not wait until all aspects of the problem have become apparent to everyone. Winston Churchill was available as a leader for Britain in World War II because his political identity had been forged through years of opposition to the policies of those he sought to replace— policies which had enjoyed popular support until their inadequacy became apparent to all. Slavery was not abolished in Britain or in the United States, nor such public-health measures as compulsory vaccination introduced, by women and men who took their cues from public-opinion polls.

The political leader is necessarily a visionary and a creator. As former Puerto Rican governor Luis Muñoz Marin once put it, "The successful politician like the poet needs to possess imagination. He needs to have both the ability to imagine non-existing states of affairs and the ability to influence other people to bring them about."[30] Leadership is not simply a quality of character or personality; politicians can have what is loosely called "charisma" without being leaders. Leadership is an emergent property of a political process in which future-oriented men and women interact with a future-oriented citizenry in such a way as to make coherent collective social action possible. The legitimacy of leadership is based on both the legal powers it has within a given political system and acceptance by the public of the results of its actions, but not necessarily on the complete agreement of all concerned parties to the desirability of these actions at the time they are taken. Some liberal theorists have argued that the only fully legitimate basis for political action is unanimity.[31] This is logical in theory—reflecting as it does the liberal belief in individual autonomy and its rejection of the existence of a public interest—but it is absurd in practice, as the fate of the late medieval Polish parliament and the League of Nations which embodied the principle of the "liberum veto" illustrates. Leadership serves the public interest by doing what men of years later would wish had been done.

The role of leadership is all-important because no government can act effectively simply by mirroring the public opinion of any given moment. People hold opinions with varying degrees of intensity. No theoretical solution has ever been found at the ethical level to the problem of whether 51 percent who are casually in favor of a certain measure should prevail over 35 percent who bitterly oppose it.[32] Public opinion is highly volatile on day-to-day issues (though not on fundamental values) and subject to influence by the media and, of course, by dramatic actions of the leadership itself. The only test of a truly democratic leadership is the extent to which all those affected by its actions have some way of making their views heard and the availability of peaceful means by which the society can reject its leaders if it finds their actions unacceptable. People's perceptions of their interests change in the course of political and social life; their perceptions are affected by, among other things, the words and actions of their leaders. The relationship between leaders and followers in any society, regardless of its political forms, is a subtle dialectic between trust and final judgment.

Political Power

Political intelligence is the perception of the world by society functioning as a political system. Political will is the determination of how to relate to that perceived world in order to create some desired future. Political power exists when society acts to implement its decisions.

Within living systems there is an overall pattern of order deriving from the goals of the system combined with freedom of action for its constituent elements. Within political systems composed of living, self-actualizing human beings who are the elements of a society which seeks to create a collectively chosen future, there is a similar synthesis of apparent contraries. Political leadership acts in accordance with the general will of the society as a whole. All of society's agents—the bureaucrats and functionaries of government—must seek to attain the social goals already chosen. At the same time, the maximum amount of freedom must be left to individuals, both citizens and officials, to act in accordance with their own perceptions of how the general goals of the society may best be realized.[33] Responsible government, in which political leadership is accountable to the electorate as a whole through regular processes, must be combined with participatory democracy. Systemwide planning designed to deal holistically with problems must be combined with decentralized action which permits each individual to make a contribution to determining the state of the total system.

The kinds of administrative structures best suited to operationalize these two goals will vary with time and place. In general the pattern will be one in which basic decisions are made at the most inclusive, the most holistic level possible, while the implementation of these decisions takes place at the lowest level possible. This is the approach implicit in a recent congressional committee report which recommended a reorganization of American government in which "the creation and coordination of national policy must be centralized at the highest level, in the President, the Cabinet and the Congress to insure that it is under effective control of elected officials, but at the same time administration must be decentralized so that decisions within the policy guidelines are made close to the people where administrators can know how to adapt national programs to local needs."[34]

Ecosystems are composed of mutual, interactive systems of controls.

The same thing is true of the American political system. The executive controls its agents by law and through the day-to-day operations of the Bureau of the Budget. Congress seeks to ensure that the laws are being faithfully executed by empowering the General Accounting Office to cross-check expenditures after they have been made. Citizens can protest actions of government to their elected officials, executive or legislative. The courts have the power to intervene to make sure that particular actions conform to the system-goals as set forth in legal statutes and in the Constitution. It has been suggested that to fill gaps in this process of mutual control the United States should adopt the Scandinavian institution of the ombudsman, a person empowered to make inquiries and interventions in order to ensure that governmental actions which improperly or unjustly affect individual citizens are redressed or prevented. (The congressional committee report previously cited includes a recommendation for a "People's Intervenor" with similar functions).[35] The press serves both as a countercontrol to government and—though this is sometimes forgotten—as a means of publicizing and promoting governmental objectives.

The many particular questions about how government should be organized and how its functions may best be carried out cannot and need not be answered here. Questions of jurisdiction, of methods of securing compliance with agreed-upon goals, and of administrative and legislative organization, cannot be answered in advance in any definitive fashion, because certain answers to some of them may preclude certain equally justifiable answers to others. Balance, it must be remembered, can be created in a number of ways. What precise balance will be arrived at will be determined by an immanent process involving the interaction of all the elements inherent in the present situation—it is an "emergent property" of the present. One thing, though, is clear: Whatever powers government—however organized—needs in order to deal with the current crisis in the relationship of man to his natural environment and to his technology, it must have. To deny these powers to the political order in the name of freedom would be to destroy that ability to create a human future which is the very essence of human freedom.

16

Necessary Utopia II:
The Economics of Ecological
Humanism

A central tenet of liberalism is the primacy of economics (especially of economic growth) and the autonomy of economics vis-à-vis the political order. Central to the philosophy of ecological humanism is the subordination of economics to politics and the creation of a society in which growth yields primacy to balance.

Like all words imported into discussions of society from the physical sciences, balance is a term which is difficult to define in a social context. It is a metaphor based on a physical concept which is in itself elusive.[1] To speak of a balanced society as an aim or aspect of the political process is to evoke images of a static, rigid, and even uniform world. But such connotations are misleading.

A balanced society is actually one in which no aspect of social activity excludes any other activity, and in which the sum of the activities can be maintained over time, as in the "balance of nature" of which ecologists speak. Balance can be spread over time or space or groups of individuals. When we speak of a balanced diet, we mean that a person must have a certain proportion of a number of nutrients not at each and every meal but over a certain limited but flexible period of time. If we talk of a balanced job-market, we are speaking about a situation in which over a particular period no more teachers are trained by colleges than schools can hire, while at the same time houses do not deteriorate for lack of plumbers or other craftsmen. By a balanced population we mean one in which the total rate of growth reaches zero even if one couple has four children and another none.

Balance, therefore, does not mean uniformity. Individual meals differ even though the diet is balanced; different individuals may—indeed must —specialize in different skills in a balanced job market; different families may take different shapes in a balanced population. Nor does balance necessarily mean direct control of those actions of individuals which contribute to balance or imbalance. Though it is sometimes difficult, a housewife concerned with her family's diet can usually ensure that those who consume the food prepared will, over the course of time, get the nutrients they require even if they do not eat everything on their plates. The relationship between availability of training and economic incentives can be organized in such a way that people who want jobs can find acceptable work without anyone being compelled to take a particular job or being denied the opportunity to acquire or use a particular skill. Though it is a much more difficult achievement, a combination of social pressures and economic incentives can influence population growth patterns in the direction of stability. All the decisions which lead to social balance—or imbalance—are the decisions of individuals or groups responding to structured choices. The role of government is to structure the choices in such a way that overall social balance is achieved or maintained. This can usually be done by manipulating positive incentives and only rarely requires personal coercion.,

Just as a balanced society need not be rigid and uniform, it need not be static. Balance is synonymous with equilibrium, and as we noted earlier, an equilibrium can be a moving one. The content of a balanced society changes as its elements change. Ecologists speak of "ecological climaxes," situations in which a forest or a plain or a lake reaches relative stability, but they know that over time nothing in nature is unchanging. Lakes become meadows, species evolve, new mountains thrust upward from the earth's crust. A balanced political system or economy is not one which is unchanging but one which changes within the limits of ordinary human adaptability and the possibility of human control. We can all tell the difference between the behavior of the American economy in the 1950s and the German inflation of the 1920s, between the evolution of the American Constitution and the postindependence political history of the Congo.

What are the elements of a balanced economy? What must government do to bring about balance in society? Before political will and power can be brought to bear on this question, we must have recourse to political intelligence. But this is a problem which at present cannot be readily dealt with by political intelligence because of the failure of the social sciences, on which this intelligence must rest, to treat the relevant phenomena adequately because of the weakness of the existing economic model of reality.

The Need for a New Economic Science

To say that existing economic science has failed the political needs of society is to disparage not the intelligence or integrity of economists but only their implicit claims to omniscience when it comes to dealing with the material bases of human social life. Economics is divorced from reality—just as liberal political philosophy is divorced from reality—not because, as some allege, it is too materialistic, but in large measure because it is not materialistic enough.[2]

The physical life of humanity depends on the interrelationship of material factors—food, clothing, shelter, energy, space, etc.—all ultimately forms of energy in particular patterns. A society in equilibrium within itself and with its natural environment is involved in an almost infinitely complex set of relationships among material factors. Economics has claimed to be able to describe, and has been accepted by liberal society as in fact describing, these relationships. But it does not. It can only describe those material factors of life which enter into market transactions—scarce "economic" goods which are in some fashion consciously exchanged among persons. Moreover, it can only value them in terms of the price at which they are bought and sold. Utility for the economist is measured by market evaluation.

Material factors necessary to human existence, such as water, sunlight, air, and space, have not normally entered into economic calculations because they have been "free" goods in the terms of economics. Nor do the material consequences of economic activity ordinarily enter into the calculations of economics except in unusual circumstances; for instance, the labor it takes to wash a car in danger of corrosion from chemical pollutants in the air or the costs of dealing with the juvenile delinquency caused by unemployment. If washing the car involves the purchase of detergents, it enters into economic data somewhere, as do the salaries of guards at juvenile detention homes—indeed, both are counted as contributions to Gross National Product (GNP) and thus ostensibly to human welfare. But the nexus between cause and effect is obscured in economic calculations and usually in law as well; the cost of washing the car does not enter into the prices paid for the products of polluting factories, and the car-owner finds it impractical to sue the offending firm which produces the pollutants, a firm which he can rarely identify, much less bring to judgment.

The real flow of material elements in society is lost to economics. This is not to say that some of the formal tools of economic science are not potentially capable of dealing with these flows if the data is available. In large measure they can, and a few economists are concerning themselves with environmental problems under the concept of "negative externalities."[3] But economists as a group have not dealt with the real material substructure of human society, because the relevant natural factors do not ordinarily enter into the formal processes of economic exchange among individual (even corporate or governmental) buyers and sellers. Various models of the American and other economies exist, but they do not include the most fundamental material bases of human life.

There is no standard of value, even in purely material terms, for many important elements of our physical lives, since they carry no price tags, existing as they do outside of market relations.[4] Need for such a system of valuation is immediately apparent in any attempt to discuss the relationship between economic growth and living standards even in a material sense (leaving aside any of the purely psychological components of happiness). Growth in GNP may or may not be an adequate measure of growth even in purely material satisfaction; but we simply do not have the intellectual tools—the data or the theories—to discuss the matter intelligently. It may be that if we could include material factors now left out, we would still be far from measuring how goods contribute to happiness, but at least our accounts would balance. What we need, therefore, is a more realistic science of economics—based on an alliance with such disciplines as ecology and engineering—which will actually trace the total flow of the material transactions which sustain human life and attach valuations to these elements, an economics which, in Boulding's phrase, will enable us to move from a "cowboy" into a "spaceship" economy.[5]

A more realistic economic theory will take into account the fact that in a finite planet with a fragile biosphere there are no longer any "free" goods, that everything has a cost and, therefore, must have a price. It will have to deal not only with second-order externalities in the present but with the impact of present economic decisions on future costs over an extended period of time. Doing things cheaply or "efficiently" today may cost our grandchildren dearly. Present liberal economics is based on a bias in favor of growth which automatically leaves bills for the future to pay. The very notion of interest implies future growth.[6] Modern finance capitalism is based on the rejection of the traditional cultural and theological prohibitions against usury.[7] The ecological approach to human material life suggests that perhaps Aristotle and the theologians were right after all.

A new science of this sort—meta economics or ecophysics, if you will—becomes increasingly feasible as computers make it possible to manipulate vast complex bodies of data and to simulate present and future relationships among them, but it will become a reality only if it becomes politically necessary. It will come into existence not simply because a political leadership seeking more adequate political intelligence arbitrarily decrees it shall exist or because it follows from ecological humanism in the way Marxian economics follows logically from Marxism, but because political systems which attempt to take into account the total effects of economic decisions, for the present and in the future will require its existence.

Such a discipline will restore the old concept of political economy (the Greek meaning of *economics* is "the science of housekeeping") to its rightful place and give it a new relevance. Given its lack of development at present, we can only roughly estimate what changes in social and economic policy might result; for a reoriented society based on ecological humanism will change its policies in accordance with increased knowledge and experience; it will be a learning society. But the general outlines of some of the economic implications of ecological humanism are already clear.[8]

If, as I have postulated, the two greatest challenges facing humanity today are the relationship between man and nature and between man and the machines he has created, it is vital that economic arrangements in society be so ordered as to curb those economic forces which destroy the balance between man and nature and to counter the economic forces which make it difficult to control the development and use of technology.

The primary task of human society in the decades ahead will be to create a balance between a growing population and its claims on economic goods on the one hand, and the limits on such claims imposed by our finite resources of energy and space on the other. Future societies will have to create this balance within cultures which allow human beings to coexist with nature and to maintain sufficient continuity with their cultural past so as to retain their identity. To accomplish this will require control by society as a whole—through its political institutions—of the total pattern of economic life.

Traditional societies can assimilate economics into a larger cultural pattern. Decisions as to who hunts and who consumes the hunters' kill, which crops are planted by whom and who harvests and consumes them, can be treated as matters of custom or religion. But in all large-scale modern societies, conscious processes of economic organization—of ownership and exchange—exist. Even small communes must allocate tasks and rewards, even socialist economies have pricing mechanisms and managers responsible for particular factories and offices. The economic system of a society based on the principles of ecological humanism would not necessarily differ from others in all its forms of organization or legal control, but it would differ from them in its subordination of economic criteria (above all of the desire to maximize productivity for its own sake) to social criteria —that is, to human needs and the paramount necessity of creating and maintaining ecological balance and controlling technology.

The economic, political, and legal techniques needed to shape economic activity to the service of socially agreed-upon ends will necessarily vary from one national economy to another, and from one time period to another. Just as we cannot predict or prescribe how the governmental structure of a reformed society will look in detail, so we cannot—nor should we try to—prescribe in advance which industries should be nationalized and which left in private hands, to what extent prohibitions and sanctions rather than economic incentives should be used to reduce pollution or control technology or population. These decisions will have to be made by future societies on the basis of their knowledge and experience as part of the immanent process of social change.

We need not wait for any fundamental change in human nature in order to create an economy based on ecological humanism. Selfishness and greed will still exist, even though human beings are basically good rather than evil and human nature is progressive and capable of changing for the better. No transformation of ordinary human beings into saints is called for. What is required is a recognition that human beings are often impelled to act in ways detrimental to the public good and to their own futures because of faulty social arrangements.

Garrett Hardin, in writing of "the tragedy of the commons," points out that each herdsman is virtually compelled to graze his cattle to the maximum extent possible even though all know that this will lead to the permanent destruction of the pasture because, unless everyone cuts down his grazing at the same time, the pasture will be destroyed anyway, and those who hold back will be worse off than if they had participated fully in the socially detrimental behavior.[9] This is an example of what John Platt calls "social traps"—positive feedback loops in which socially destructive behavior is reinforced and society becomes "locked in" to destructive behavior patterns.[10] The kind of behavior triggered by inflation is an example of this kind of social trap; people buy goods they would not otherwise buy at a given time in order to avoid paying higher prices, thus causing prices to rise. Social arrangements can be redesigned to avoid such situations, and human beings can act in more objectively moral ways even though they remain just as selfish or shortsighted or as inclined to take the easy way out. Incentives and penalties can be built into the economic system which will discourage rather than encourage ordinary human beings acting in ways which damage the community, and themselves in the long run.

The Economic Goals of Ecological Humanism

Present-day society is locked into four positive feedback loops which need to be broken: economic growth which feeds on itself, population growth which feeds on itself, technological change which feeds on itself, and a pattern of income inequality which seems to be self-sustaining and which tends to spur growth in the other three areas. Ecological humanism must create an economy in which economic and population growth is halted, technology is controlled, and gross inequalities of income are done away with.

CONTROLLING ECONOMIC GROWTH

The primary goal of a society based on ecological humanism will be to put an end to economic growth for its own sake. Ending economic growth may or may not mean an increase or reduction in the GNP as measured in conventional economic terms. What it does mean is the creation of a steady-state economy in physical terms—what the contemporary economist E. J. Mishan calls "constant-physical-product" economic growth.[11] John Stuart Mill defined a "steady-state" economy as a condition in which capital was not increased but only replaced as capital goods wore out,[12] but this is only a crude measure. An adequate definition of a steady-state economy in operational terms is one of the tasks of the emerging science of ecophysics.[13] But what it clearly implies is an economic system in which productivity is limited to levels which can be sustained over a long period of time, given available resources and the indirect costs of using these resources. This does not necessarily mean that there will be no increase in the satisfaction of material wants. Technological progress may make it possible to increase productivity, to satisfy human wants, with less use of energy or space—what Buckminster Fuller has referred to as "miniaturization" and "ephemeralization."[14] But where economic growth necessarily involves an increase in the consumption of finite resources of materials or space, it must be slowed and eventually brought to a halt.

While there is a general positive correlation between levels of income and consumption of goods, different spending patterns (technically called the *composition of demand*) are to be found among different individuals, classes, and societies. A society based on the principles of ecological humanism will not be concerned with reducing incomes as such any more than with cutting down the GNP (essentially the total of all incomes) as such. It will be interested in decreasing the "physical-product" consumption of individuals and groups, using sophisticated social and economic planning so as to alter the composition of demand in such a way that while incomes may remain stable or even increase in monetary terms, the total growth of the economy in ecophysical terms is damped down.

Putting an end to growth is the only way to control the total impact of pollution on society. Most measures taken to decrease pollution are in themselves contributions to putting an end to growth in a physical-product sense, since they usually involve limitations on the generation of waste products. Recycling has the same dual function of being both a measure and a means of limiting physical-product growth. The concept of "husbandry," which has been the essence of traditional agriculture in much of the world and which means that what otherwise would be pollution and waste is recycled into socially useful products, must be extended to the whole industrial system of technological society.

Central to curbing growth is social (ultimately political) control of the use of space; land-use planning must be a basic element of comprehensive national and international planning.[15] Such questions as whether or not we should turn to synthetic food production to save wilderness from being consumed by agricultural uses (as some have envisioned),[16] and whether present urban concentrations should be decentralized into new cities, are particular issues which will have to be solved in terms of their merits as these become more apparent with the increase in our social and political intelligence. But they will have to be decided by political society as a whole, and they should be decided in a way that maximizes the extent to which man can relate to his natural environment on its terms rather than his. What one leading theorist has said of regional planning must be true of all land-use planning: It must "uncover, reveal, and visualize" not only the ideas of planners but nature's, "not merely . . . formulate the desire of man, but . . . reveal the limits imposed thereto by a greater power."[17]

Committed liberals are likely to argue that to attempt to build an economy as socially controlled as the one here advocated is to launch a frontal assault on human freedom. But, as I have tried to demonstrate, such a charge is meaningless. No one of us is a free agent; we all live within an economic system beyond our individual control, and what Sir Geoffrey Vickers says of human freedom in general is especially true in the economic sphere: "The idea that liberty means freedom from limitation rather than freedom to choose our limitations is a particularly dangerous delusion for the overcrowded inhabitants of a rather small planet."[18]

The charge that an ecologically sound economy would diminish freedom is meaningless on empirical as well as theoretical grounds. The contemporary American economy is not free. Even leaving aside the Nixon Administration's Phases I, II, III, and IV, the federal government has long played a major role in shaping the economy. Government has built canals and roads and given public lands to the railroads. It has constructed irrigation dams and subsidized farmers and manufacturers, airlines and shipowners. Power to take land from private owners for highways and public utilities and so-called urban renewal is well established under the traditional common-law concept of eminent domain. The contemporary American economy, like that of all modern nations, is a maze of governmental regulations, tariffs, and subsidies, and of legally enshrined monopolies and oligopolies. Present political controls are a chaos of ad hoc responses to pressures from special interests, held together only by a bias in favor of economic growth, slightly modified by fears of inflation.[19]

Mention of "total" controls usually triggers a visceral negative reaction. But all social controls based on general laws are in a sense total. All murders at any place and time are subject to sanctions; indeed, all willful injuries are. Every item one buys is subject to tax at some point in the economic process. But the existence of total control does not mean there is a policeman or a tax collector at everyone's elbow. It simply means that certain socially determined patterns of behavior are enforced without exception in all situations in which they are relevant. Social management of the American economy would not substitute unfreedom for freedom, nor increase the amount of control to which anyone is subject or of which most people are conscious. It would simply attempt to rationalize the controls which now exist and orient them toward the common good.

But economic techniques—incentives, penalties, and regulations—aimed at curbing economic growth in general and such especially noxious aspects of it as air and water pollution and the destruction of open space, cannot succeed if they focus solely on the processes of exchange and distribution, either through detailed regulation of day-to-day economic activities or through control over fiscal and monetary policies. If the present imbalance between man and nature is to be replaced by an equilibrium, technological change must be controlled as well; for changes in technology affect the balance between man and nature both positively and negatively[20] and have a direct effect on man's own nature and identity.

CONTROL OF TECHNOLOGY

"Death to the Technocrats," read wall slogans during the French student revolt of 1968. Although this is carrying things a bit far, "death to technocracy" makes good sense. Technology has become not only the "opiate of the intellectuals"[21] but that of all of us. It provides the rationale for the existing economic system and for the system of political authority which buttresses it.[22] If humanity is to regain the freedom to be human, the spell of technology must be broken and technology must again become our servant rather than our master. A socially managed economy will not only involve control over wages, prices, and the use of capital and land, as is theoretically the case in socialist economies. It will also undertake a vital task which socialist societies normally neglect: social control of technology. At the height of the liberal era, technology was raised to the status of a deity. "SCIENCE FINDS—INDUSTRY APPLIES—MAN CONFORMS" was the motto of the Century of Progress International Exposition held in Chicago in 1933.[23] But technology is a false god, whose devotees must be overthrown and whose pretensions must be rejected. Both the introduction of new technologies and the continuance of existing ones must be made subject to social control.

Control of technology must be distinguished from control of science. Science is ordered knowledge, and knowledge is a good in its own right. From the standpoint of the human race as a whole the growth of human knowledge and understanding necessarily represents an unfolding of human potentialities and an evolutionary step toward human fulfillment.

This is not to say that the pursuit of knowledge is an absolute value and that scientific endeavor has a claim to absolute autonomy, independent of any social control. There may be circumstances in which the truth is dangerous to society, and a plausible—if not compelling—argument can be made that certain lines of research—into, say, genetic differences among the races—should not be encouraged. There quite obviously are forms of research which violate ethical norms because of their contempt for human dignity: The sadistic research of Nazi concentration-camp physicians is an obvious case in point, and some types of research today involving live animals, live human fetuses, or even consenting human beings could perhaps be so classified. Society can forbid such research in order to maintain its integrity as a cultural system.

Freedom of scientific research is no more an absolute than freedom of speech. Science, like communication, is a human activity which takes place within a larger system which has superior needs and purposes. Even at present, governments do in fact have "science policies," and they attempt to support and direct science toward national goals, though what the means and ends of such policies can and should be is hotly debated.[24] But science in general can and should be free, unlike technology. Today (in contrast to the situation in previous historical epochs) science and technology are closely conjoined, which makes the distinction between them difficult to operationalize in practice. But it is clear in principle: Knowledge is one thing, application another. Simply because I may be a skilled surgeon and know how to kill you with a single thrust of a knife does not mean that I may use my knowledge whenever I will.

Technology is a human instrument for human needs and—unlike nature—cannot be legitimately spoken of as having rights of its own. It exists for human purposes. But whose? There is a sense in which technology as such can be spoken of in terms of its relationship to humanity as a whole. This is both necessary and legitimate. But we have to be careful in doing so. Though liberal society teaches its members to worship technology, this false god is an idol behind which are often hidden particular social forces and groups which derive special benefit from particular technologies. The "domination of nature" is to a large extent a cloak for the domination of some men by other men who use science and technology as their instrument.[25]

Technology has never been autonomous. It is a product of social forces and is modified by them, just as it also modifies them. This is especially true in the contemporary world. One fundamental characteristic of modern technological society is what Alfred North Whitehead calls "the invention of invention."[26] Our technology is largely created by a consciously organized social process, in corporate or government-managed or subsidized research laboratories. Research and development—"R and D"—is financed and directed by social groups, public and private, for various ends—usually private profit or national military power. The political control of technology, like the political control of economic growth, would substitute not social control for individual choice nor unfreedom for freedom, but rationality for chaos, the public interest for private interests.[27]

Nor does technological control mean an end to technological progress. Different technologies have different consequences. Some aid human development, some hinder it. For every technology that we may decide to abort because its introduction cannot serve the common good, there are other technologies which might not otherwise exist which we will want to create and encourage. Indeed, technology assessment and control will call for new extensions of scientific and technological research and skills. One problem in controlling technology is that new developments are usually upon us—involving irrevocable or seemingly irrevocable choices and investments, difficult to reorder or rescind—before we are aware of their existence. Controlling the future of technological society means anticipating new technologies through future research so that, through experimentation and simulation, we can directly or indirectly pretest changes before we become committed to them. Such an exploration of alternatives will necessarily involve an increase rather than a decrease in scientific and technological development and creativity.

Technology assessment involves difficult intellectual problems. It will be necessary to anticipate and measure the second-order effects of existing or proposed technologies: Will they promote centralization or decentralization, help one economic interest in society or another, bring man closer to nature or further alienate him from it, lead to greater consumption or better resource-conservation? We must also be able to measure the effects of the interaction of technologies upon each other and the effects of the interaction of their consequences. But just as difficult, we must make basic social decisions about the values we are going to use in measuring and judging technology. No matter how much information we can obtain about the possible consequences of various technologies, such information is useless unless we know which consequences we consider good and which we consider evil, which we want to encourage and which we wish to avert.

The assessment and control of technology present structural problems similar to those involved in controlling growth generally. Agencies for technology assessment can be located in either or both the executive and legislative branches of government, and they will have available economic incentives and disincentives as well as legal sanctions to use in shaping technological change toward politically chosen social goals. As in the case of the basic structures of the government or the economy, the structures of technological assessment and control in a society based on the principles of ecological humanism will vary with time and place, and will come into being as part of an immanent process of social change.

But technology assessment and control, however institutionalized, must be capable of helping us escape "the tyranny of small decisions,"[28] which causes us by imperceptible steps to become locked into an undesirable future. Technology assessment must see all our problems as interrelated. In the words of the report of the National Academy of Sciences to the American Congress, we must "replace tunnel vision with a more holistic view."[29] In addition, though every choice in some sense precludes other choices, whenever possible society must avoid making technological decisions which have irrevocable consequences and must above all keep the future as open as possible in the name of freedom.

Technology exists to assist man in relating to nature, but it cannot be allowed to upset the balance of nature and destroy humanity in the process. Human beings are the highest form of life the earth has produced, but they hold their mother planet not as property in the liberal sense, but in trust to the cosmos, in stewardship for human generations to come and for the forces of life which some call God.

Technology geared to unlimited economic growth can make the future subsistence of the human race impossible. "Dominion over nature," John Black has written, "is incompatible with long term sustenance."[30] Unbridled technological dominion over nature is also incompatible with the preservation of human cultural and biological identity; the uncontrolled development of the technology of genetic engineering and the technologies which kill off other species and put an end to the last vestiges of wilderness can destroy the identity and purpose of the human race. As one sociologist puts it, we must change the Marxian formulation and recognize that "the devaluation of the human world increases in direct relation to the decrease in value of the world of things."[31] A technology which sets man at war with his physical environment sets him at war with himself. Humanity must control technology on behalf both of nature and of its own future.

But creating a balance between humanity and nature, and bending technology to human purposes (the ultimate ends and justifications for ecological humanism), will be impossible without the prior or simultaneous accomplishment of two logically intermediate ends. Population growth must be checked and greater economic equality must be instituted. Unless these goals are reached, there will be little chance of halting runaway economic growth or unbridled technological change, with the dire consequences these portend for all mankind.

POPULATION CONTROL

Continuing high rates of planetary population growth threaten both the subsistence of the human race and any possibility of order and stability.[32] Little is known about the effects of crowding on human beings, but it is known to lead to abnormal behavior and social breakdown among other animal populations.[33] Differing rates of population growth among nations create the possibility of war between the less densely populated richer nations and the more crowded poor ones, especially as nuclear weapons become more readily available. More immediately, population pressures on subsistence not only create demands for increased economic growth, which cannot be sustained without placing insupportable burdens on the environment; they also create a demand for technological changes which threaten human culture and identity. It is easy to conceive of a world in which even several times the planet's present population could maintain the minimal life processes of nutrition, respiration, and so on; but it would have to be a completely mechanized, regimented world in which man would have lost all contact with nature and with his own cultural past—a world fundamentally inhuman.

Control of population growth on a planetary scale is therefore a necessity for the preservation of human civilization. Indeed, in some countries, and on a worldwide basis, population has probably already passed optimum levels.

But population control raises basic ethical questions, both as to ends and means. Since life in itself is a good, both in terms of the cosmos as a whole and of its individual possessors, how can it be argued that the number of lives should not be maximized? One can hold such a position only if one postulates that the quality of life is as important as the quantity of life. Nature itself gives support to this proposition. Biological processes are careless of life. Primitive creatures lay untold numbers of eggs, most of which are wasted. Lower animals have large litters, few members of which survive. Nature's aims appear to be twofold: the survival of species and the development of species, the upward thrust of life toward ever higher levels of self-consciousness and self-actualization. Whatever ethical force is inherent in the universe seems to be concerned with quality as well as quantity.

Each individual who comes into existence seeks to survive, choosing life over death save in exceptional cases where it "chooses" death to escape from pain. Human beings on occasion deliberately choose death in preference to physical or psychological conditions where survival appears to be more dehumanizing than release from life. Humans also choose to sacrifice their lives for others or for ideals, thus bearing witness to the belief that the individual and collective quality of life is a primary value.

But what right does any individual creature have to choose his life—including its quality—over the lives of other beings? Creatures other than man instinctively kill in order to preserve their lives or those of others of their species. Human beings also kill in the name of self-defense or social order. We are only slowly coming to grips with the ethical problem of whether we should kill members of other sentient species in order to maintain our diets or other cultural qualities of human life, and are only slowly recognizing our obligation to keep other species in existence. We are more rapidly, though haltingly, starting to confront the problem of the extent to which we are justified in ever deliberately taking other existing human lives.

But veneration for life—a fundamental premise of any natural morality—cannot in practice imply an obligation to create new life ad infinitum, for all life exists in a context of finitude. Our reluctance to kill other animals does not mean that we feel an obligation to increase the world's population of dogs, cats, starlings, or cattle without limits.[34] Nor does it mean we have an obligation to create—as we increasingly can—new animal species which have not so far existed. Much life that is within our power to create we will never create.

This principle applies to human life as well. There are still those who argue, against reason and common sense, that married couples should not practice birth control. But such a proposition obviously cannot be based on the premise of the more life the better. To be consistent, those who hold such a position would have to argue that celibacy should be outlawed, sterile couples separated, and sterile females only allowed to marry sterile males so as to maximize fecundity; that all females should begin procreating immediately upon sexual maturation; and that where separation from usual mates reduced the possibility of procreation, alternative temporary mates should be found lest potential life be wasted. The children produced in such a system would be just as unique and lovable as those born today. Why deny them life? The only possible reason is that some principle—social order, personal freedom, human integrity—is given precedence over the creation of life per se.

The position of those who believe in maximizing life without limit becomes absolutely ridiculous now that we are on the verge of artificial conception and "prenatal" nurture in artificial wombs. So-called test-tube babies will have personalities—souls—just as do traditionally conceived humans, however scarred these personalities may be. Do we have a moral obligation to give priority to perfecting such techniques and devoting a major share of our resources to creating assembly lines for the production of more people by such a method? If life is an absolute value, it is difficult to see why we should not do so. It is possible to understand if not necessarily to accept the position of those who base their opposition to birth control on an adverse technological assessment of some of the means toward that end, but a generally "pronatalist" position is ethically absurd.[35]

Every human being born is a unique individual who, having entered the social system, is by definition part of it with claims upon it for existence and development. Uncreated beings not having existence cannot have "rights." But is it not a pity that some child who might have been born and been a receiver and giver of love, a contributor of his or her uniqueness to mankind, is not born? In a sense it is. But we routinely prevent the existence of such persons. The child who would have existed had Jane married Sam—the result of their unique mixture of genes (actually the chance mixture of some among many of their genes only)—will never come into existence because Jane married Tom instead. This child is lost forever, just as is the particular mixture of genes of the child that she might have borne Tom had they mated on one occasion rather than another. But this is the way of life in a finite universe. Every choice individuals make in the course of their potentially procreative lives denies existence to many potential human beings.

How population should be kept at a level consistent with the common good of individual societies and of humanity as a whole raises more difficult ethical questions. War, famine, pestilence, infanticide, abortion, homosexuality, voluntary or involuntary celibacy for various periods of time, sterilization, and contraception are all means of population control. Animals limit their populations by various automatic means.[36] Birth control is a natural phenomenon. Human societies throughout premodern history have always controlled the size of their populations.[37] Human beings must choose their means in accordance with their personal and societal ethical norms. Such means as war, famine, and pestilence can be dismissed out of hand as inconsistent with the requirement for fulfilling basic human needs and so as contrary to nature. But insofar as rationality and technology are natural attributes of man, limiting birth is natural, and opinions can logically differ on the difficult questions of which means, on balance, are more in accordance with the requirements of human nature.

A future society based on the principles of ecological humanism will choose among means of population control in the same way that it chooses means of limiting economic growth or of controlling technology: by an immanent process in which decisions are the result of debate, experimentation, scientific investigation, and particular historical circumstances. But if control of population is necessary to fulfill the common good of subsistence, order, and purpose, social policy can legitimately aim at its encouragement. No individual or group or nation has an absolute right to bring into existence additional human beings whose nurture and sustenance necessarily create claims against the whole of human society, despite whatever attempts may be made to justify such behavior on the grounds of personal ideology or ethnic or national strivings for increased political or economic power. In exceptional circumstances where individual or group behavior interferes with the achievement of the common good, negative legal sanctions may be necessary to supplement persuasion and economic and social pressures. As in any other aspect of maintaining social balance, coercion is an undesirable and intrinsically limited means of social control, but it cannot be ruled out on ethical grounds. The assertion of the United Nations Declaration on Human Rights that choice of family size is a basic human right is an expression of liberal philosophy incompatible with ecological humanism, because it justifies conduct which infringes on the rights of others and of society as a whole.[38]

Population control is herein conceived of as relating to the gross size of population. Despite the claims of some eugenicists, there is no firm evidence that the biological quality of human beings is declining and no warrant for being particularly concerned about encouraging the breeding of certain individuals or groups and discouraging that of others in order to improve the quality of the human race.

If human society as a whole has a common good which all of its members must help attain, population control is a problem for all societies. Even leaving aside the extent to which population pressures exist within individual developed nations as such, it is important to recognize that a child born into a rich society, even one born to members of a disadvantaged group within such a society, has a much greater impact upon world ecological balance than one born into a poorer country. In her lifetime she will consume far more resources and create far more waste and pollution. Therefore, developed societies have an obligation to control their populations for the sake of the common good, not merely to make worldwide programs of population control more creditable but also because of the direct consequences of their actions on others.[39] The problems of redistribution of economic assets between rich and poor countries are bad enough as it is, without increasing the populations of the economically powerful societies.

If the subsistence needs of humanity in the long run require a leveling off of economic growth, this will further acerbate already existing problems of inequality within and among nations. Liberal society has always been based on wide disparities in economic well-being among its members. Today in the United States the most well-off 20 percent of the population receives about 40 percent of the national income, while the poorest 20 percent of the population gets little more than 4 percent, and no change seems imminent.[40] Roughly similar patterns exist in other western industrial nations.[41] Why is this tolerated? In some capitalist societies this situation represents the perpetuation of patterns of class domination which existed in precapitalist society; in socialist nations it evidences the ability of a bureaucratic "new class" to take care of itself even under allegedly egalitarian conditions. In the United States disparities in income are buttressed by belief in the "work ethic," which implies that all reward originates in merit and by the desire even of those who are not among the winners in the race for success that there continue to be large prizes to strive for.[42]

But the most important reason why these vast differences in income are tolerated is that, while the relative share of the poorer segment of society has remained constant, its absolute material standard as conventionally perceived has increased dramatically in recent generations. Even most members of minority groups in the United States are relatively well off, as measured by money income, compared with the average person in most developed nations, and are rich by the standards of the developing nations. This pattern is the result of overall economic growth achieved through the use of technology and the mortgaging of the future of the species. Liberal society has solved its social and political problems of redistribution by ignoring them and increasing the size of the pie rather than changing the way it is sliced.

Given these patterns of income distribution, what would the results be of a slowing down or cessation of economic growth within advanced societies? All other things being equal, this would mean at the very least that the aspirations of the less-well-off for higher absolute living standards (as measured in purchasing power at least) would be thwarted. It would also probably mean, given the current relationship between economic and political power in liberal society, that the burden of lessened production would fall more heavily on the poor than on the rich, and that the poor might end up less well off absolutely as well as relatively. A high tax on gasoline to conserve energy or for the purpose of cutting down smog obviously has its biggest impact on the driving habits of the poor. There is no need to wonder, therefore, why the more economically sophisticated leaders and members of lower income groups are suspicious of measures aimed at environmental protection that involve placing limits on economic growth and material consumption.[43]

However politically difficult the solution to this problem, the economic policies it mandates are obvious. In order to meet the ethical imperative for basic human equality, as well as to be politically feasible in democratic societies, any measures aimed at curtailing economic growth must be accompanied by a massive restructuring of the existing patterns of income and living standards within all societies, developed and underdeveloped alike. As in the case of the other goals of ecological humanism, it is impossible to specify particular means and structures for accomplishing these goals in advance, but certain possibilities are logically obvious. If all persons are to have the maximum possible opportunity for self-actualization, then changes will have to be made in the access of the least advantaged to the economic product of society. Whether this would be most effectively accomplished by improving opportunities for participation in the productive process itself,[44] or through income-redistribution by means of the tax system, or through the direct provision of benefits such as housing, medical care, educational and recreational facilities, and other services, or through some combination of these, will have to be determined on the basis of the best available information and of subsequent experience. Whatever changes are instituted, one caution will almost certainly have to be observed: Most of the population must perceive themselves as benefiting to some degree (even if not to the same extent as the least-well-off) if the changes are to be accepted and if the new institutions are to have a constituency politically effective enough to ensure their efficient and successful operation.[45]

Implicit in any planned economy is what economists often refer to as "incomes policy." Wages and salaries of individuals in liberal societies are the result of a variety of factors: supply and demand, custom, prestige, length and availability of training for various jobs, special skills, personal aptitudes and characteristics, danger and unpleasantness of the work, and so on. Utopian communes find one of their most difficult problems to be the allotment of tasks among supposedly equal members: Who will work in the library and who will take out the slop?[46] The French utopian socialist Charles Fourier solved the problem to his satisfaction by assigning the dirtiest jobs to small boys, since they like being dirty anyway.[47]

A planned economy must necessarily determine the relative economic positions of various occupational groups, whether it wishes to do so or not, just as a liberal economy does by default. A planned economy with a no-growth goal would soon have to face up to the question of making social and cultural judgments about the worth of differing jobs outside the marketplace—asking, for example, what the ratio of an average physician's income should be to that of a high school teacher or a truck-driver. Creation of an incomes policy will be an intellectually and politically difficult task, an exercise in economic and ultimately moral judgment, but will be unavoidable, since freezing existing income patterns would in itself be a de facto judgment about the relative social worth of occupations.

The need to make these kinds of decisions might well lead to a drastic reevaluation of the need or the utility of gross inequalities in income. As in the case of other kinds of social decision-making, this would be a learning process and in itself an evolutionary force, and what the ultimate outcome would be is impossible to predict in advance with any certainty. But what is perfectly clear is that these decisions are much too important to remain outside the control of the community acting through the political process.

The problem of equalizing economic well-being on the international level is as complicated as trying to create greater economic equality at the national level. It will undoubtedly entail some redistribution of economic benefits which will inevitably reduce living standards in the richer countries. Here, too, care will have to be taken in economic planning so that these reductions do not fall primarily on the poorer part of the population of the richer nations rather than on their populations as a whole. The imperative of halting growth is even more obvious on the international than on the domestic level, since it is possible to conceive of the poorer classes of the developed nations enjoying—in the short run at least—living standards as high as those of the rich, but it is simply impossible for the total population of the globe in the twenty-first century to enjoy standards of material well-being, in physical-product terms, equal to those now enjoyed by the more favored classes in the more favored nations.

17

Politics and Culture in
a Humane Society

THE kinds of changes in the political and economic life of society required
to implement the philosophy of ecological humanism must be preceded by
changes in the cultural values of society. Any reorientation of government
toward the common good presupposes major changes in cultural norms and
in individual behavior patterns. But obviously this is not a one-way street.
Governmental actions necessarily affect as well as reflect culture.

How much internal cultural divergence different societies will choose
to tolerate or encourage in the future will depend on their particular self-
image. Varying degrees of cultural pluralism are compatible with ecological stability. Ecological stability does not demand cultural uniformity or
stagnation; quite the contrary.

Cultural Pluralism

In a society based on the principles of ecological humanism, human
beings would be able to enjoy more freedom in individual cultural behavior than exists in most societies today. In contemporary America we are
involved in an anomalous situation in which individualistic chaos dominates the areas of life which determine the future of society as a whole,
while in our private lives, which can and should be mostly free, we are
subject to an unnecessary degree of control and uniformity. Laws forbid us

to consume certain allegedly dangerous drugs such as marijuana (or alcohol in the skies above Kansas), and most states by legal statute regulate our most intimate sexual behavior. Education is a relatively uniform lockstep process in the United States, though freer than in many other nations. Censorship still exists in many communities, officially or unofficially. Surveillance, as the Watergate conspiracy has emphasized, is rampant. A society based on ecological humanism would on balance increase rather than decrease personal freedom or, strictly speaking (since freedom in the abstract is an illusion), would increase diversity in our personal lives. No act which is affected by or affects others (and all do) can ever be completely "free" either in a physical or ethical sense, but the premise of ecological humanism is that even acts which affect others need be "controlled," by positive or negative measures, only to the extent necessary to maintain the equilibrium of the total society.

A society based on the principles of ecological humanism would seek to maximize real cultural pluralism, since social as well as biological diversity is a source of strength in the evolutionary process. Norman Mailer in his demi-serious campaign for mayor of New York a few years ago suggested that the city be divided into multiple districts, each of which could choose its own way of life. In one neighborhood divorce and marijuana might be outlawed; in another "free love" could be mandatory and marijuana distributed free. However difficult it might be to manage such a polity administratively, behind the obvious absurdity there is a sound premise. In fact, local police have traditionally enforced laws according to local community standards; what is juvenile delinquency or disturbing the peace in one neighborhood may not be in another; and many of our difficulties in social control today stem from the fact that the police often do not understand or cannot accept the norms of the communities in which they work. Communalism has a long history in human society; during certain historic periods members of different ethnic groups in parts of the Near East, Africa, and Asia have been subject to different laws based on their own communal standards. Cultural pluralism poses difficult practical problems, of course. It is not only veneration for the abstract concepts of uniformity of laws and national sovereignty which has led modern liberal societies to insist on social controls based on uniform standards. Nonetheless there is no reason why subgroups in large populations cannot be allowed to live according to their own norms whenever possible. The Amish can be permitted, as they sometimes are today, not to send their children to high school; nudists can be allowed stretches of public beach for their own use; motorcyclists can be provided with bike paths; ethnic groups can have their own neighborhoods and schools which conform to their own customs and cultural norms.

Cultural pluralism should be limited in a society aiming at ecological balance and human self-actualization only by three constraints. One is that physical living spaces for subgroups be equitably distributed in terms of size and location, and that, when uses of space conflict, the activities be segregated to the extent necessary for them to coexist. Cultural zoning should not always give the best land to the same subcultures. Motorboating should not be permitted to make safe or quiet swimming or sailing impossible; wilderness areas should be protected against motorcycles and snowmobiles; apartment buildings or city blocks could be zoned against dogs, small children, or loud late parties.

Another requirement is that cultural subgroups should be open and fluid, no one should be forced to remain in a particular subculture, and those wishing to break away from their communities or communes or ways of life should be able to do so. Free choice of lifestyles should be guaranteed by government to the extent humanly possible. In addition, all activities of subgroups must be kept within the physical boundaries necessary for ecological balance. This may require that the sum total of activities be restricted; not all land can be given over either to recreation or to industry, nor will it be possible to meet the energy costs of all potential activities.

Finally, any society will place some limits on the futures individuals are allowed to choose for themselves, based on its sense of identity and purpose. The British régime in India took a laissez-faire attitude toward local cultures, but it did, over objections even of some Englishmen, curtail suttee, the practice of widows immolating themselves on the funeral pyres of their husbands. No humane society today is likely to allow any subculture to reinstitute it, just as any human society would probably prevent the introduction of palpably dangerous and addictive drugs. The boundaries between cultural pluralism and social control will always be fluid and difficult to define in isolation from particular cases, but certainly a society such as ours which permits private individuals to degrade our air, water, and diet, and yet has laws which make it possible for students to be thrown out of school for wearing their hair below their collars, has its priorities upside down.[1]

Politics and Communication

An essential concomitant of ecological humanism is freedom of speech in relation to public policies. Such freedom is justifiable not as an inherent right of the individual or a minority but as an absolutely necessary part of the social process by means of which a majority makes a rational choice. Freedom of political communication is essential to a future-oriented, learning society. Contemporary technology is increasingly making available new means for decentralized cultural and political expression, giving individuals more effective means for influencing future-creation through that form of social interaction we call communication. The Xerox machine, the tape-recorder, the Polaroid camera, small and cheap motion-picture cameras, and videotape recorders increase the possibilities of effective communication among individuals and within communities. So significant are the potentialities of these media for challenging the control of existing power structures over ideas, that they have been labeled by their enthusiasts as "radical software."[2] Government should encourage their proliferation and use.

But at the same time, decentralization of communication should not be allowed to imperil the common consciousness of national and world society. The physical interdependence of modern nations and of the total global society makes them organic wholes, and they need the equivalent of a central nervous system in order to be able to function above the dinosaur level of activity. The United States has never had national newspapers as do Britain and Japan; its mass-circulation general magazines are rapidly dying off, and even national TV programming is in peril. Future American governments may have to take positive action to ensure the existence of national media of some kind. In order for social systems to have the self-consciousness necessary for future-creation, common media of communication as well as specialized and local media must exist.

But speech and political activity are not ends in themselves, and like other human "rights" cannot be thought of as absolutes existing apart from the social context in which they function.[3] While government in a human society will allow its citizens the maximum personal freedom and the maximum choice among lifestyles compatible with social equilibrium, it should not and indeed cannot avoid exercising a positive influence on culture in accordance with the premises on which it rests. Liberal societies have operated under the pretense that government could be completely neutral in cultural matters, and have often taken the position that all cultural activity which is not illegal should be equally encouraged. In point of fact, liberal theory in this respect has been largely ignored by liberal societies in historical practice, and culture in liberal societies has been in thrall to the tyranny of public opinion.[4]

Complete government neutrality in relation to the communications process is simply incompatible with a politics whose goal is the common good. Advertising should be controlled so as to increase the consumer's opportunities to exercise rational choice (as in regulation of labeling), to decrease consumption of socially undesirable products (such as cigarettes), or simply as a brake on the stimulation of demand for particular products (such as energy) or in general. Television and radio stations can be forced to devote time to public-service broadcasting. Aesthetic zoning, varying with local needs and circumstances, can control the sensory input originating in the public environment. Arts and sciences and education can be encouraged and subsidized whether or not they are carried on under direct public auspices. And—this is all-important—public energy must be devoted to increasing the level of scientific and technological training and sophistication of the citizenry so that they will be better able to understand and thus evaluate these forces which are central to the lives of humans in a technological society.

There is nothing startling about any of these suggestions as individual matters. They have been proposed or carried on in many liberal societies. But they have been largely invisible to political theory and beyond public debate because everyone has taken certain cultural standards, derived from past beliefs, for granted; liberals, inconsistently with their philosophical premises, have always assumed that public support of museums and parks was preferable to public support of bear-baiting or demolition-derbies. But to say that government has a legitimate role in shaping culture toward higher standards and in enlightening its citizens, bringing them closer to the beautiful and the true, becomes controversial when the nature of the standards and the substance of truth and beauty are in doubt, and during periods of cultural and political revolution such activities take on a new importance and political sensitivity. In the late twentieth century, after three centuries of liberalism, we can no longer take for granted, as did the liberals of John Stuart Mill's time, that all decent and well-bred men and women know the difference between right and wrong and between the beautiful and the ugly.

A society based on the principles of ecological humanism will carry on propaganda activities in favor of the principles upon which the total system rests—the need for curbing economic growth and population, for eliminating racial and economic inequality, for subordinating individual interests when necessary to the common good, and for reconciling man with nature—even if not every citizen accepts these principles. Liberal societies carry on propaganda in favor of basic social norms today. American government agencies propagandize against racial discrimination; Canada exhorts its citizens to "Stand together. Understand together." Every healthy social system seeks to maintain itself and to advance its particular cultural and racial goals. This is not only inevitable but ethically desirable. The liberal premise that truth is unknowable is a counsel of intellectual despair which makes human life impossible. Absolute truth is beyond finite human beings, but we are compelled to live—as individuals and as societies—according to the best approximation that is available to us. However one might disagree with some of his particular applications of the principle, the philosopher Marcuse is right in insisting that "The *telos* of tolerance is truth."[5] Irrationality, ignorance, violence, greed, and sloth have no right to equal time in social communication, however much some degree of their expression may be tolerated as a matter of convenience or necessity. Nature, which is the touchstone of morality, aims at life and development, and a society which chooses to embrace nature's norms need not accord status to the forces of degradation, disorder, and death.

Patriotism

The ancient virtue of patriotism needs to be remembered and fostered.[6] In some nations such as Canada, patriotism may take the form of nationalistic objections to the despoiling of the society's patrimony to serve foreign interests. In other countries, such as the United States, the real enemy will be found in the nation's own liberal traditions. If a foreign enemy had done to America what its "developers" have done to it over the centuries there would long since have been a revolution, with collaborators of the forces of destruction hanging from every lamppost. In an interdependent world, resources must be exchanged, and the past, natural or manmade, can never be completely preserved. But there are limits to change dictated by the need to preserve the cultural integrity of communities. A society based on the principles of ecological humanism will be a society not of egoistic individuals but of patriots—one in which citizens are brought up to love their native land, to cherish its beauties, its past, and its life processes; one in which patriotism will be based not on jingoism or competition with other societies but on a veneration for the social community which gives its members their existence and identity and which sustains and enhances their life within its own.

Patriotism of this kind is not only compatible with but is an essential part of devotion to global humanity, the future of which is dependent upon the cherishing and prevervation of its planetary home. The keystone of ecological humanism is a planetary patriotism focused on the creation of a social system which will maintain the balance between man and nature. This kind of patriotism received eloquent expression in an address by Adlai Stevenson to the United Nations Economic and Social Council in 1965:

> We all travel together, passengers on a little space ship, dependent on its vulnerable supplies of air and soil; all committed for our safety to its security and peace, preserved from annihilation only by the care, the work, and I will say the love we give our fragile craft.[7]

The Limits of Politics

Liberal society is based on a distrust of the political and a willingness to accept second-best political arrangements rather than run the risk of fanaticism and regimentation inherent in the programs of many utopian social reformers. Liberals recognize that, as Robespierre was quite content to acknowledge, virtue and terror may go together. So, fearful of terror, liberals reject virtue. Any attempt to reach the goals of ecological humanism, they argue, would lead to social breakdown and/or tyranny.

But a measure of the degree to which liberalism is an inappropriate political system for any time and place is the extent to which chaos or terror exists anyway, even in the absence of virtue. The need to go beyond liberalism in the coming decades is dictated by the depths of the crisis facing the world, not simply by abstract philosophical theory. The British industrialist-philosopher, Sir Geoffrey Vickers, warns us that "the post-liberal era will depend absolutely on adequate means to make and implement political choices of extreme difficulty."[8] Anarchist Barry Weisberg suggests that "the critical importance of ecology as a developing source of political opposition in America stems from the realization that politics in our age has acquired an absolute character."[9]

To take the current crises of man's relationship with nature and the machine seriously is necessarily to abandon liberalism. Liberalism denies the existence of absolutes in theory and in practice, save for the perverse absolute of unlimited growth. Liberalism is outmoded in contemporary society because of the need for living together in an ever-smaller world and for controlling runaway growth and technological change. This means that the alternative to political virtue is not a tolerable civic slovenliness but a chaos in which order and subsistence will be impossible to attain, or a state which is not a society of self-actualizing human beings but a new Leviathan.[10] As Paul Goodman wrote a few years before his death, "For green grass and clean rivers, children with bright eyes and good color . . . people safe from being pushed around so that they can be themselves—for a few things like these. . . . I am pretty ready to think away all other political, economic, and technological advantages."[11] Insofar as liberalism stands in the way of the defense of these things—of a restored earth and the integrity of human civilization—it must be thought away and soon.

Yet we should not be overly sanguine about the possibility of using government as a means to human survival. Strong government, however modified by popular feedback, will not necessarily work for the good of mankind. Not only may even well-intentioned governments pose threats to individual freedom,[12] but they can become the captives of special interests and intensify rather than alleviate the threats which individualism and mechanization pose to the maintenance of human civilization. Government can institutionalize, rather than fight, the destruction of a human world. In recent years the American federal government has too often sought to negate rather than reinforce the attempts of states to protect their populations against environmental degradation, has supported such socially and environmentally undesirable policies as the SST and the Alaska Pipeline, and has used modern electronic technology to spy on its citizens and to destroy the political enemies of those in power. Strong governments—with popular support—thrive on wars and imperialisms which strike at the heart of human solidarity and planetary survival. Strong governments often seek to perpetuate outmoded cultural patterns, the more so as they feel themselves threatened by the forces of innovation and renewal.

To advocate strengthening the ability of society through its political structures to control its own future is therefore to take terrible risks, to gamble with the future. In the last analysis such a step can only be justified as a strategy by the conviction that, while strong governments may not solve our problems and may make some of them worse—injuring some individuals in the process—there is no alternative. Dropping out, withdrawing into the hills or urban communes, making of community control a panacea rather than a technique, will not salvage the future for humanity. Technological society is already here, and it can no more be left without someone at the controls than can an automobile racing for a cliff. Without governments able to govern, human society, at both national and global levels, has little chance of avoiding a twenty-first century racked by insupportable poverty and suffering, national, class, and racial warfare, the steady deterioration of the biosphere, and the collapse of human civilization as we have known it throughout the past millennia.

The only chance that future human beings have of being able to live lives of dignity and grace lies in the quality of the life of the community as a whole. And the only hope for the quality of life of the community lies in the will of the community. Government can be an instrument whereby the effective majority of the community sets and enforces the norms for the whole social system, but it cannot alone do all that is necessary to create a decent human future. Political decisions leading to governmental action are a necessary condition for the creation of a decent human future, but not a sufficient condition. There are no systems of surveillance and sanction powerful enough to enforce norms of conduct which have not already been overwhelmingly internalized within a community, while if such norms are sufficiently widespread direct political control will be unnecessary. If people wish to live in societies where their humanity can thrive, they will have to create them through each choice they make in their daily lives. The revolution of ecological humanism will lead to and require a new politics and a new economics, but if it is to be successful it will have to be a revolution which penetrates every aspect of human thought and action.

IV

The Emergent Future

18

Getting There from Here:
The Immanent Revolution

POLITICS is power. To change the shape of the future means to change the patterns of interaction among men and things. Since the human persons involved in social processes are self-conscious beings, they are capable of initiating actions which will alter the impact they have on others and hence the character of the whole system of which they are parts. They can take actions which have consequences. But how should they act?

Ecology has been defined as a resistance movement.[1] So it is, a movement of resistance to changes in the relationship of man and nature and man and the machine which threaten the survival of human culture and identity. But resistance movements always fail if they remain such. Resistance can be a necessary and useful tactic for survival, but it is rarely a strategy for victory. Sooner or later, the forces of resistance must mount a counterattack which seizes the initiative and wrests control of the direction of events from the opposing forces. In the face of a concerted counterattack, some invaders lose heart and simply fade away. But the forces of environmental degradation and technological change will not do so. They must be not only repulsed but pursued and defeated on their home ground; the breeding grounds from which the aliens within sally forth must be seized and destroyed. Premature counterattack can expose the forces of resistance to complete destruction, but simply hanging on cannot bring victory against a determined and powerful foe. Resistance can only be a prelude to mobilizing for an assault upon the citadels of enemy power.

At the beginning of our discussion of the nature of political philosophy

the point was made that all serious political philosophies include a theory
of social change, implicit or explicit, which identifies the forces for change
and the means by which change takes place. The great intellectual strength
of Marxism is that it not only describes what is wrong with capitalist
society from a Marxist perspective and prescribes, however sketchily, the
utopia which will follow the overthrow of capitalism but, in the concept of
the proletarian revolution, it specifies the agent and mechanism of change.
Furthermore, Marxism's analysis of the agents and process of change flows
logically from its overall philosophy of man and society. The time has come
to ask how the philosophy of ecological humanism proposes to restore
humanity to its proper role in the universe. To fail to deal with this question would reduce ecological humanism to an exercise in utopianism in the
most pejorative sense—to the building of a dream castle which could serve
as a standard for a critique of the real world but could never become an
instrument for its transformation.

Any proposal for change must be reasonably specific about how to get
to where we want to go from where we are. Recently some social scientists,
discussing the problem of change, recalled the fable of the owl and the
grasshopper. Suffering from the winter's cold, the grasshopper asked the
wise old owl what to do. "Simple," the owl replied, "become a cricket and
hibernate." "But how?" asked the grasshopper after his unguided attempts
at transformation failed. The owl was enraged. "Look," he said, "I gave you
the principle. It's up to you to work out the details."[2] The process of
reordering society so as to achieve the promise of tomorrow will take place
in a future of contingencies no one can foresee, but the political philosopher must do more than suggest an alternative future and tell anyone who
accepts it as desirable simply to will it into existence. We must at least try
to understand the nature of the metamorphosis and how it takes place.

Revolution and Social Change

A revolution will be necessary to put an end to liberal society and to
institute a society based on the principles of ecological humanism. By
revolution I do not mean an act of violent overthrow of existing governmental or social structures. By revolution I mean a radical transformation
of the guiding principles of social and political life which will so alter the
form of society that it will be recognizably different from its previous
shape. The Industrial Revolution was a revolution in this sense, involving a
change in ethos, in the structure of economic life, in the physicial contours
of society, and ultimately in the distribution of economic and political
power. The human revolution—by which man was transformed from ape
into human—involved changes in form and behavior.[3] These revolutions
were the outcome of changes in the processes which constitute societies

and species which were so profound, pervasive, and cumulative in their effects that a new society or a new species came into existence. Feudalism was not overthrown by capitalism through a coup d'état or civil war—though the transformation, which took centuries, involved conflict and sometimes violence. Nascent man did not violently overthrow his simian ancestors, though there probably were bitter and perhaps bloody struggles between the bands of protohumans and their nondeveloping cousins.

New societies and new species arise out of the old in an evolutionary process, one which builds with the materials of the old but reorders them in new ways. The new society or species is an emergent property of the old. But though the new grows out of the old in evolutionary fashion, there are points of radical disequilibrium and disjunction, when the permanence of the change hangs in the balance, and the new struggles to maintain and develop its identity in confrontation with the old.

Political revolution differs from other evolutionary processes not simply in the magnitude of the changes it involves but in its self-conscious nature. The men whose economic activities created industrial capitalism may have recognized the magnitude of the change their new way of doing things occasioned, but the early capitalists did not undertake to revolutionize society by a coordinated conscious act of will.[4] Political revolution is a specifically human act in that it is a fully conscious process. Revolutionaries are engaged in deliberate, willed actions aimed at creating radical social change.

Change does take place in society and change can be discontinuous.[5] Societies do not simply proceed along the same paths, with the future only a projection of the past. To some extent change is the result of the fact that no society is ever completely integrated, that every society includes within it conflicting forces and conflicting values, but for the most part it is necessary to look for the cause of changes within society to changes which occur outside of society. Change is possible and inevitable because societies are acted upon by outside forces. Some external forces are social, originating in the activity of other societies: The contemporary societies of Asia, Africa, and the rest of the "developing" world are today undergoing the advanced stages of a traumatic process of modernization which is not simply a continuation of their own indigenous social processes but reflects the impact of western colonialism and imperialism, the voluntary or forced assimilation of new cultural norms, technologies, and legal and political forms derived from abroad.[6] Throughout history, such external forces have made new inputs into social systems, which have radically altered the balance among internal forces and changed the shape of the whole.

But even western societies have undergone drastic change. In the case of the western world the forces of change were ordinarily population growth or the introduction of new technologies or ideas.[7] In a certain sense growth in population comes from within; it is the people already living in society who give birth to new human beings. So too technological and cultural innovations are internally produced; they are the creations of the men and women of the existing societies. James Watt, the inventor of the modern steam engine, which revolutionized western society, was an Englishman; Einstein was born in Germany. But, from the standpoint of the preexisting social systems and their dynamics, such forces as population growth, technology, and cultural innovation are new, outside forces. They act like mutations in genetic patterns, which, if successful, force their radical restructuring.

Finally, the physical environment can change. Biological evolution is a process through which species evolve in response to environmental challenges as well as, and in conjunction with, response to pressures from other species. Throughout human history the physical environment has been a given. Although changes in climate have affected societies within the period covered by recorded history (the desiccation of the Sahara during the last two thousand years—a process which is still continuing—is one example), on the whole man has been able to develop in association with a constant or at least almost imperceptibly changing environment. Now the human species has reached a point in history where to the slow, almost imperceptible, naturally induced changes in his environment have been added violent and drastic changes of his own making, currently accelerating in speed and significance. Man's own activity reacts back upon him, and he must adjust to it as to an outside force.

The setting for change in the next century, and the need for political action to control change, is the result of the concatenation of all of these forces which now impinge on liberal society: population growth, technology, cultural innovation, and humanly created environmental instability.

How do societies change internally in response to such outside pressures? Change creates crisis. Sociologist Robert Nisbet, following W. I. Thomas, describes the nature of crisis as "a relationship between human being and environment precipitated by the inability of the human being (or social group or organization) to continue any longer in some accustomed way of behavior."[8] Crisis forces men and societies to become aware of change as old forms of behavior cease to be adequate to achieve their goals. Most human beings seek to deny the existence of crisis—the response of liberal society to the problems raised by technology, population growth, and man's impact on his environment is a prime example of this approach. But crisis has a positive aspect in that it permits new leadership, conscious of the crisis and willing to deal with it, to come to the fore.

When societies become fully conscious of crisis—a point which unfortunately is sometimes not reached until the crisis attains such proportions that it is almost impossible to deal with effectively—the possibility of radical social transformation exists. External events upset the balance of the system, making radical change not only possible but necessary. The anthropologist Anthony Wallace has studied this process in tribal societies confronted with the need for dealing with the challenges presented by contact with more advanced societies. The process he describes, in which leadership takes advantage of the failures of the existing system to meet members' needs and, through the formulation of new ideals and the creation of new movements, brings the society to a new equilibrium, can occur in advanced societies as well.[9]

Change need not take place according to such a scenario—indeed change may not take place at all, except for change that results in the disintegration and disappearance of a system too rigid to change. But it *can* take place in this way. Any change—on whatever model—can take place with great rapidity, as is testified to by the spread of innovations in the contemporary world. The income tax and the withholding tax spread around the world in a few years.[10] Automobiles and television are accepted by peoples whose parents lived in Stone Age conditions. The spread of oral contraception within western society has taken only a few years. Ten years from the time it was first suggested at a Pugwash conference, an atomic test-ban treaty became a reality. Change in the basic character of cultural and social systems does not come quite as easily, but it too occurs quite rapidly once the foundations have been laid by such basic events as population growth, technological innovation, or new environmental pressures. What is all-important is the perception of crisis and the willingness to deal with it.

Crisis, then, provides the basis for cultural revitalization and for legitimation of change and movement toward new goals. Crisis also provides a basis for legitimizing new leadership. Liberal society is today in crisis. Most people, including many intellectual leaders, deny this, as is predictable from what we know about how human beings respond to crisis generally. But the objective elements in the crisis are real, however some may seek to deny them: the growing shortages of resources, dramatized by the "energy crisis," the degradation of the environment as exemplified in frequent "smog alerts" in large cities, crime and social unrest, growing cynicism about government, cultural conflicts, population pressures, the constant threat of global war. The only real question is whether the needed leadership will arise and will find enough acceptance and support in time to revitalize society.

The Revolution of Technological Man

Students of history have long argued about the relative importance of "great men" and broad social forces in the generation of social change. Robert A. Nisbet contends that great men and social forces combine to produce change when the historic moment is ripe. "Major changes," he writes, "are incomprehensible save in terms of superlatively endowed individuals, or effectively marshalled elites, working within social circumstances, usually those of crisis."[11]

No one can predict whether the great men we need will be available to lead liberal society out of its present difficulties, who they will be, or where they might come from. But we do know that crisis is upon us. Where then can we look for those who might compose an "effectively marshalled elite"? There is one obvious source. In modern western societies the revolution of ecological humanism will have to be led by technological man.

Who is technological man? In an earlier work I defined technological man as an emerging human type "in control of his own development within the context of a meaningful philosophy of the role of technology in human evolution . . . a new cultural type that will leaven all the leadership echelons of society." Technological man, I argued, "will be man at home with science and technology for he will dominate them rather than be dominated by them."[12] Technological man will be inspired by a philosophy based on naturalism, holism and immanentism—the philosophy of ecological humanism.[13]

Social change occurs when conditions in society are altered so that new forms of behavior are necessary if valued needs and goals are to continue to be met. Those who recognize that this is so become the leaders of the new society.

The ability to recognize the need for change has a dual origin. Some individuals are capable of greater rationality and imagination than others and have a greater devotion to the well-being of humanity as a whole. They compose a self-constituted, open elite definable by no other characteristic than their historic role itself. Their actions in times of crises have an overwhelmingly voluntaristic aspect in that they are able to rise above their social origins, and sometimes their material interests as well, to act on behalf of society as a whole. Other individuals and groups, more bound by habits of mind and social position, are nonetheless willing to alter their behavior in times of crisis because of the logic of the changes affecting their position in society. All social movements combine voluntaristic and quasi-deterministic elements among both leaders and followers.

The revolution of the bourgeoisie which created modern industrial capitalism resulted from the combined influence of such leaders as John Locke and Adam Smith and Alexander Hamilton, and of the struggle for increased economic and political power on the part of the merchants and industrialists and professionals throughout western society who were not moved primarily by ideology but by a sense of where society was going and how it affected them and their personal futures. The socialist movement was the creation of leaders of varying—often nonproletarian—origin, men such as Proudhon and Lasalle and Bakunin and Marx and Lenin, and also of millions of workers and peasants who were aggrieved about the conditions and course of capitalist society. The revolution of ecological humanism will likewise be the result of the combined efforts and activities of scientists and statesmen and millions of otherwise ordinary persons who are simply disturbed about the conditions of contemporary technological society and anxious about the future of the human race.

Ordinarily revolutions come about not because conditions continue to be as bad as they have long been—this is more likely to lead to fatalism and apathy than to revolution—but because conditions fail to improve fast enough or because conditions are deteriorating.[14] The revolution of ecological humanism will come about because of a combination of unsatisfied utopian hopes and growing conservative fears of disaster. This is the unique historical situation described by psychologist Robert J. Lifton: We see technology and science as simultaneously presenting us with the possibility of liberation from humanity's age-old burden of oppression, arduous labor, and want and at the same time threatening us with nuclear holocaust and ecocide.[15] Or, as a political scientist puts it, it is a paradox that "at the very time many people are coming to an awareness of the possibilities for creative freedom, fear is spreading that humanity will be reduced to an army of robots."[16] The revolution will come into being when a sufficient number of people recognize that the future cannot be a simple projection of the past but requires a conscious choice between liberation and destruction.

As in any social revolution, some groups will be more disposed to meet the challenge of change than others because of the factors which condition their ability to perceive change and the factors which enhance or inhibit their ability to accept change. Many members of the upper classes recognize intellectually that the industrial system on which their power and privilege rests carries within it the seeds of its own destruction but are unable to break their profitable ties to it. Most members of the industrial working class are still unable to grasp the true nature of the crisis facing humanity and pin their hopes for the future on an increase or redistribution of the fruits of continuing technological change and economic growth. Thus, while technological man represents a new cultural type based on perception and knowledge rather than class position—a person identifiable by his cultural rather than his economic or social position—most technological men and women will necessarily come from the ranks of what is usually referred to in American usage as the middle class, defined in terms not simply of income level but also of education, occupation, and lifestyle.[17] They will be people who are sufficiently educated or skilled to understand how technological society works and the problems it confronts, but not so powerful that they enjoy or profit greatly from playing the games of war and expansion and change which threaten its viability. They will be people who are sufficiently secure economically not to have to struggle for subsistence but not so well off they can hope to use personal wealth as a means of escaping from the unpleasant and dangerous side-effects of technology.

Every developed industrial society produces individuals and groups who are in a position to ask whether the society is really working or not—to question its ends (its "consummatory values") as well as its means or instrumental values.[18] The potential revolutionaries in technologically advanced societies consist typically both of representatives of older social forces—intellectuals and professionals with norms derived from older cultural modes, humanists in the traditional sense—and representatives of newer forces generated by technological civilization itself—managers, scientists, and technicians who see the difficulties of a decadent liberal society from within and perceive its inherent irrationality from the point of view of technological and scientific standards of efficiency and long-range survival capabilities. (Some individuals, of course, combine both the humanist and technical outlooks.) The better-educated young play a special role in the perception of crisis because they are sometimes able to see the developing future more clearly than their elders, since it is the society into which they were born and they know no other,[19] and since they are engaged in a process of creating their personal identities and goals which necessarily involves the questioning of the meaning of life, including the value of the system in which they are growing up and which they must choose to serve or alter.

The social matrix out of which technological man is emerging is world-wide, but it is most clearly visible in the United States. The United States has been the world's leading technological society, and, despite its rapidly declining position economically and technologically, it continues to be so. It is the cutting edge of the human future, and American developments tend to prefigure developments elsewhere. American society has on the whole offered greater perceived social mobility than exists in any other nation, and both the economic elite and the working class are more oriented toward middle-class values than is true elsewhere. In addition, the United States has from its founding had a special sense of national destiny. America was to become not only a great and powerful nation but the scene of the creation of a good society capable of being a model for all mankind. The dream of a brave new world on American soil has always been a fundamental element in American national identity, and even after Vietnam and Watergate and in the face of the waning of material plenty, Americans will not easily surrender that dream.

Technological man is therefore especially likely to be the product of the American middle class. As one sociologist claims, "The American middle class and its ethics are at the center of American moral traditions."[20] The cultural and political struggle for the reordering of American society is in large part a war for the soul of the American middle class and a struggle over the continuance of its dominant role in American culture. Critics of the so-called New Politics of the late 1960s have tended to recognize this, though they state the problem in divergent ways. Herman Kahn sees "upper middle class elitists" imposing their values on the masses,[21] while Irving Kristol argues that a "mass intelligentsia" is engaged in a war with the "business class" for the control of American society.[22] What in fact they dimly perceive are the birth pangs of technological man, of a mass movement of women and men who question not the existence and necessity of technology itself, but the desirability and survival value of a society in which technology is used in accordance with liberal norms. The revolt against liberalism is real and deep-rooted because it represents not simply the philosophical views of a few intellectuals but the conviction of large numbers of the very women and men who keep technological society in operation that there is something wrong, that we have to go beyond liberalism if we are to build the American dream.

The struggle against uncontrolled growth and technological change and in behalf of a society which will seek full self-actualization for all its members is therefore not simply a reactionary, romantic nostalgia for the past, real or imagined, on the part of social groups displaced and downgraded by industrial society, as some apologists of things as they are and are becoming allege or imply. Rather, it is primarily a revolt on the part of the most intellectually open and technically sophisticated elements of the populations of modern societies.

The revolution of ecological humanism looks both backward and forward: backward to man's biological identity and cultural history as a guide to what being human means, and forward to a world in which a humanely controlled technology will permit man's identity and cosmic purpose to manifest itself fully rather than becoming lost in some nonhuman form of existence. The first step in the revolution is already beginning to be taken; this is what the historian Crane Brinton has called "the transfer of allegiance of the intellectuals," in which the most educated members of the society, normally members and supporters of the "establishment," question the intellectual and ethical bases of the existing order, as intellectuals did prior to the American, French, and Russian revolutions.[23] But if the revolution is to succeed it must be democratized, and carried on at every level of society. This democratization will occur in the course of the spreading revolution itself, as the housewife concerned about the future safety of her children joins the atomic physicist in protesting inadequate safeguards at nuclear power plants, or the homeowner joins the urban planner in struggling to keep a new freeway from destroying a viable neighborhood, and in the process each learns that the evil he or she is fighting is only a symptom of the increasingly antihuman and dysfunctional nature of the existing social system as a whole.

The Tactics and Strategy of Revolution

Political philosophies reflect particular images of nature and of man, and this is as true of their theories of change as of their normative prescriptions. Liberal political philosophy conceived of nature and man in terms of a mechanistic Newtonian view of the world. Liberal theories of revolution therefore saw revolution as a situation in which the levers of power in the social machine were to be grasped and society steered in a different direction, with new social forces in the driver's seat. Marx conceived of history as a deterministic process in which consciousness was a mere derivative of historical forces, and the political and cultural victory of socialism was an inevitable byproduct of the physical socialization of the means of production as a result of the development of large-scale industry. Lenin combined this determinism, somewhat inconsistently, with a return to a liberal, voluntaristic conception of revolution, a revolution which would be carried on by an elite when the time for revolution was historically ripe.

Ecological humanism views the universe and society as a process of interaction in which changes in the activity and consciousness of the elements of a living system come together in a learning process in such a way as to create a new whole with a structure different from that of the organism or social system as it existed in the past. Revolution for ecological humanism is an immanent process. It is the result of the intermeshing of the conscious and unconscious movements of all the elements of a social system. Primates became humans and primate society became human society through such a process of immanent change. Discontinuities occurred when environmental challenges created crises which required or elicited new adaptations; there are "macro-jumps" in evolution. But the new arose out of the old not because "centers" of power were seized by certain cells or individuals, but because certain kinds of cells and individuals with certain characteristics became more prevalent and significant and system-defining as a result of the nature of externally triggered change and crisis. At a certain point in time, new configurations were recognizable; man could be said to have replaced ape; agricultural or industrial society could be said to have come into existence. But this was not a centrally directed process of change, nor was it one achieved through violence. The new patterns developed as a result of piecemeal changes in the old, as modifications in some elements of the system interacted with modifications in others to reshape the whole.[24]

The triumph of ecological humanism, if humanity is to prevail, will have to be this kind of piecemeal, incremental process of creeping, convergent revolution. But it will not be a mostly unconscious apolitical process such as that prematurely heralded as "the greening of America." It will be a process of social learning—of self-conscious evolution of society toward a new form. Ecological revolution will be a conscious struggle, somewhat like a guerrilla war, the kind of process the New Left German student leader Rudi Dutsche spoke of in a metaphor based on the Chinese Communist experience as "the long march through all the institutions of society."[25]

Because of its immanent character, the revolution of ecological humanism will not and cannot be centrally directed. Some advocates of utopian change have argued that what we need is a new political party, international in scope, aiming at the capture of political power in all nations by an elite dedicated to the creation of a viable future.[26] The debacle of the kamikaze charge of the "New Politics" forces in the McGovern campaign for the American Presidency is 1972 illustrates the futility of such a strategy. The first requisite of successful guerrilla warfare is decentralized action, rather than combining and thus isolating revolutionary forces so that they can be easily identified, discovered, and destroyed. Another requisite is flexibility, an ability to hit and run rather than to stand still and engage in the traditional warfare of massed opposing groups of combatants. A single political party, even a single centralized movement, dedicated to ecological humanism would be an invitation to annihilation by the alien forces which control the citadels of wealth and power in modern society.

But the emphasis of the forces of ecological humanism must remain political in the broadest sense. While they will strive in their personal lives in whatever ways they can to bring a new society into being within the womb of the old, they will not be seduced by those who believe or pretend to believe that the world can be saved by Boy Scouts picking up tin cans when nonreturnable containers should be outlawed, or by individuals giving up automobiles in communities where adequate public transportation is not available, or by families refusing to live in freestanding homes with backyards or swimming pools in cities where decent communal housing or recreational facilities do not exist, or by people turning down their furnaces when building codes do not require new homes to be properly insulated to conserve energy. Campaigns to pick up litter or ride bicycles can be a useful means of consciousness-raising both for those involved and for as-yet-uncommitted spectators, but they should not be allowed to divert energy and attention from the need for overall, socially mandated and supported changes in the laws and lifestyles of the whole community. They can be devices for diverting attention from the real sources of social problems and for placing the burdens of social mismanagement exclusively on the shoulders of those who have sufficient public spirit to care about the whole community.

In any event, the results of such activities are at best marginal contributions to necessary change. Individual witness is no substitute for revolution in an era of total crisis. What is required at such times is not the purgation of individual feelings of guilt, but effective action to reform those social dynamics which make the antisocial action virtually inescapable for most men and women, no matter how much good will they have toward their fellow citizens.[27]

What applies to individual action applies to local and community action as well. Local governmental jurisdictions and local voluntary associations can be laboratories for social change and redoubts for mobilizing opinion, but however useful they may be as training grounds or staging areas, their activities toward the creation of a better society can only be the beginning of the struggle. Ultimately it is national governments that control the most important tax structures, the technology used in the products which go into the marketplace, and the relationships among national economies. Local governments and voluntary groups can be used as bases for forays into the jungle of national policy, especially in as open and uncoordinated a political system as that of the United States, and combinations of local actions and pressures can radically affect the shape and direction of federal policy, but there is no possibility of escaping the need to directly influence government at the national level.[28]

The revolution of ecological humanism will be an "open conspiracy," an unstructured, unorganized movement for reform. Such a movement will have its organizations and its leaders, but they will come and go. There will never be a single body or a single head. The adherents of ecological humanism will join its ranks by individual acts of intelligence and will be united only in the possession of a common consciousness. Insofar as they share common convictions about the nature of man and society and the needs of the human future, they will act in accordance with these convictions in every facet of their daily lives. People will work for conservation or tax reform or consumer interests through already-existing or new environmental organizations or public-interest lobbies. They will form groups for political lobbying, legal action, or the ad hoc support of political candidates as the need arises. Ecological humanists will work within existing political parties and churches and schools and media and community and voluntary associations as they individually can and see fit to do so. The only thing uniting them will be their common perceptions and their common dedication. The first Christians, persecuted by the Roman Empire, recognized each other by the symbolism of the fish; ecological humanists will recognize each other by their own symbols, some of which are already emerging—the specially charged use of familiar words such as love and peace, the veneration of greenness, the shining blue globe of earth set against the dark depths of space. The immanent revolution of ecological humanism will be above all a revolution of consciousness.

The most basic element of the new consciousness will be the rejection of the dualism which is at the heart of liberal society. Ecological humanists will reject both the "single vision" of rationalistic science and technology and the inchoate romanticism of the counterculture by combining them in a new synthesis reflecting the essential unity of matter and spirit, mind and body, the rational and the affective. As political activists they will be mystics who know how to do precinct work; as administrators they will be dreamers who can program computers.

This new consciousness will be heightened by books and journals and newsletters and conferences and "invisible colleges"—communications networks of like-minded individuals—through which revolutionaries can come to recognize their friends and their enemies. This is the vital area where leadership and organization will play a crucial role in mobilizing and guiding by persuasion and example. Hard cores of activists and even professional political and cultural revolutionaries will emerge, but they will not be so much generals mobilizing armies as prophets and scholars rallying the forces for change by providing information and inspiration.

The revolution of ecological humanism will be a *convergent revolution* rather than a single, centrally directed assault against liberal society. But it will be a single effort despite the fact that it will use a variety of means and will have no central leadership. It will be a single movement because it will have a coherent, common goal which will require simultaneously changing the priorities of American social and economic life, the operative value-structure, and the distribution of costs and benefits within the society. The concept of convergent revolution recognizes that any social system is a complex moving equilibrium which can and must be restructured while it still continues to function, rather like rebuilding a spaceship while in flight —which, after all, is essentially the task confronting ecological humanism on a planetary basis. Happily, even in the most intricately interrelated system there are enough redundancies of structure and function, enough short-range discontinuities, enough contradictions and stresses, to make this possible. A ship at sea can seal off part of its hold while the water is being pumped out and a ruptured bulkhead repaired. Human beings can function with a single kidney or lung, and in time surgical techniques may make it possible to repair the injured organ or replace it while the patient continues to live normally. Different structures can take over the functions of those damaged or destroyed, in the same way as the blind use canes to "see." What the revolutionaries of ecological humanism must do is difficult but not impossible: They must devise a strategy for maintaining the political legitimacy and economic stability of the system while drastically reorienting its purposes and the outcome of its activities.

Ecological humanists will not follow the example of many supporters of the "counterculture" by "dropping out," for they are part of the system they must reorder, and it is only through a change in the nature of their activities within the system and that of those whom their activities influence that the system can be transformed. They will therefore (as even such a proponent of counterculture values as poet Allen Ginsburg now urges his followers to do) "drop in" or "drop up"; they will take active roles in existing social institutions and change them from within.[29] They will refuse the counsel of despair of men such as Ellul; they will refuse to leave the world to the snake.[30]

A strategy of changing society from within, of altering it without being altered by it, will require both dedication and sophistication. For those who enlist in the ranks of ecological humanism, dedication will be expressed not simply by traditional political activity but by asking of every action in life: Is what I am doing—my support for this organization or public figure or activity, my choosing this job or recreation or product—conducive to restructuring society or not? The crucial weakness of the existing liberal system is its focus on the particular and the short-range, its leaders' lack of historical perspective and scientific knowledge, and its consequent inability to focus its powers of repression on forces of revolution which operate simultaneously and in a sense anonymously in all areas of social and cultural life. The advantage of those committed to ecological humanism will be their superior perception of the significance of what is happening and their superior awareness of how what they and others are doing is shaping the future, and therefore their greater ability to distinguish between friends and enemies.

This dedication will have to be informed by a social sophistication unequalled in human history to date. Partisans of ecological humanism will have to learn how to promote their ends indirectly as well as directly. They will have to take advantage of differences among their opponents in order to undermine the position of first one and then the other, or in order to build relationships of trust or obligation for later use, even in contexts which may seem unrelated to their major objectives.[31] Ecological humanists will have to wage cultural war against nationalism and militarism and the cultural patterns which reinforce bourgeois society, such as philosophical dualism, addiction to consumer goods, psychological rigidity, and individualism, taking advantage of the conflicts which already exist in all societies.

Despite setbacks and reactions, a cultural revolution is proceeding in the United States as in all present-day technological societies. Most of its elements—the "revolt of the body," the growing importance of affective life and "physical mysticism," the "sexual revolution" and "women's lib," the renewed pride and assertiveness of various ethnic minorities, renewed interest in spiritualism and eastern religions—are healthy reactions against the dualism of liberal society and its emphasis on technological norms. Ecological humanists will foster the positive aspects of these movements and reject the negative, assimilating them into their own synthesis as part of the basis of a new, postliberal civilization.[32]

Ecological humanists, while hewing steadily to their own goals, will not seek to form an exclusive sect of the elect, taking pride in their own rectitude as many reform movements do, but will work with whomever they can in whatever context they can. They will recognize that cultural revolution is a prelude to and part of political revolution but cannot be a substitute for it, that consciousness-raising can be a useful technique for mobilizing people but cannot be allowed to become an end in itself—in short, *they will aim at results rather than emotional catharsis.* They will above all else be futurists, in the best sense of the word—not seeking novelty for its own sake or seeking refuge in the future from unsolvable problems of the present nor playing oracle to mystify the multitudes. They will be futurists because they will recognize that only those who seek the future will find it, that only those who think constantly about the consequences of actions in the present can control what the future will become, and that only through consciously working to create the future can humans exercise their freedom.

Revolution Against Violence

The revolution of ecological humanism is an absolute revolution. Ecological humanism differs from liberalism in assuming that there is such a thing as the common good, that its content is knowable, and that it takes precedence over all private goods. Ecological humanists are not simply another interest group with demands to be met within the structure of the liberal system but a movement which speaks on behalf of interests common to all human beings, not only those of its members but of generations yet unborn, a new perspective from which the total system must be judged and restructured. Ecological humanism does not offer new answers to old questions but is instead a set of new questions which challenge the validity of the total intellectual and political framework of liberal society. The claims of ecological humanism are uncompromisable because they constitute a new framework for the resolution of particular claims. What is at stake in the revolution of ecological humanism is not the position and interests of particular groups of individuals, but the total direction of human society.

Yet despite its absolute character the revolution of ecological humanism cannot and will not be a violent revolution. This is not because violence could not be justified in the name of the human future.[33] Violence as such is not necessarily evil: That "the violent shall bear it away" was spoken in the New Testament of the kingdom of heaven. Physical force is a fact of life, and of the nonliving universe as well. Within the context of human life, there are situations where violence and evil can only be countered by violence.[34] To struggle against the forces of death is a law of life.

But violence, though sometimes ethically justified, has its own inner dynamics. Violence is common throughout nature, but intraspecies violence is not. Whatever may have been its role in prehistory, violence in the current stage of human evolution is a regressive, if deepseated, propensity. Liberal industrial society is based on violence toward nature; all forms of political oppression are based on violence toward members of the human species. Ecological humanism must reject both.

There are times when violent protest can serve a useful function in calling attention to otherwise ignored grievances, or when violence may be necessary in counterrevolutionary situations before human consciousness has evolved to full acceptance of the new ways of behavior. Any political coercion involves the potential use of violence, and violence will always be the last resort of political authority—however remote, guarded, or disguised it may be. But means condition ends, and this makes it difficult if not impossible to build a decent human society on the basis of violence. The passions which accompany violence are difficult to put aside, as every settlement after a revolution or civil war attests. Beyond this, violence is by its very nature more congenial to the social and cultural forces opposed to ecological humanism than to the forces of human fulfillment.

The triumph of western liberal capitalism throughout the world was achieved primarily through violence, as warrior and merchant combined their efforts to force the liberal economic system on other peoples.[35] In large part as the result of its pioneer heritage, America has long been the most violent nation in the developed world. The large proportion of the American budget which goes for military purposes not only represents a considered judgment about the world strategic situation, but also symbolizes the central role that violence and its institutions play in American life. In no industrialized nation in the world do as many private citizens have firearms, and in none is the possibility of violent private conflict nearer the surface in ordinary social relations. As a militant black leader once said, "Violence is as American as cherry pie." Nothing more clearly illustrates the intellectual and political puerility of the New Left than their attempt to use violence to change the course of a society rooted in violence. To attempt to defeat the American establishment through violence is to engage the enemy in battle on his own chosen terrain; it is a confrontation the forces of liberalism and their allies would be sure to win.

The strength of ecological humanism does not lie in its ability to win pitched battles but in its ability to understand and harness the diverse forces of social change and to oppose to the haphazard actions of the agencies of destruction the creative power of pressures aimed at achieving a consciously sought common goal. Each change in consciousness has a potential for altering the actions of self and others. If enough of the interacting elements which make up a system perceive what the future shape of the system ought to be, they can restructure it by consciously altering the direction and velocity of their own movements and the responses they make to the movements of others. It is as though a large number of the atomic particles which constitute a caterpillar were consciously to decide to turn it into a butterfly. Social change becomes a process of self-conscious metamorphosis.

To believe in the possibility of the immanent revolution of ecological humanism is an act of faith in the possibility of freedom, in the power of the human mind to predict the consequences of individual actions, and in the potential of the human individual to act so as to shape the future. To many this will appear the purest fantasy and self-delusion. But the only alternative is to believe that human beings are no different from electrons or the unconscious cells in a plant or animal, that human actions have no significance, and that human history is utterly without meaning.

19

Ecological Humanism and Planetary Society: The Restoration of Earth and Beyond

HUMAN beings are members of societies which must be reshaped by the immanent revolution of ecological humanism. But all human beings are also members of one planetary human society. The crisis in man's relationship with nature and the machine is global; it involves the relationship of the human race as a whole with the planet Earth, and its ultimate resolution must be global. The common good of world society is simply the common good of individual national societies writ large. It consists of subsistence for the species as a whole, controlling population growth and technological change on a global basis, and attaining a greater degree of economic equality for women and men of all nations. Like the common good at the level of individual national societies, it can only be achieved through political action, which means through the creation of political intelligence, will, and power on a global scale. But whereas the primary task of the revolution of ecological humanism within nations is the reordering of the objectives of existing political and social systems, at the global level we confront a dual task: The global community must be reoriented at the same time a global political system is being created.

Mankind's first ventures into space and the landing on the moon were not only a triumph of physical and social technology but a supreme irony. For many human beings these space adventures marked the beginning of humanity's abandonment of its native planet and first home, Earth.[1] "We're winding up on earth and we're heading back to the stars. . . . Our children's children will be immortal. That's what space travel is all about," rhapsodized Ray Bradbury, the well-known science-fiction writer, giving voice to the hubris implicit in much of the space program.[2] But this is not likely to happen in the foreseeable future. The space programs of the United States and the Soviet Union are more of an end than a beginning, given the inhospitability of other planets to human life and the impossibility of journeying beyond the solar system by means of any known technology while still maintaining contact with earth.[3] Recent space programs may well be the last episode in man's exploration and conquest of Earth and its accessible environs, and the results will not be an explosion of new opportunities such as accompanied the discovery and colonization of the western hemisphere by Europeans, but a new sense of closure, of limits and finitude.

Indeed man's journeyings to the moon have provided the first dramatic symbolism for an ecological approach to the future of the human race. Photos of Earth taken from space show humanity as living on a small globe, the only part of the solar system capable of sustaining the human race, its mantle of white clouds, green hills, and blue oceans a unique cosmic legacy.

A major use of earth satellites is environmental monitoring, both mapping Earth's soils and mineral deposits and recording the spread of pollution. Even the military and communications uses of satellites point inward toward Earth rather than out toward the stars. Not only have many of the astronauts become ardent conservationists but some have, in their personal lives, already gone beyond the conquest of nature through technology to the expansion of human powers through the revolution of consciousness, with one even conducting experiments in parapsychology from outer space.[4] What has in recent years been called the conquest of space may be the final necessary step toward the human species's recognition that the "stargate" is not to be found by means of machines but in and through our own bodies and minds.[5]

The Oneness of the Earth

This earth, it is now obvious, is one system consisting of the physical planet, the living biosphere, and the humans who inhabit it. What is true of particular subsystems is also true of this one comprehensive system. It is kept in being through a process of interaction among its component parts, and must be kept in balance through the principles of ecological humanism. The consciousness of Earth as a planetary society gives the final lie to liberalism.

On a planetary level liberalism has operated on the basis of two postulates: that man could treat Earth as an object of his desires and break it to his will, and that human political societies could live on the earth's surface in a Hobbesian state of nature (or war) with each other.[6] Both of these postulates are false. Humankind must learn to live in balance with the other elements of the planetary ecosystem, and human societies must learn to live in harmony with one another, as elements of a planetary society. The sovereignty of man over nature and the sovereignity of political states vis-à-vis each other must both yield to the realities of human existence. Just as individuals have no rights which contravene the common good, so individual nation states have no rights against the common good of the world community. The basic needs of human beings for identity and purposeful activity, for security, self-esteem, and variety, must be fulfilled at the planetary as well as at the national level.

The new planetary society will be an emergent property of the old—the result, like the new ecologically oriented local societies, of an immanent revolution in which a myriad of changes in behavior and consciousness converge to create a new system. A world society cannot be created by a process of traditional-style political revolutions throughout the world, on the model of the Communist movement or some of the other current recipes for international reconstruction. Neither can it be created by a single formal act of federation of existing sovereignties, as has been true of the American and other federal systems.[7] The revolution of ecological humanism on a planetary scale will parallel the revolutions on the national level; it will be a creeping, convergent process leading to the emergence of a new entity. As more and more individuals and nations are enmeshed in international patterns of trade, law, communications, and mutual undertakings of many kinds, effective power over these areas of life, and their interrelationships, will pass to international bodies, and a new political order of control over human relationships will come into being. This clearly implies that national sovereignty, which is the basis of the current nation-state system and the cause and expression of international disorder, can be whittled away.

Many distinguished political theorists would deny that this is possible.[8] But to do so is to reify national sovereignty and give it a metaphysical quality which is false to reality. Sovereignty is simply the legal symbol of actual or claimed independence. A nation or an individual is sovereign only to the extent to which it is free to do as it wishes. But freedom, defined as absolute, is an illusion insofar as human individuals are concerned. We are all affected by the actions of others and affect them in turn; we are free only to choose the character of our interactions.

As a result of the existence of a worldwide technological society, freedom defined as absolute autonomy is today becoming as illusory on the international level as it is within states. The only truly independent nation would be one whose physical and cultural processes could not be affected by other states or their members. On the basis of such a criterion, it is obvious that most states are not free. Such a freedom could conceivably be purchased by individual nations at the cost of giving up all economic and cultural contacts with those outside their borders. But the course of western imperialism has long made this impossible for most lesser nations, while in the nineteenth century even as large a state as Japan was forced into relations with the outside by Perry's warships, and in the twentieth century remote Tibet has come completely under the domination of China.

All states today of necessity participate in economic and cultural relations with the "outside world"; the great powers are no more independent than the small. The United States, Western Europe, and Japan are dependent, in varying degrees, on Middle Eastern oil, and the time has passed when they can extract it at gunpoint at prices they dictate. At the same time, underdeveloped nations cannot hold back their raw materials from the developed because of the demand of their growing populations for the manufactured goods the latter produce. Erstwhile enemies, such as the United States and the Soviet Union, are entering upon a relationship which could leave America dependent on Siberian natural gas while the Soviet Union would be dependent on the United States for agricultural products. The crisis of the dollar and the recent rises in food and energy costs in North America illustrate the intimate symbiotic relationship of contemporary economies. All nations today are parasites on one another, and cannot live independently of their hosts.

As economic relationships are becoming both more intensive and more extensive, so are cultural relations. Despite the attempt of Communist states to curb foreign ideas while welcoming foreign trade in material goods, ideas cannot be kept out. Communications satellites are a technological reality just as international styles in music, dress, and the arts are a cultural reality. The global village—or global metropolis[9]—is coming into existence, slowly but surely.

But for the revolution of ecological humanism to take place on a global level, the present inchoate movements toward the creation of a planetary system must be consciously guided so as to result in a planetary political order capable of dealing with the current crises of human existence, a planetary order created and buttressed by a planetary culture.

World Order

The necessary prerequisite for any world political system able to secure the common good of humanity is the elimination of the private use of force among national societies—the ending of the state of nature among nations. Nuclear war would be the greatest ecological catastrophe imaginable, and peaceful settlement of disputes among major powers is a necessity for the survival of the human race.[10] Peaceful settlement of disputes among smaller powers is necessary to prevent the intervention of great powers and the possible subsequent clash of the latter. For the foreseeable future, relations among the great powers will be a matter of direct negotiation and traditional diplomacy, such as have led to the precarious and inherently dangerous balance of nuclear terror between the United States and the Soviet Union, or the regional balances of power in Europe and the Far East. The United Nations and regional international organizations can, however, act as agencies for dealing with disputes among lesser nations and for keeping such conflicts from engulfing the major powers. Eventually, of course, the United Nations or some successor organization will have to evolve into a true world government, but this will occur only after both the absolute and relative military might of the great powers has declined and the world has moved further away from the bipolar balance of power between the United States and the U.S.S.R. which characterized the Cold War era. The staggering costs of large military establishments and modern weapons-systems will generate economic pressures toward disarmament, just as economic ties among nations will lessen the danger of war, but disarmament will not come easily, and working to reduce the level of armaments is a major priority for all forces concerned with making world order or survival possible.[11]

A nascent world political system will have to do more than prevent wars, however. It will have to deal directly with the problems of the relationship of man and his physical environment in the total sense, and in such a way that national interests or decisions become subordinated to and overruled by the needs of humanity as a whole.[12] This will be a necessity not only for the survival of humanity in the long run, but for the preservation of peace in the short run. As the world becomes smaller and the margin of subsistence becomes narrower, economic problems may increasingly threaten world peace. Famines can lead to instability and to local wars which involve greater powers, industrialized nations may fight over the last scraps of dwindling sources of energy. The world political system will have to perform the same tasks on the international plane that reoriented political systems will be called upon to perform at the national level: curb growth (in the sense of physical product), curb population growth, control technological innovation, and equalize economic conditions throughout the globe. Already steps are being taken in some of these directions by various international bodies—specialized agencies of the United Nations, the U. N. Economic and Social Council, the World Bank, and regional organizations such as E.E.C.—and through multilateral treaties. Attempts are being made to control pollution and curb population growth on an international basis and to set up a regime for the governance of the oceans, mankind's last terrestrial frontier.[13] These efforts face a double handicap, the vested interests of richer nations in maintaining the economic dominance they have long enjoyed and the fears of poorer nations that any attempts to stem growth are a plot on the part of the better-off to relegate them permanently to a status of poverty and inferiority. The need to find new definitions of human welfare and development is one which both the rich and poor nations share.[14]

Ultimately all the problems of planetary balance of man and nature are interrelated, both physically and psychologically. When, a generation ago, a group of international idealists working at the University of Chicago under the inspiration of Robert M. Hutchins drafted a world constitution, its preamble read, "the four elements of life—earth, water, air, energy—are the common property of the human race," and stated that their ownership should be assigned in accordance with the common good.[15] A world political system, however structured, would have to be capable of dealing with these elements of life on an interconnected basis simply because of their physical interrelatedness as aspects of the common human environment. It would also have to deal with economic and demographic growth, with technology, and with mitigating economic differences among and within societies in an integrated way. Equalizing economic levels is necessary in order to induce—perhaps even to make possible—the curbing of population growth in less developed nations and to enable them to forego paths to economic growth which would add to the burden already placed on the biosphere by the economic activities of the developed nations.[16] Controlling economic growth, and redistributing assets on a world scale, obviously involves technological assessment, control, and development on a world scale.[17]

A world political order—whether a formal world state or, as is much more likely, a de facto world state emerging out of the intensification and intermeshing of the existing international processes of social control—would have to have the ability to plan for the future, in however decentralized, complex, and overlapping a fashion. Curbing growth, controlling technology, and redistributing economic benefits and burdens will necessarily involve international coordination and adjustment of tariffs, fiscal policies, and national economic planning, as well as a system of de facto international taxation. Such a planned world society would have the same principles and objectives and face the same obstacles of class and cultural and ideological differences that a planned society aiming at balance and justice would at the domestic level, but these difficulties would be magnified by being expressed through political subsystems possessing considerable legal autonomy and moved by nationalistic passions. A world political system—whether a formal world state or a de facto union—would require some degree of administrative decentralization and therefore would still have to operate in part through local subdivisions, which could become the instruments of special interests or aberrant visions of the public interest.

For a long time to come, many of its subsystems would be undemocratic, and it would face the problem of eliciting the will of humanity through governments ruled by narrow oligarchies in developed and nondeveloped nations alike.[18] To overcome these obstacles poses a challenge to the inherent possibility of political leadership and cultural regeneration unparalleled in human history, but one which ecological humanists cannot refuse.

Planetary Culture

If a world political order capable of dealing with crises of global society is to come into being, its creation will require, as well as contribute to, the creation of a planetary culture. The world political revolution of ecological humanism will have its counterpart in a global cultural revolution.

Throughout human history, political systems which were effectively world states were united by a common culture. The Roman Empire was dominated by Hellenic civilization, medieval Christendom by the Catholic church. Islam at its height was society and culture in one, literally as well as etymologically "the way," and the Chinese Empire was based on common language and religion as well as race in its triumphant periods. In the nineteenth century, when western European nations and their descendants overseas dominated the world, they spread a culture based on liberalism—rationalistic in philosophy, capitalist in economics—across the planet.

Today technological society is still spreading throughout the underdeveloped world, even though liberalism is increasingly under attack in industrial society. The "republic of science" is as international in scope as are the world religions.[19] What is called for now is not a new, artificially created world culture—cultures do not come into being in such a fashion—but a growing universal syncretism. Such a universal world culture is already in the process of formation and will become increasingly widespread as technology and trade promote the growth of communication throughout the world. The new world culture will be cosmopolitan and eclectic, drawing alike on universalistic science and technology, on the traditional spiritual and psychological insights of both East and West, and on the claims of the body and the natural world which have been accepted by the most primitive societies and are now being reasserted in the most advanced.[20]

Belief systems provide answers to questions. Human beings—having a common nature, living in interdependence on the same small planet, possessed of a modicum of intelligence and good will—can find common answers to their common problems, and it is these answers which will become the basis of the belief-system necessary to the creation and success of a world government.[21] The common problems of planetary society will provide the basis for the new ecological humanism, based not only on philosophy and religion in the traditional sense but on science and art as well, which will replace liberalism, transcending capitalism and Communism (both of which are based on liberal premises), ideologies which are both equally irrelevant to the world of today and tomorrow. Dying seas, poisoned soils, malnutrition and ignorance, the tyranny of mechanization, plagues both old and new, the alienation of man from nature, his own body, and his fellow men—all are problems which transcend traditional national and ideological boundaries. Men may cling to much of the language and symbolism of old creeds—secular and religious—but, unless a new faith based on man's burden and glory as the carrier of morality and self-consciousness in the world overcomes the old ideologies and creates a new planetary synthesis, world government is doomed. Metanoia must be global.

This new world culture, despite its varied traditional components, will end the dichotomony between science and humanism, mind and body, matter and spirit, East and West, tradition and modernity. In so doing it will help heal the split between man and nature, humanity and earth. In the words of one utopian theorist, "the real world is itself transcendent, when it is seen integrally, as a holistic and dynamic system."[22] Piety toward planet Earth will thus simultaneously promote consciousness of the community between man and nature and consciousness of the transnational community among men. Ecological humanism, in postulating the harmony of individual human needs in the common good and the congruence of human and natural values, provides the necessary basis for peace and order not only within national societies but on a planetary basis. Conversely, no scheme for political reconciliation among men on an international basis can succeed unless humanity as a whole is reconciled with its mother planet.

Prominent among the carriers of this new revolution in consciousness on a world scale and among the leaders promoting the economic and political reforms necessary to implement it, will be the scientists and technicians of all nations. Much of the informed and effective impetus for disarmament has come from within the international scientific community, and it is the scientists and managers of nations as disparate as the United States, the Soviet Union, Japan and Western Europe who have persistently called attention to the ecological problems facing all the world's peoples.[23] This is to be expected, because while culture is based on the differing worldviews of individual societies, nature is everywhere the same and comprises one worldwide system of relationships. Those who perceive it as scientists and technologists necessarily perceive the reality and universality of its laws.

The world of the future can be managed and threats to the physical survival of the human species dealt with on a purely technical basis if political leadership allows this to happen. But such a situation—while preferable to irreversible global catastrophe—would be both unstable and unworthy of humanity. Technological man, if he is to control his own destiny, must go beyond technology to create a common world philosophy and culture based on the governing principles of naturalism, holism, and immanentism, a culture in which all can participate. The alternative is world rule by a narrow elite of experts, a gnostic priesthood of specialists undemocratically controlling an ignorant "technological peasant."[24]

A new world culture does not need to mean cultural uniformity among the world's peoples. Nationalism in the sense of group identity and loyalty is deeply rooted in human consciousness, but it does not have to be incarnated in independent "sovereign" political entities; it can exist within functioning political systems as well as among them. Any political system must have a common culture in some vital respects or it will not long survive, yet even within nation states a plurality of subcultures can flourish. So also within a world political system based on certain necessary shared understandings. Indeed, one possible effect of a world culture underpinning a de facto world political system would be that the present alliance between the nation-state and culture which prevails in much of the world would be broken, and people could more easily afford to be Welshmen, Singapore Chinese, French-Canadians, black Americans, or Ukrainians than at present. Religions and quasi religions will be able to exist in ecumenical relationship or in new syntheses, as in recent years individual spiritual explorers like the late Thomas Merton have sought to combine Zen or Hinduism with Christianity.

Differing lifestyles can coexist internationally as well as within nations as long as ecological balance is maintained. Groups of the world's population should, for a long time at least, be able to choose traditional ways rather than better health for their members, or even perhaps in some cases elect to have larger families rather than higher living standards. What alone is necessary to create a humane world society is that all subcultures be, in terms of some interpretation, compatible with the world society's view of itself and of nature, and of the unity of humanity and the earth. To the extent that any cultural tendency denied this it would be aberrant; to the extent that such a tendency was expressed in action it would be subversive; and to the extent that it might become powerful it would have to be fought. The purest expression of denial of the premises of ecological humanism would be belief in expansion and proliferation at the expense of the harmony of the total system, the phenomenon known in political philosophy as liberalism and in medicine as cancer.

In the political system of the world community, individuals would be citizens, but the word *citizen*, despite its hallowed origins, has a connotation of membership in one among contending political entities, and would not be definitive of human identity. Women and men would simply think of themselves as human—children of the universe, animals and kin to all beasts and birds, heirs of Earth and creators and masters of all machines, as simple as the first protein molecule and therefore as marvelous and complex as the nebulae.[25]

Beyond Equilibrium

But suppose that a steady-state world is created—population stabilized, ecological balance restored, technology brought under conscious social direction, war and poverty abolished. Who would want to live in such a world? Is not ecological humanism a philosophy which means an end to human history just as much as would the triumph of the machine—a snuffing out of the divine spark of creativity and adventure in man? Does not it advocate purchasing balance and security at the cost of future development and progress? Does not a stable world mean, as the liberal economist Henry Wallach argues, one in which people become "routine minded, with no independent thought and very little freedom, each generation doing exactly what the last did?"[26]

A generation ago an aristocratic-minded European socialist leader abandoned his party to become a fascist, branding socialism as a "shopkeeper's paradise." Would not a steady-state world be nothing more than a middle-class schoolteacher's or accountant's paradise, much like the avantgarde's caricature of suburbia, a world without conflict or change, without challenge or hope, doomed to collapse into boredom and then into anomic violence and despair? Have not some of the vagaries of the youth culture of the late sixties—especially the misuse of drugs—shown us the perils of a world of peace and economic security? Would not our necessary utopia be just as dull as the traditional popular concept of heaven: a place where wantless, conflictless, riskless creatures wander about staring at God? Does not the popularity of space flight among the masses prove that humanity needs a collective adventure that everyone can share at least vicariously? Even if change presents problems of imbalance, would it not be better, the proponents of technological messianism and unrestricted growth ask, to do whatever can be done, to take our chances on what the future may hold rather than go through our collective planetary life like clerks waiting for their pensions?

There is much to be said for such arguments. Once we cease to adventure and create as individuals, we are already beginning to die. Nations must continue to strive for greatness or accept decay, and America especially, as a nation built on hopes for the future, would find it difficult to accept a social order based on conservation and stability. Is this not true of the species as a whole? The human race came into being because our ancestors were not afraid to leave the sea for the land, and later to come down from the trees to accept the challenges of life on the ground. Some insist that the attempt to "conquer space" is necessary for the psychic health of humanity. Others see humanity finding collective adventure in following the dolphin and returning to the sea. Creating "underwater man" would be perhaps both more humane and practical than conquering space, but both groups agree on the need for continued conquest of the physical universe in the name of human development.

Evolution cannot come to a deliberate end. The human race cannot opt for stagnation. Given the possibilities of science and technology and the uncertainty that attends all human affairs, is it not better to take the risk of future ecological disaster, doing what we can to alleviate day-to-day difficulties and solving particular limited problems as they arise, than to try to create a changeless Platonic model of the good life and bring the human adventure to an end? One scientist has even speculated that progress or invention in both science and the arts will soon necessarily come to an end, and the world will become on a vast scale what Polynesia was before European incursion and conquest. He calls it a Golden Age.[27] Others might call it stagnation or degeneration.

Of course the only possible human answer to the future is that of Molly Bloom at the close of Joyce's *Ulysses*: "Yes, yes, yes." We are achieving animals. But we do have an option as to the kind of future we wish to strive to achieve. For some the preferred future is to go on doing what we have been doing in the past: increasing material consumption, extending our domination over physical nature, increasing our population, doing what we have been doing for centuries, only doing it bigger and better, in a cosmic orgy of boosterism. Closely allied with this mentality is that which, through a commitment to technology as an autonomous force not subject to human control, is willing to consider the present state of development of our minds and bodies as final and regards the future evolution of man as destined to take place through genetic manipulation—breeding monsters for utility or whim—man-beast hybrids or cloned copies of the socially acclaimed. Others with the same basic attitude would channel man's evolution toward symbiotic relationship with the machine—creating cyborgs composed of human brains in mechanical bodies for the long voyages they project for future space-travelers, or as a means toward immortality on this planet—thus constructing a future world, whatever their particular fancies, in which man becomes a parasite to the machine, not only dependent upon but overshadowed by his host. Both these groups have their logical genesis in the antinatural, antihumanist vandal-ideology, with its dichotomy between man and nature.

But suppose the vandal ideology of liberalism is based on a false picture of the universe? Suppose the nexus between mind and matter, body and spirit, is much closer than liberalism lets us admit? Suppose that, as I have suggested here, mind and matter are really the same thing. If so, the creation of an ecologically balanced planetary society does not mean the end of human progress any more than a secure income means the end of adventure or spiritual growth for an individual. Throughout history some men have sought wealth for its own sake, or for the joy of the quest. But others have sought it for the freedom—including the freedom from wealth-seeking itself—that it can bring. Progress is identified in liberal society with the material conquest of nature. But this is not the only or the best way of defining progress. In a humane world-order, suggests a contemporary philosopher, "Progress in human affairs will shift from the realm of externals— the realm of social, economic and political institutions—to the interior life of man, the life of the mind and spirit."[28] If we could create a steady-state world, a secure, just, and humane earth, it could be the basis for wilder adventures in human development than most of us—mired in dualism and enamored of material growth—have yet dreamed of. Once we have broken the chains of poverty for the mass of mankind, we can enter into the kingdom of freedom. A steady-state world is the radical, not the stand-pat, option. It clears the human agenda for the voyage through the "star-gate," for the real breaking of the bonds of space and time.

Decades ago a popular book by a distinguished medical scientist was titled *Man the Unknown*.[29] Man is still unknown to himself. We do know, however, that mind-body dualism is an artifact of how we learn about ourselves at a certain stage of our life as individuals and as a species, not a true reflection of reality. The human body and mind are one. The distinction between voluntary and involuntary muscles is increasingly coming to be seen as a cultural creation, not a condition of nature. Not only have most of us forgotten how to wiggle our ears; we have forgotten how to control our heartbeat. In the future we should be able—on a mass scale rather than in esoteric cults—to learn how to gain control over our bodies through the use of our minds, eliminating virtually all illness in the process, possibly even learning how to generate new limbs and organs as lower animals do. We can foresee, as in the vision of the late geneticist and Marxist J. B. S. Haldane, an era in which sciences will arise that are to yoga what contemporary chemistry is to medieval alchemy, and that will open up to us almost inconceivable vistas of individual and collective self-control and development.[30]

Just what human intelligence can do we do not know, but the evidence is strong that we use only the smallest fraction of our mental capacities. Without genetic tampering we can perhaps produce a species of geniuses, a match even for the computers yet to come, ensuring our continued mastery over them.

But more than increased intelligence or control over our bodies is at stake in our lack of knowledge of the human mind and body. We are at present able to do little more than speculate about what potentialities for personal growth and interpersonal communication are inherent in human sexuality once it has been divorced from its cultural and physical relationship to procreation. We know little of the nature or potentialities of telepathy, clairvoyance, psychokinesis, and similar psychic phenomena. Ironically, only in the "materialist" Soviet Union—where by coming down on one side of the question they have eliminated the mind-body, matter-spirit dualism—has much serious research been done in these areas. One important contribution of the counterculture to American life has been the opening up of these issues, bringing them closer to the mainstream of western thought.

Who knows what adventures a humane, physically secure future may make possible? Perhaps our real exploration of the universe—the infinitesimal as well as the infinite—may be made not through technology but through mind-body transfiguration and transmigration. Possibilities in this area may be as limited as current-day established science and common sense hold, but we simply do not know. All we know is that the more we seek to identify the basic building blocks of the universe, the more they become faint and far away, like the smile on a ghostly cosmic Cheshire cat. We obviously need a more pluralistic, eclectic approach to the search for reality than that provided by the established physical sciences.

If we remain enmeshed in our present social trap, where growth breeds poverty and problems which we assume can only be solved by more of the science and technology which created the problems in the first place, we shall never know what possibilities are open to humanity. The desperately poor cannot worry about art or science or mysticism or abstract ideas, except as they contribute to the immediate goals of psychic or physical survival. Nor can those bemused by dreams of material accumulation. For the human race to break through the stargate will require the time and energy we will never have if we are forever trapped in attempts to stay alive or to run faster in order to stand still. We will never have an opportunity to take the next step in human development if we destroy the human being's genetic identity or turn him into a machine. Such evolutionary discontinuity would mean that humanity no longer existed. If we are going through the stargate, we must do it as integral human beings.

But even if all speculation about breaking through the barriers of time and space and matter and spirit by building on our present genetic endowment as human beings is idle fantasy, there is no question that vast new possibilities of human intercommunication based on currently available social and physical technology already exist and could be developed if only we were not kept so busy setting the table and taking out the trash on a planetary scale. Throughout history philosophers have held that the fullest human development was possible only within a human community, that happiness was to be sought through the polis. We have, if we will restrict our numbers and our infatuation with things, the power to create such a community on a planetary scale for the first—and perhaps the last—time in human history. Is not movement toward the creation of such a community adventure and progress? Is it not challenge enough to create a world in which we can ask ourselves the only important questions, those aimed at telling us who and what we really are? Humanity can be freed from the chains of liberalism and the spiritual dangers of an impoverished anthill world and liberated to seek its true heritage if we want it to be. The choice is as immediate as it is real. We must become more fully human and make "the step to man" now,[31] or we will be writing finis to the adventure which began when the first protohuman's eyes lit up in a flash of self-recognition no animal before him had ever known.

Epilogue

BEHIND the desk at which the final draft of this book was written is a window which overlooks People's Park. People's Park in Berkeley has become world-famous; it is for many both a political and a religious symbol, a place where blood was shed and life lost in the cause of radical ecology and the struggle against the American political and economic establishment.[1] Blind nature conspired against her own defenders in the winter of 1972–1973, which was unusually rainy and cold, driving off most of the remaining squatters. People's Park became a muddy island, on which student cars, insatiable in their hunger for parking space, increasingly encroached. Only the financial problems of its owner, the Regents of the University of California, have delayed the construction of the student housing for which the land is destined.

The fate of People's Park is a parable of the nation and the world in which these words are written. The revolt of the late sixties against the liberal establishment, epitomized by the existence of the New Left and the counterculture, is now mostly memory. The first flush of interest in ecology which culminated in Earth Day 1970 has waned, and environmentalists are on the defensive before a determined counterattack by big business, the major labor unions, the federal government, and their intellectual apologists. They are under indictment as "elitists" who are the enemies of the poor, of racial minorities, of the underdeveloped nations, of the democratic process, of national security, of the American standard of living, and of America's position in the world.[2]

Leading statesmen respond to warnings about the dangers of growth by conjuring up a vision of the destruction of our liberties and freedom.[3] Environmental standards already enacted into law are whittled down; the long-range problem of energy scarcity has been converted into a largely fake short-run gasoline shortage, making it possible for oil companies to increase their profits, for the Alaska pipeline to be rammed through Congress, and for environmentalists to be discredited.[4]

Rightwing forces are resurgent in the Democratic party. Students return to their books. Even the mass media and higher education—despite the vast social investment they represent—are losing prestige and power, in large part because of their roles as conduits for social protest and as vehicles of cultural change resented by partisans of the past. The first wave of the second American Revolution is clearly over, and reaction is everywhere triumphant, despite the handwriting on the walls of liberal society in the form of food and energy shortages and increasingly rampant inflation.

The last days of People's Park coincided with the overwhelming re-election of Richard Nixon, the standardbearer of liberalism, to the American Presidency. And with Watergate. Yet, though the depths of corruption revealed by the Watergate and associated scandals corroborate some of the wildest charges about "America" of the bedraggled protesters of the sixties, thus turning paranoia into prophecy, most Americans assume that the President is guilty of crime, consider it normal, and are apathetic. The American electorate hardly seems like the kind of soil from which a future-redeeming politics might spring. At a planetary level, despite the Stockholm conference on the future of the global environment,[5] pollution continues to pour into the earth's atmosphere and oceans; and the world's population grows.

What hope can one legitimately have for the future of humanity in such circumstances? The Martians may not be coming, but the alien invaders—the mutants of our own flesh and spirit—are everywhere triumphant. Humanity is increasingly threatened by the invasion of the post-human future, and to struggle against the world of our nightmares seems more and more a sign of madness, a work of futility, and an invitation to personal disaster. To believe in a green tomorrow is an act of faith which forces one out of the world of philosophy and politics and into the world of epic fables which have so significantly enchanted a generation; it requires one to believe with Tolkien that the dark hordes of Mordor will somehow not prevail against the Shire, with C. S. Lewis that the Lion will return and redeem his kingdom. A matricidal humanity, under the spell of liberal ideology and technological messianism, has its Mother Earth tied to the stake and makes ready to strike the last blows that will signal its own death, the end of human history.

Yet the universe still abides, and its life processes still go on. Somewhere in deepest space stars transmute energy in patterns beyond our understanding. The earth's crust remains restless and its movements mock human pretensions to dominance over nature. Somewhere hawks still wheel in the sky, lovers' pulses quicken at the sight of the beloved, men and women still feel awe at the sacred, children still marvel at the sea and the sky. The straggling army of the human cause lifts its ragged banners yet again, regroups its broken legions, and prepares for its final battles to preserve its patrimony and keep the stargate open, serene in the knowledge that whatever the future holds, to be human means to keep faith with the cosmic processes which made man. The partisans of humanity know in their bones that in a world where doom portends, resistance and life are identical, and the odds against the survival of human existence can hardly be greater than those against its creation. They sing to themselves as they go about their tasks—merging their silent song with that of every buried seed struggling toward the sun and of the earth as it spins around its star.

NOTES *

1.   The Invasion of the Human World

1. Quoted in *Not Man Apart* 2 (August 1972), p. 1.
2. My position regarding the world's ecological problems is less pessimistic than that of the Club of Rome (see Meadows et al., *The Limits to Growth*) or of the Ehrlichs (*Population—Resources—Environment*) but more pessimistic than that of Lester Brown, *World Without Borders* or of Barbara Ward and Rene Dubos, *Only One Earth*. For a viewpoint diametrically opposed to mine see Grayson and Shepard, *The Disaster Lobby*, or Maddox, *The Doomsday Syndrome*.
3. Ellul, *The Technological Society*.
4. On contemporary threats to human biological integrity see especially Leach, *The Biocrats*, and Taylor, *The Biological Time Bomb*. On threats to human psychological integrity see Delgado, *Physical Control of the Mind*, and Young, *A New World in the Making*.

2.   The End of the Modern Era

1. For a description of Locke's world see Clark, *The Seventeenth Century*; Cowie, *Seventeenth-Century Europe*; Friedrich, *The Age of the Baroque, 1610–1660*; and Hill, *The Century of Revolution, 1603–1714*.
2. Population estimates are from Bates, *The Prevalence of People*, and Cippola, *The Economic History of World Population*.
3. See Toffler, *Future Shock*.
4. The phrase is that of the philosopher Alfred North Whitehead, *Science in the Modern World*, p. 98.
5. Clark, *Seventeenth Century*, p. xvii.
6. *Second Treatise on Civil Government*, sec. 82.
7. For a survey of these beliefs see Baillie, *The Belief in Progress*.
8. On the study of the future see McHale, *The Future of the Future*, and Toffler, *The Futurists*.
9. For a radically contrary view see Velikovsky, *Earth in Upheaval* and *Worlds in Collision*.

* Books and articles are cited by titles and by authors' last names. After the first citation, titles are, wherever possible, given in abbreviated form. Full citations will be found in the two bibliographical sections, "Books" and "Periodicals." News stories are not included in the bibliography and are therefore cited in full in the notes. Individual articles in anthologies are not listed separately in the bibliography, so authors' full names are given in the notes; anthologies are listed in the bibliography under the name of the senior editor.

10. On the problem of distinguishing significant trends see Boulding, *Beyond Economics*, pp. 158–163.

11. Kahn and Weiner, *The Year* 2000.

12. The phrase is that of Denis Gabor, *Inventing the Future*.

13. See McLuhan and Fiore, *War and Peace in the Global Village*.

14. On the consequences of even moderate growth see McConnell, *Britain in the Year* 2000.

15. See "Population: The U.S. . . . A Problem; The World . . . A Crisis," supplement to the *New York Times*, 30 April, 1972.

16. This distinction is an important theme in Kuhns, *The Post-Industrial Prophets*.

17. White, *Machina Ex Deo*, pp. 11–31.

18. Fuller, "Planetary Planning," p. 56.

3. Liberal Ideology and the Assault on Nature

1. On the development of liberalism see Orton, *The Liberal Tradition*; de Ruggiero, *The History of European Liberalism*; Watkins, *The Political Tradition of the West*; and Wolin, *Politics and Vision*, pp. 286–351. On American liberalism see Girvetz, *From Wealth to Welfare*, and Hartz, *The Liberal Tradition in America*. On the problem of defining liberalism in the American context see especially Dolbeare, *Directions in American Political Thought*, pp. 15–24.

2. See *Libertarianism* by philosopher John Haspers, who received one electoral vote for President of the United States from a disgruntled former Nixon supporter in 1972. For a modern exposition of the classic liberal position see Hayek, *The Constitution of Liberty*; also Miller, *Individualism*. For a defense of the liberal tenet of competition see Knight, *The Ethics of Competition*.

3. Wills, *Nixon Agonistes*, p. 528. Nixon was described as being more nearly a "classic liberal" than his opponents in 1968 by a Yale Law professor, Robert Bork, whom Nixon later appointed Solicitor General; see John Pierson, "New Solicitor General Called a Conservative in Radical's Clothing," *Wall Street Journal*, 8 March, 1973.

4. Minar, *Ideas and Politics*, p. 43. On Hobbes as the founder of liberalism see also Mansfield, "Disguised Liberalism," p. 621.

5. McNeilley, *The Anatomy of Leviathan*, p. 25. On Hobbes's philosophy generally see Macpherson, *The Political Theory of Possessive Individualism*, pp. 9–106; Strauss, *Natural Right and History*, pp. 166–202, and *The Political Philosophy of Hobbes*; and Wolin, *Politics*, pp. 239–286.

6. *Leviathan*, pt. 2, chap. 13.

7. On the social context of Locke's teachings see Macpherson, "The Social Bearing of Locke's Political Theory," in Kramnick, *Essays in the History of Political Thought*.

8. On Locke's philosophy see Macpherson, *Possessive Individualism*, pp. 194–262; Strauss and Cropsey, *History of Political Philosophy*, pp. 433–468; Strauss, *Natural Right*, pp. 202–251; Seliger, *The Liberal Politics of John Locke*; and Wolin, *Politics*, pp. 286–351. Seliger's study offers a useful critique of various schools of Lockean interpretation, but I am not persuaded by his attack on the positions of Macpherson and Strauss, which I follow here.

9. *The Meeting of East and West*, pp. 66–164.

10. Ibid.

11. Minar, *Ideas*, p. 47.

12. See Easton, *The Political System*, p. 146.

13. *Second Treatise*, sec. 42, as rendered by Goldwin in Strauss and Cropsey, *Political Philosophy*, p. 468.

14. *Natural Right*, p. 245.

15. Goldwin in Strauss and Cropsey, *Political Philosophy*, p. 450.

16. For a typical expression of the contemporary liberal attitude toward nature see Brandy, "Moral Presuppositions of the Free Enterprise Economy."

17. *Second Treatise*, sec. 41.

18. Ibid., sec. 37.

19. Strauss, *Natural Right*, p. 241.

20. *Second Treatise*, sec. 42.

21. Strauss, *Natural Right*, p. 258.

22. Ibid., pp. 250–251.

23. See Macpherson, *Possessive Individualism*, pp. 222–258.

24. Strauss and Cropsey, *Political Philosophy*, p. 452.

25. Macpherson, *Possessive Individualism*, p. 2.

26. See Leiss, *The Domination of Nature*.

27. The phrase is from Macpherson, *Possessive Individualism*.

28. Paradise, "The Vandal Ideology."

4. American Liberalism in Triumph and Decay

1. Goldwin in Strauss and Cropsey, *Political Philosophy*, p. 467.

2. Jones, *O Strange New World*, p. 70.

3. Didion, *Slouching Toward Bethlehem*, p. 72.

4. Hartz, *Liberal Tradition*, p. 62.

5. Hacker, *The End of the American Era*, p. 128.

6. Potter, *People of Plenty*.

7. See Naipaul, *The Loss of El Dorado*.

8. See Marcuse, "Aggressiveness in Advanced Industrial Society," in Cutler, *The Religious Situation: 1969*.

9. On the history of American attitudes and policies toward nature see Burch, *Daydreams and Nightmares*; Huth, *Nature and the American*; Marx, "American Institutions and Ecological Ideals"; and Murphy, *Governing Nature*. On recent American policies toward the environment see Marine, *America, the Raped*; Sundquist, *Politics and Policy*, pp. 322–381; and Rathelsberger, *Nixon and the Environment*.

10. Murphy, *Governing Nature*, p. 7.

11. See Beard, *An Economic Interpretation of the Constitution of the United States*; Brown, *Charles Beard and the Constitution*; and McDonald, *We the People*.

12. Hartz, *Liberal Tradition*, pp. 67–113.

13. Minar, *Ideas*, pp. 169–170.

14. "Report on Manufactures" (1791), reprinted in Dolbeare, *Directions*, pp. 151–165.

15. "Thoughts on Government in a Letter from a Gentleman to His Friends," reprinted in ibid., p. 127.

16. Ibid., p. 124.

17. Quoted in Marx, *The Machine in the Garden*, p. 144.

18. Wolin, *Politics*, p. 485.

19. Letter to John Adams, 1813; reprinted in Dolbeare, *Directions*, p. 77.

20. On the role of science in early American development see Price, *Scientific Estate*.

21. Quoted in Ekrich, *Man and Nature in America*, p. 211.

22. Quoted in ibid., p. 46.

23. "A Bill Concerning Slaves" (1779), excerpted in Dolbeare, *Directions*, pp. 50–51, and discussed in Jefferson's *Autobiography*, excerpted in ibid., p. 50.

24. Hsu, *Psychological Anthropology*, pp. 216–228.

25. "On Politics," reprinted in Dolbeare, *Directions*, p. 196.

26. Murphy, *Governing Nature*, p. 6.

27. Calhoun, *Disquisition on Government*, pp. 1–2.

28. The spread of liberal industrial society to Latin America is currently having much the same results there as in the American West. See Bodard, *Green Hell*.

29. Murphy, *Governing Nature*, pp. 210–211.

30. On the relationship between subsidized capitalistic growth, "free enterprise," and corruption see Sale, "The World Behind Watergate."

31. On the origins of the conservation movement see Hays, *Conservation and the Gospel of Efficiency*.

32. See Galbraith, *The New Industrial State*.

33. For an exposition of pluralism see Dahl, *Pluralist Democracy in the United States*.

34. See Wolff, *The Poverty of Liberalism*, pp. 38, 159.

35. Habermas, *Toward a Rational Society*, pp. 100–106.

36. On the economic institutions of neoliberal society and their development see Heilbroner, "Phase II of the Capitalist System"; Kolko, *The Triumph of Conservatism*; Lockhard, *The Perverted Priorities of American Politics*; Mintz, *America, Inc.*; Reagan, *The Managed Economy*; Weinstein, *The Corporate Ideal in the Liberal State*; and Ziegler, *The Vested Interests*.

37. On reform liberalism see Dolbeare, *American Ideologies*, pp. 50–106.

38. See Ferkiss, *Technological Man*, pp. 117–153.

39. Wills, *Nixon*, p. 528.

40. "The Confrontation of the Two Americas."

41. On the distribution of economic power see Kolko, *Wealth and Power in America*; Parker, *The Myth of the Middle Class*. On political power see Domhoff, *The Higher Circles* and *Who Rules America?*; and Dye, *The Irony of Democracy*. On the consequences of class differentials see Sennett, *The Hidden Injuries of Class*.

42. "Poll Finds Crime Exceeds Reports," *New York Times*, 14 Jan. 1973.

43. See Cook, *The Corrupted Land*.

44. One of the astronauts reprimanded for his role was recently (1973) appointed deputy director of the U.S. Space Agency's flight research center at Edwards, California.

45. On political alienation see James L. Kirkpatrick, "The Crisis in Public Confidence," Oakland (Calif.) *Tribune*, 4 December 1972; and Burnham, "Crisis of American Political Legitimacy."

46. Meggysey, *Out of Their League*.

47. An example of this mentality is Lyndon Johnson's statement, "Danger and sacrifice built this land, and today we are the Number One nation. And we are going to stay the Number One nation" (quoted by Tom Wicker,

"The Ideology of War," *New York Times*, 2 May 1972). But see "Survey Shows Nixon Is Right: Many Americans Don't Care to Be No. 1," *Wall Street Journal*, 16 November 1971.

48. See Ford, "Casualties of Our Time"; Glass, "Urban Fog Dims the Brain"; McCaull, "Building a Shorter Life." Also "Developed Nations' Death Rates May Be Ending 150-Year Decline," Washington *Post*, 28 Oct. 1970.

49. Hacker, *End of the American Era*, p. 136.

50. "Poll Finds Most Content in Life," *New York Times*, 26 Sept. 1971.

51. See Murphy, *Governing Nature*, pp. 53–56.

52. Harwood, "We Are Killing the Sea around Us."

53. The United States is now the dumping ground for radioactive wastes produced by the Atoms for Peace program throughout the world, notably in Japan, Canada, and Italy. Lee Dye, "U.S. Becoming Dumping Ground for World's Radioactive Wastes," Vancouver (B.C.) *Sun*, 2 August 1973.

54. See Nader, *Unsafe at Any Speed*; Marine and Van Allen, *Food Pollution*; Wellford, *Sowing the Wind*; and Zwerdling, "Food Pollution."

55. "Pesticide Killing the Nation's Symbol," Washington *Post*, 8 June 1970.

56. Betty Ross, "Pity the Poor Lincoln Memorial: It Has a Very Bad Case of the Blahs," Washington *Post*, 6 February 1972.

57. Even the desert is threatened. See Stephen V. Roberts, "Coast Desert a Vast, Littered Playground for Millions," *New York Times*, 11 April 1971. The Ford Motor Company has been especially active in destroying wilderness through its aggressive promotion of off-road motor vehicles. See John Fialka, "New View of Ecology," Oakland *Tribune*, 30 November 1972, and Fialka, "Vehicle Threat to Wildlife Aired," ibid., 7 December 1972.

58. The extent to which and the precise manner in which American prosperity is the result of the exploitation of other peoples is a complex question on which the extant literature is inadequate for reaching any firm conclusions. For varying views on this issue see Baran, *The Political Economy of Growth*; Hudson, *Super Imperialism*; Jalee, *The Pillage of the Third World* and *The Third World in World Economy*; Julien, *America's Empire;* Kindleberger, *The International Corporation*; and Magdoff, *The Age of Imperialism*. See also Borgstrom, "The World Food Crisis"; and "Symposium: Does the U. S. Economy Require Imperialism?"

59. Ward and Dubos, *Only One Earth*, p. 119.

60. Ibid., p. 120.

5.   Technology and the End of Liberalism

1. For a political scientist's fictionalized account of the manipulation of modern electorates see Burdick, *The 480*.

2. On the dependence of the liberal economic system on underlying social norms see Parsons, *The Structure of Social Action*, pp. 715–717.

3. According to Kenneth and Patricia Dolbeare, for liberalism "the chief goals of political activity are stability and the avoidance of disruptive conflict" (*Ideologies*, p. 67); "Liberalism regards all absolutes with profound skepticism, including both moral imperatives and solutions produced through rational analysis" (ibid., p. 69). Wolin argues that liberalism "possessed a theory which made objective social and political judgments impossible" (*Politics*, pp. 319–324). Wolff asserts that liberalism depends on people acting selfishly instead of altruistically or even rationally (*Liberalism*, pp. 35–38).

4. Wolin, pp. 319–324; Hoselitz, *Economics and the Idea of Mankind*, p. 31.

5. *Federalist*, no. 22. Hamilton would have scorned Calhoun's devices to create deadlock.

6.   Roads to Nowhere: Conservatism, Socialism, Anarchism, and the Antipolitics of Despair

1. Dolbeare, *Ideologies*, pp. 17–18. On reform liberalism, ibid., pp. 81–95; also Kaufman, *The Radical Liberal*.

2. Dolbeare, *Ideologies*, p. 81.

3. This seems to be the suggestion of Lowi, *The End of Liberalism*.

4. For a general overview of American conservatism see Rossiter, *Conservatism in America*. For an influential statement of the conservative position see Kirk, *The Conservative Mind* and *A Program for Conservatives*. *Conservatism Revisited* is a critique of contemporary American conservatism by poet-historian Peter Viereck, who regards himself as a conservative.

5. Weiss, *Nature and Man*, p. xiii.

6. Grant, *Technology and Empire*, p. 66.

7. Ibid., p. 69.

8. Strauss, *Thoughts on Machiavelli*, p. 298. See also Strauss, *What Is Political Philosophy?*, pp. 310–311.

9. Grant, *Technology and Empire*, p. 17.

10. Strauss and Cropsey, *Political Philosophy*, p. 834.

11. Pfaff, *Condemned to Freedom*, p. 133.

12. Buckley, *Did You Ever See a Dream Walking?*, p. 82.

13. Ibid., p. 83.

14. Ibid., pp. xx–xxii. Ayn Rand's views are to be found in *The Anti Industrial Revolution* and *For the New Intellectual*, as well as in her novels.

15. Kendall, *The Conservative Affirmation*.

16. Meyer, "The Recrudescent American Conservatism," in Buckley, *Dream*, p. 81.

17. "The Convenient State," in Buckley, *Dream*. For similar views see Friedrich, "The Deification of the State" and "The Greek Political Heritage and Totalitarianism."

18. Buckley, *Dream*, p. 93. Oakeshott holds that "the world is the best of all possible worlds and *everything* in it is a necessary evil" (*Rationalism in Politics*, p. 136). Belief in man's inability to change the world is a basic element in the conservative ideology and is gaining increasing currency among social scientists. See for example, Kristol, "A Foolish American Ism—Utopianism," and Etzioni, "Human Beings Are Not Very Easy to Change after All."

19. Hart, *The American Dissent*, p. 252.

20. William V. Shannon, "The Highwaymen," *New York Times*, 16 September 1972.

21. Ropke, *A Humane Economy*, pp. 39–52.

22. "Population Explosion: A Conservative Reacts."

23. Buckley, *Dream*, p. xxxvi.

24. Ibid., p. 1.

25. Grant, *Technology*, p. 32. A recent public-opinion poll taken in California showed those who considered themselves "conservatives" were much less concerned about the impact of technology on human life than those

who considered themselves "liberal." La Porte and Metlay, *They Watch and Wonder*, p. 12.

26. Goodman, *The New Reformation and Notes of a Neolithic Conservative*, p. 191.

27. Howe, *Steady Work*, p. 326.

28. Dolbeare, *Ideologies*, pp. 225–226.

29. Harrington, *Socialism*, p. 238.

30. Weisberg, *Beyond Repair*, p. 161. See also Schmidt, *The Concept of Nature in Marx*.

31. See Meek, *Marx and Engels on Malthus*.

32. *Literature and Revolution*, pp. 251–256.

33. See Sibley, "Social Order and Human Ends" in Spitz, *Political Theory and Social Change*, p. 253.

34. Engels, *Dialectics of Nature*, pp. 291–292.

35. But see Volkov, *Era of Man or Robot?* and the illegally circulated *Progress, Coexistence and Intellectual Freedom* by Andrei Sakharov. Also Mandel, "The Soviet Ecology Movement."

36. See especially Goldman, *The Spoils of Progress*, and Pryde, "Victors Are Not Judged."

37. A Soviet scheme threatening major ecological perils for Europe has been temporarily shelved, apparently for economic reasons: "Siberian Rivers to Be Reversed," *New York Times*, 26 March 1970; Claire Sterling, "A New Peril Eased for Polar Icecap," Washington *Post*, 10 September 1971.

38. Weisberg, *Beyond Repair*, pp. 149–156.

39. Weisberg does this. Ibid., p. 155.

40. Murphy, *Governing Nature*, p. 281.

41. Theodore Roszak reaches a similar conclusion in *Where the Wasteland Ends*, p. 47. For Chinese attitudes and policies see Orleans and Suttmeier, "The Mao Ethic and Environmental Quality"; also William D. Hartley, "Chinese Make Progress in Cutting Pollution by Making Multiple Use of Industrial Excess," *Wall Street Journal*, 21 November 1972.

42. See Lichtheim in Howe, *The Radical Papers*, pp. 55–72; also Harrington, *Socialism*, p. 207.

43. See Lowenthal, "What Prospects for Socialism?"

44. Williams, *The Great Evasion*, p. 12.

45. Ibid., p. 173.

46. For the American Communist party position see Hall, *Ecology: Can We Survive under Capitalism?*

47. Harrington, *Socialism*, p. 344.

48. See their contributions in Fromm, *Socialist Humanism*.

49. Harrington, *Socialism*, p. 108.

50. Ibid., especially pp. 347–351.

51. Rosenberg in Howe, *Radical Papers*, p. 84.

52. Harrington, *Socialism*, especially pp. 353–362. On the role of labor in support of American foreign policy see Berger, "Organized Labor and Imperial Policy."

53. See "Harrington Quits as Socialist Head," *New York Times*, 25 October 1972, and James King Adams, "Battle Royal among the Socialists," *Wall Street Journal*, 8 December 1972.

54. On the New Left see Goodman, *The Movement for a New America*; Jacobs and Landau, *The New Radicals*; Oglesby, *The New Left Reader*; Sale, *SDS*; and Teodori, *The New Left: A Documentary History*. For a useful short critique see Dolbeare, *Ideologies*, pp. 145–184. For a leader's retrospective

analysis see Findley, "Tom Hayden." On the intellectual background of the New Left see King, *The Party of Eros*; Megill, *The New Democratic Theory*; Raskin, *Being and Doing*; and Waskow, *Running Riot*.

55. Reprinted in Teodori, *New Left*, p. 166.

56. Ibid., p. 167.

57. Ibid., p. 163.

58. Ibid., pp. 412–418.

59. For a summary and critique of mass society theories see Ferkiss, *Technological Man*, pp. 69–76.

60. See Habermas, *Rational Society*, p. 87 circa. For Marcuse's thought see his *Essay on Liberation*, *Five Lectures*, and *One-Dimensional Man*; also Keen and Raser, "Conversation with Herbert Marcuse."

61. For a recent illustration see Oglesby, "The Crisis Is Political."

62. On the legacy of the New Left see Sale, "The New Left—What's Left?"

63. See Kelley, "Blissed Out with the Perfect Master," and Kopkind, "Mystic Politics: Refugees from the New Left."

64. On anarchism see Goodman, *Seeds of Liberation*; Joll, *The Anarchists*; Krimerman and Perry, *Patterns of Anarchy*; and Woodcock, *Anarchism*.

65. Bookchin, *Post-Scarcity Anarchism* and "Toward an Ecological Solution"; Slater, *The Earth Belongs to the People—Ecology and Power;* and Weisberg, *Beyond Repair*.

66. *Post-Scarcity Anarchism*, p. 41.

67. Weisberg, *Beyond Repair*, p. 68

68. Bookchin, *Post-Scarcity Anarchism*, p. 22.

69. Ibid., p. 59.

70. Ibid., p. 22.

71. Ibid., p. 119.

72. Ibid., passim; Bookchin's essay "Toward a Liberatory Democracy" also appears in Benello and Roussopoulos, *The Case for Participatory Democracy*.

73. This theme runs through much of Goodman's work, especially *Notes of a Neolithic Conservative*. See also Sibley in Spitz, *Political Theory and Social Change*.

74. The nineteenth-century anarchist leader Benjamin R. Tucker once said, "Yes, genuine anarchism is consistent Manchesterism." Quoted in Krimerman and Perry, *Patterns*, p. 34.

75. Bookchin, *Anarchism*, p. 51.

76. Ibid., p. 293.

77. Goodman, *New Reformation*, p. 145.

78. For instance, ibid., passim.

79. Mumford, *The Myth of the Machine*, vol. 2, p. 435.

80. Reich, *The Greening of America*. For commentary see Nobile, *The Con III Controversy*.

81. Revel, *Without Marx or Jesus*.

82. For an appraisal of the youth rebellion see Ferkiss, "American Youth and the Process of Social Change."

83. Berger, "The Blueing of America."

84. Marcuse, *One-Dimensional Man*.

85. See Thompson, *At the Edge of History*, p. 121.

86. See Roszak, *Wasteland* and *The Making of a Counter Culture*. At the intellectual level Roszak tends to admit the need for technology and ration-

ality, but his rhetoric and emphases suggest his heart isn't in it. His clearest statement about action for survival is found in "Can Jack Get Out of His Box?" For a discussion of his ideas see Wade, "Theodore Roszak." For a position even more apocalyptically antitechnological and politically quietist see Schwartz, *Overskill*.

87. Pfaff, *Freedom*, pp. 121–122.

88. For an extreme but useful appraisal of Luther see Brown, *Life against Death*.

89. Ellul's negativism is apparent not only in *The Technological Society* but in *The Political Illusion* and, in a more specifically Christian context, in *False Presence in the Kingdom*.

90. See Kegley and Breitall, *Reinhold Niebuhr*.

91. Niebuhr, *Moral Man and Immoral Society*, p. xii.

92. Lippmann, *The Public Philosophy*, p. 109.

7.   The Real and the Ideal: Philosophy and Politics

1. *The Causes of World War III*, p. 81.

2. *New York Times*, 12 June 1962.

3. Ferkiss, *Technological Man*.

4. Fuller, *Utopia or Oblivion*.

5. Ferkiss, *Technological Man*, pp. 232–242.

6. Rothman, "The Revival of Classical Political Philosophy," and Cropsey, "A Reply to Rothman."

7. One example of oversimplified analogy between the physical and social worlds is Cannon, "The Body Physiologic and the Body Politic."

8. Margenau, *Open Vistas*, p. 40.

9. Quoted in ibid., p. 44.

10. Boulding, "Philosophy, Behavioral Science and the Nature of Man"; Frank, "The Need for a New Political Theory"; Rodman, "The Ecological Perspective and Political Theory"; and Thomas L. Thorson, "The Biological Foundations of Political Science," in Graham, *The Post Behavioral Era*.

11. J. B. S. Haldane, quoted in Collingwood, *The Idea of Nature*, p. 24.

12. But see at least Edel, "The Relation of Facts and Values: A Reassessment," in Lieb, *Experience, Existence, and the Good*; Leavenworth, "On Integrating Fact and Value"; and Searle, "How to Derive 'Ought' from 'Is.'"

13. *Ethics*, bk. 1, sec. 4.

14. Skinner, *Beyond Freedom and Dignity*, p. 94.

15. See, for instance, some of the essays in Wolfe and Surkin, *An End to Political Science*.

16. Strauss, *Natural Right*, pp. 35–80.

17. Becker, *The Structure of Evil*, p. xiii.

18. Berger and Luckmann, *The Social Construction of Reality*. For an extreme position on the nonobjectivity of reality see Pearce, *The Crack in the Cosmic Egg*.

19. Margenau, *Open Vistas*, p. 101.

20. See David Bohm, "Some Remarks on the Notion of Order," in Waddington, *Towards a Theoretical Biology*.

21. Lovejoy, *The Great Chain of Being*.

22. See for example Nogar, *The Lord of the Absurd*. For more integrative views see Balthasar, *God Within Process*; Carothers, *The Pusher and the Pulled*; Hartshorne, *Beyond Humanism* and *A Natural Theology for Our*

Time. See also Francoeur, *Evolving World, Converging Man* and *Perspectives in Evolution.*

23. On natural law see Center for the Study of Democratic Institutions, *Natural Law and Modern Society*; Passerin D'Entreves, *Natural Law*; Rommen, *The Natural Law*; Sigmund, *Natural Law in Political Thought*; and Strauss, *Natural Right.*

24. On the attitudes of various cultures toward nature see Black, *The Dominion of Man*; Glacken, *Traces on the Rhodian Shore*; Murphy, *Governing Nature*; Northrop, *Philosophical Anthropology and Practical Politics*, pp. 238–257; and White, *Machina ex Deo*, pp. 75–94. On Muslim culture, Nasr, *The Encounter of Man and Nature.* On Indian culture, Hart, "The Natural Environment in Indian Tradition." On American Indian attitudes, Petersen, "Lessons from the Indian Soul." But on the actual activities of various cultures see Moncrief, "The Cultural Basis of Our Environmental Crisis."

25. See Martin and Wright, *Pleistocene Extinctions*, and Martin, "The Discovery of America." See also Walter Sullivan, "How Man Became a Mass Killer 11,000 Years Ago," *New York Times* 22 March 1970, and "Overkill of Animals Laid to Huntsmen in 9000 B.C.," *New York Times*, 11 February 1972.

26. See Yankelovitch, *The Changing Values on Campus* and "The New Naturalism."

8.   Man in Nature and the Nature of Man

1. Francoeur, *Evolving World*, p. 94.

2. This position underlies such contemporary works as Cox, *The Secular City.*

3. On the Romantic movement see Roszak, *Wasteland.* For a different perspective see de Rougement, *Love in the Western World.*

4. Quoted in Dubos, *So Human an Animal*, p. 61.

5. These lines provide the name of the periodical *Not Man Apart*, organ of the activist environmental organization, Friends of the Earth, underlining the essentially religious dimensions of the movement.

6. Laszlo, *The Systems View of the World*, pp. 104–107.

7. Polak, *The Image of the Future*, vol. 2, p. 117.

8. See Ferkiss, "Technology and the Future of Man."

9. Collingwood, *Idea*, p. 3.

10. Jonas, *The Phenomenon of Life*, p. 14, circa. On medieval cosmology see also Guardini, *The End of the Modern World.*

11. See Eliade, *The Myth of the Eternal Return.*

12. For a discussion of current thinking on the physical origin and age of the universe see Calder, *The Violent Universe*, pp. 113–118; also Metz, "The Decline of the Hubble Constant."

13. Barnett, *The Universe and Dr. Einstein*, p. 100.

14. On the prerequisites of life as we know it see Henderson, *The Fitness of the Environment.* For a skeptical view regarding extraterrestrial life see Dobzhansky, *The Biology of Ultimate Concern*, p. 49; and for skepticism about extraterrestrial life anything like our own see Hardin, *Nature and Man's Fate*, p. 292. Among those who hold that life as such is a much more common phenomenon than we imagine is Cowen, "Life on Earth: The Beginning of the Beginning."

15. The time interval between the sending and receipt of messages would

span the lifetimes of whole societies or even species. Clarke, *Profiles of the Future*, pp. 118–119.

16. Margenau, *Open Vistas*, p. 127.

17. Ibid.

18. Koestler, *The Ghost in the Machine*, p. 205. On contemporary physics see Matson, *The Broken Image*, pp. 113–140; Park, "Complementarity without Paradox"; and Taylor, *The New Physics*. Not all scientists accept the quantum theory, of course; see Blissett, *Politics in Science*, pp. 146–151.

19. Krutch, *The Great Chain of Life*, p. 210. See also Waddington, *Theoretical Biology*, p. 1; W. H. Thorpe in Koestler and Smythies, *Beyond Reductionism*, pp. 433–434, and Ludwig von Bertalanffy in the same collection, pp. 57–58.

20. Bertalanffy, *Problems of Life*, p. 134.

21. Collingwood, *Idea*, p. 147.

22. Margenau, *Open Vistas*, p. 156.

23. Quoted in Collingwood, *Idea*, p. 149.

24. Brooks, *The Government of Science*, p. 213.

25. Collingwood, *Idea*, p. 152.

26. On the primacy of form in defining reality see ibid., pp. 150–152; David Bohm, "Some Remarks on the Notion of Order," in Waddington, *Theoretical Biology*; and Whyte, *Accent on Form*.

27. Y. P. Mei, "The Basis of Social, Ethical, and Spiritual Values in Chinese Philosophy," in Moore, *Essays in East-West Philosophy*, p. 302.

28. The term was originally popularized by the anthropologist Lloyd Morgan and later taken up by philosophers such as Jan Christian Smuts and Samuel Alexander. On emergence see Nagel, *The Structure of Science*, pp. 366–380; Needham, *Time, the Refreshing River*; Polanyi, *The Tacit Dimension*, pp. 29–52; and Simon, *The Sciences of the Artificial*, p. 86, circa. Also Platt, "Properties of Large Molecules That Go beyond the Properties of Their Chemical Sub-groups."

29. On hierarchical organization of natural systems see Bohm in Waddington, *Theoretical Biology*; Koestler, *Ghost*; Simon, *The Artificial*, pp. 108–111; also Koestler, "Beyond Atomism and Holism—the Concept of the Holon"; Bertalanffy, "Chance or Law," in Koestler and Smythies, *Reductionism*; and Platt, "Hierarchical Growth."

30. Quoted in Collingwood, *Idea*, p. 160.

31. Ibid., p. 167. For Whitehead's thought generally see his *Concept of Nature* and Anshen, *Alfred North Whitehead*.

32. "The Real World."

33. Barnett, *Universe*, p. 110.

34. For a naturalist's views on "anthropomorphism" see Krutch, *Chain of Life*, pp. 226–227. For a definition of metaphysical pathos see Lovejoy, *Chain of Being*, p. 11.

35. R. W. Sperry, quoted in Koestler, *Ghost*, p. 211.

36. On Teilhard's thought in this context see Zaehner, *Matter and Spirit: Their Convergence in Eastern Religions, Marx and Teilhard de Chardin*; and Francoeur, *Perspectives*, pp. 129–131.

37. Weiss, *Nature and Man*, pp. xviii, 71–76.

38. Koestler, *Ghost*, p. 64.

39. Bertalanffy, in Koestler and Smythies, *Reductionism*, p. 70.

40. "The hand of man . . . *is* made for grasping. Darwin said so, and then provided a natural scientific explanation for the fact" (Simpson, *This View of Life*, p. 102).

41. On recent evolutionary theory see especially Hardin, *Nature;* Koestler and Smythies, *Reductionism*, pp. 357–395; Sol Tax, *Evolution after Darwin*; and Hoagland and Burhoe, "Evolution and Man's Progress."

42. Koestler, *Ghost.*

43. Ibid., pp. 68–69.

44. Ibid.

45. See Beck, *Modern Science and the Nature of Life*, pp. 184–186; N. W. Pierce, "The Meaning of the Terms 'Life' and 'Living,'" in Needham and Green, *Perspective in Biochemistry*; W. H. Thorpe, "Vitalism and Organicism," in Roslansky, *Uniqueness of Man*; and Watts, *Does It Matter?*

46. Bertalanffy, *Problems*, p. 129.

47. In Waddington, *Theoretical Biology*, p. 27.

48. *Man on His Nature*, p. 85.

49. "An Essay on Religion, Death, and Evolutionary Biology," p. 331.

50. Sherrington, *Man on His Nature*, p. 149; Dubos, *So Human*, p. 71.

51. Ibid., p. 73. On man's animal heritage see Bleibtreu, *The Parable of the Beast.*

52. Sherrington, *Man on His Nature*, p. 164.

53. W. H. Thorpe, quoted in Koestler, *Ghost*, p. 208.

54. Aristotle complicated the issue by postulating an "intellectus agens" —the locus of some of man's reasoning powers—which apparently participated in collective immortality.

55. Francoeur, *Evolving World*, p. 120.

56. Carrell, *Man the Unknown*, especially pp. 117–120.

57. Simpson, *Biology and Man*, p. 69.

58. R. W. Sperry, "Mind, Brain and Humanist Values," in Platt, *New Views of the Nature of Man*, p. 78.

59. On the evolution of the mind see Thorpe, *Biology and the Nature of Man*, pp. 20–59; Roe and Simpson, *Behavior and Evolution*; and Leslie A. White, "Four Stages in the Evolution of Minding," in Tax, *Evolution after Darwin*, vol. 2; also the discussion, ibid., pp. 175–206.

60. This position does much to vitiate Adler's otherwise useful book, *The Difference of Man and the Difference It Makes.*

61. Simpson in Roe and Simpson, *Behavior*, p. 519.

62. Dobzhansky, *Biology*, p. 68.

63. Koestler, *Ghost*, pp. 267–296; Koestler and Smythies, *Reductionism*, pp. 258–275.

64. Krutch, *Chain of Life*, p. 127.

65. See Wooldridge, *Mechanical Man* and Ferkiss et al., "Mechanical Man: Review Symposium."

66. Donald M. Mackay in Wolstenhome, *Man and His Future*, p. 165. Also see Mackay and Sperry, in Platt, *New Views*, p. 83.

67. See Roger A. MacGowan and Frederick J. Ordway, "Artificial Thinking Automata," in Kostelanetz, *Social Speculations.*

68. *Dialectics of Nature*, p. 175.

69. *Man on His Nature*, p. 316.

70. Ibid., pp. 318–319.

71. *Evolution and Ethics*, p. 198. See also Whitehead, *Nature and Life*, p. 25; and Birch, "Purpose in the Universe."

72. Platt, "Can Determinism and Freedom Be Reconciled?"

73. Dubos, *So Human*, p. 130.

74. *Science and the Common Understanding*, pp. 79–82.

75. *Open Vistas*, p. 199.

76. "Indeterminism Is Not Enough," p. 26.

77. Sherrington, *Man on His Nature*, pp. 165–166.

78. On possible relationships between the mystical and the psychedelic see Braden, *The Private Sea*.

79. Carrell, *Man the Unknown*, pp. 259–262; Hardy, *Living Stream*, pp. 234–261; Thorpe, *Biology*, p. 108. Psychic research flourishes in some Communist nations. See Ostrander and Schroeder, *Psychic Discoveries behind the Iron Curtain*, and Krippner and Davidson, "Parapsychology in the U.S.S.R." In the United States, parapsychology is slowly gaining recognition as a legitimate science (Stuart Auerback, "ESP Officially Recognized as Science," Washington *Post*, 28 December 1970). Koestler, *Ghost*, p. 219, and Hardy, *Living Stream*, suggest the possibility that something akin to telepathy was a skill once possessed by man but lost in the evolutionary process; ironically, traditional Christian theology held that Adam knew by means of "infused" rather than acquired knowledge.

9. Nature and Human Values

1. Waddington, *The Ethical Animal*.

2. "The Ethical Basis of Science," p. 1254.

3. As Paul Weiss notes, *Nature and Man*, p. 92.

4. While "instinctual" is a term of dubious acceptance in modern psychology, I use it here for the sake of convenience rather than "preprogrammed" or other possible substitutes with essentially the same meaning. On infant dependency and the origins of society, see Corning, "The Biological Bases of Human Behavior and Their Implications for Political Theory."

5. Waddington, *The Ethical Animal*, p. 7. Ethics at this level becomes similar to what Michael Polanyi refers to as "tacit knowledge" and bears the risk of becoming unduly conservative in content. See *Dimension*, especially pp. 60–61.

6. Natural-law theories can have a stultifying effect on social thinking, as in the use of the concept of "substantive due process" by late nineteenth-century American courts. See Haines, *The Revival of Natural Law Concepts*.

7. Bidney, *Theoretical Anthropology*, pp. 140–141.

8. Ibid., p. 81.

9. *Essays in Sociology and Social Philosophy, Vol. I: On the Diversity of Morals*, p. 120.

10. Ibid., p. 39.

11. Dobzhansky suggests this, though skeptically, in *The Biological Basis of Human Freedom*, p. 119.

12. "The Impact of the Concept of Culture on the Concept of Man," in Platt, *New Views*, p. 103.

13. Ibid., p. 112.

14. Ibid., p. 106.

15. The classic work on the Dobu and Hopi for our purposes is Ruth Benedict's *Patterns of Culture*, an extremely influential work, the ethnographic accuracy of which is questionable. Even Benedict, however, recognized that there were universal standards by which cultures could be judged. See Gorney, *The Human Agenda*, pp. 171–175.

16. On the benighted Soriano see Holmberg, *Nomads of the Long Bow*.

17. Bidney, *Theoretical Anthropology*, p.81.

18. Anderson, *Politics and the Environment*, p. 336.

19. Greeley, *Unsecular Man*.

20. For a variety of recent approaches to the problem of "natural theology" see Alexander, *Space, Time, and Deity*; Hartshorne, *Natural Theology*; Overman, *Evolution and the Christian Doctrine of Creation*; and, of course, the work of Teilhard de Chardin.

21. As one example see the influential work of Alves, *A Theology of Hope*.

22. On ecotheology see especially Conrad Bonifazi, "Biblical Roots of an Ecologic Conscience," in Hamilton, *This Little Planet*; William G. Pollard, "God and His Creation," ibid.; Frederick Elder, "Two Modern Doctrines of Nature" in Cutler, *The Religious Situation, 1969*; Hefner, "Toward a New Doctrine of Man"; and Santmire, *Brother Earth*.

23. Watts, *The Way of Zen*, p. 36.

24. See F. S. C. Northrop, "Methodology and Epistemology, Oriental and Occidental," in Moore, *Essays*; also Northrop, *The Meeting of East and West*. On the relationship of eastern and western thought see, too, the essays of Y. P. Mei, T. M. P. Mahadevan, and Moore in the Moore volume; also Watts, *Zen*; Siu, *The Tao of Science*; and Needham, *Science and Civilization in China*, vol. 2.

25. *The Book*, pp. 69–70, 118.

26. Quoted in de Rougement, *Love*, p. 333.

27. Daniel Callahan, privately circulated paper.

28. Quoted in Cauthen, *Science, Secularization, and God*, p. 127.

29. As suggested by the recent controversial exeriments by polygraph expert Clyde Backster. See Tomkins, "Love among the Cabbages"; Robert Martin, "Be Kind to Your Plants or You Could Cause a Violet to Shrink," *Wall Street Journal*, 2 February 1972.

30. *Man on His Nature*, p. 375.

31. Ibid., p. 364.

32. Ibid., p. 374.

33. See Krutch, *Chain of Life*, pp. 28–36, 177. On the role of sex in nature generally see Wickler, *The Sexual Code*.

34. For differing viewpoints on evolutionary ethics see Huxley, *Evolution and Ethics*; Joshua Lederberg, "Orthobiosis: The Perfection of Man," in Tiselius, *The Place of Values in a World of Facts*; Jacques Monod, "On Values in the Age of Science," in ibid; and Waddington, *Ethical Animal*.

35. Krutch, *Chain of Life*, p. 177.

36. See Francoeur, *Utopian Motherhood*.

37. Krutch, *Chain of Life*, p. 185.

38. On the derivation of ethical norms from nature see Callan, *Ethology and Society*; Pearl, *Man the Animal*; Wickler, *The Biology of the Ten Commandments*; Allee, "Where Angels Fear to Tread"; Burhoe, "Five Steps in the Evolution of Good and Evil"; Gerard, "A Biological Basis for Ethics"; Hoagland, "Ethology and Ethics"; Romanell, "Does Biology Afford a Sufficient Basis for Ethics"; and "Symposium: On Human Values and Natural Science."

39. See Francoeur, *Perspectives*, p. 110.

40. Koestler, *Ghost*, p. 164.

41. Carothers, *Pusher*, p. 151.

42. Ibid., p. 184.

43. Ibid., p. 196.

44. For example, Hardin, *Nature*, p. 308.

45. Ibid., p. 399.

46. Quoted in Means, *The Ethical Imperative*, p. 115.

47. Cowan, "Conservation and Man's Environment," p. 1147.

48. A. Starker Leopold, "Introduction," to Shepard and McKinley, *The Subversive Science*, p. 7.

49. Quoted in Glenn T. Seaborg, "Mexican President Echeverria and Science" (editorial), *Science* 177 (1972): 1063. On Man's proper relationship to nature see Dubos, *A God Within*; Leopold, *A Sand County Almanac*; McHarg, *Design with Nature*; and McPhee, *Encounters with the Archdruid*. For antinature views see Hoffer, *First Things, Last Things*; and Krieger, "What's Wrong with Plastic Trees?" For historical perspective see Northrop, *Philosophical Anthropology*, pp. 238–257; and Clarence J. Glacken, "Man's Place in Nature in Recent Western Thought," in Hamilton, *This Little Planet*. For recent developments in attitudes toward nature see Yankelovitch, "The New Naturalism"; and Heberlein, "The Land Ethic Realized."

50. Shepard and McKinely, *Subversive Science*, p. 3.

51. Merton in Disch, *The Ecological Conscience*, p. 42. In an ironic twist of history, Congressman Wayne Aspinall, a longtime key foe of conservation interests, was espousing Leopold's philosophy in speeches after his defeat for reelection, a defeat contributed to by environmentalists. See Fred Garretson, "Defeated Rep. Aspinall in Swan Song Appearance," Oakland *Tribune*, 20 September 1972.

52. *Sand County*, p. 200.

53. Ibid., pp. 224–225.

54. Ibid., p. 214.

55. "Earth House Hold," reprinted in Disch, *Conscience*, p. 202.

56. Ibid., p. 17.

10.   The Goodness of Man and the Primacy of Politics

1. On race see Coon, *The Races of Man*; also Hardin, *Nature*.

2. For a useful discussion of this issue, see Bertalanffy, *General Systems Theory*, pp. 227–248.

3. On the relationships between cultural and genetic factors in human behavior see the articles by Gaspari, Sperry, Thompson, and Simpson in Roe and Simpson, *Behavior and Evolution*; also Montagu, *The Human Revolution*.

4. See Weldon, *The Vocabulary of Politics*.

5. Ardrey, *African Genesis*.

6. Rogow and Lasswell, *Power, Corruption, and Rectitude*.

7. But see Sorokin, *The Ways and Power of Love*.

8. See Carthy and Ebling, *The Natural History of Aggression*; Montagu, *Man and Aggression*; Gorney, *The Human Agenda*, pp. 95–111, 131–149.

9. Adler, *The Common Sense of Politics*, p. 33.

10. Engels, *Dialectics of Nature*, p. 208.

11. Fromm, *Beyond the Chains of Illusion*; King, *The Party of Eros*; Robinson, *The Freudian Left*.

12. See Morris, *The Naked Ape* and *The Human Zoo*; and Tiger and Fox, *The Imperial Animal*. For critical commentary see Willhoite, "Ethology and the Tradition of Political Thought"; and the reviews by Earl W. Counts, *American Anthropologist* 72 (1970): 869, and Morton H. Fried, *Science* 165 (1969): 883. See also Knelman, *1984 and All That*, pp. 157–160.

13. See Max Gluckman's review of Tiger and Fox, *The Imperial Animal*, in *New York Review of Books* 19 (16 November 1972): 39–41.

14. *The Biological Basis of Human Freedom*, p. 110.

15. Despite the contention of novelist Saul Bellow that "Technology is not about to create a new humanity, more rational and mature. That is only a fairy tale. We still are what we were." "Literature Amid the Noise," Washington *Post*, 19 November 1972.

16. Dubos, *Man Adapting*, p. 10.

17. Loeb, *The Biological Basis of Individuality*, pp. 7–8.

18. *What is Life?*, pp. 91–92.

19. *Politics* 3:6.

20. On the biological bases of human sociability see Montagu, *On Being Human*, pp. 29–67; and Count, "The Biological Basis of Human Sociability."

21. See Brown, *Seeds of Change*.

22. As Buckminster Fuller puts it, speaking of the planetary system as a whole: ". . . every action has not only a reaction but also a non-simultaneous but immediately subsequent resultant" ("Planetary Planning," p. 42).

23. Strauss, *The Social Psychology of George Herbert Mead*, p. 217.

24. For Mead's ideas see his *Mind, Self, and Society* and *The Philosophy of the Act*; also Strauss, *Mead*; and Winter, *Elements of a Social Ethic*. For Watts's ideas see his *The Book*.

25. Skinner makes about the same point in *Beyond Freedom*, p. 117.

26. See Fustel de Coulanges, *The Ancient City*.

27. Weber, *The Theory of Social and Economic Organization*, p. 156.

28. *The Political System*, p. 146.

29. Skinner, *Beyond Freedom*, passim.

30. Jaeger, *Paidea*, vol. 3.

31. For example, Kariel, *Open Systems*.

32. Bay, *The Structure of Freedom*, p. xiv.

11.   Beyond Liberalism to Freedom

1. On the nature of freedom see Anshen, *Freedom: Its Meaning*; Bay, *Structure*; Bidney, *The Concept of Freedom in Anthropology*; Friedrich, *Nomos IV: Liberty*; Nearing, *Freedom: Promise and Menace*; Oppenheim, *Dimensions of Freedom*; and Skinner, *Beyond Freedom*. For a contemporary liberal treatment see Hayek, *The Constitution of Liberty*. For positions closer to that taken here see Brownell, *The Human Community*; Dewey, *The Public and Its Problems*; Follett, *The New State*; Vickers, *Freedom in a Rocking Boat* and *Value Systems and Social Process*; and Schiller, "Social Controls and Individual Freedom."

2. Zijderfeld, *The Abstract Society*, p. 150.

3. *Beyond Economics*, p. 100.

4. *Leviathan*, bk. 2, chap. 21.

5. Oppenheim, *Dimensions*, pp. 111–113, for example.

6. *Power: A New Social Analysis*, p. 35.

7. George C. Homans as paraphrased in Buckley, *Sociology and Modern Systems Theory*, p. 32.

8. See Wolin, *Politics*, pp. 345–346.

9. *The New Reformation*, p. x.

10. Bay, *Structure*, p. 133.

11. Wolff, *Liberalism*, p. 45 and passim. His discussion of Mill is especially useful.

12. Cf., Bauer, *Second Order Consequences*.

13. On coercion as a tool of social control see Etzioni, *The Active Society*, pp. 320–321, 350–386.

14. On the role of discussion as a vehicle for change in perception of desires see Gross, *Action under Planning*, p. 221 circa; and White, "Social Change and Administrative Adaptation," in Marini, *Towards a New Public Administration*. Failure to recognize that social interaction changes the goals of actors is a major weakness of the game-theory model of politics exemplified in such works as Buchanan and Tullock, *The Calculus of Consent*.

15. Bay, p. 221.

16. Quoted in Carr, *The New Society*, p. 118.

12. Human Needs in Social Perspective

1. Fromm, *The Heart of Man*, p. 47.

2. For an excellent discussion of the relationship of environment and evolution see Dunn, *Economic and Social Development*, pp. 39–74.

3. Dice, *Man's Nature and Nature's Man*, p. 108.

4. On human needs generally see Montagu, *On Being Human*; and Davies, *Human Nature in Politics*.

5. Dunn, *Development*, p. 274.

6. Skinner argues that critics of S-R psychology such as Koestler are seventy years out of date and that current behavioralist theories acknowledge the importance of genetic endowment and "what is called consciousness," but he fails to explain how (*Beyond Freedom*, p. 158). On Skinner's ideas see Chein, *The Science of Behavior and the Image of Man*; Chomsky, "The Case against B. F. Skinner"; Hall, "Will Success Spoil B. F. Skinner?"; Platt, "Beyond Freedom and Dignity." On the political implications of his ideas see Champlin, "Behavior (-ism & -alism) and Theory (Political)."

7. Charlotte Buhler, quoted in Dunn, *Development*, p. 274. See also Matson, *Broken Image*, pp. 161–230.

8. *Man Adapting*, p. xviii.

9. *Structure of Evil*, p. 294.

10. *Man Adapting*, p. xviii.

11. Skinner, *Beyond Freedom*, pp. 96–110.

12. Maslow, *Toward a Psychology of Being*. Also Gordon Allport, *Becoming*; Allport makes a useful distinction between "deficit" and "growth" motives.

13. For a summary of Maslow's views see Davies, *Human Nature*, pp. 9–11 and passim.

14. On the relationship between material and psychological needs viewed in a social context see Dahrenweld, "Urban Leadership and the Appraisal of Abnormal Behavior," in Duhl, *The Urban Condition*.

15. On childhood needs for close care see Montagu, *Being Human*, pp. 53–67.

16. As Adam Smith of course recognized in *The Wealth of Nations*. For a dated but still suggestive discussion of how much government would have to do to maintain a classical free-enterprise economy see Simons, *An Economic Policy for a Free Society*. For what governments in fact did to force the market system into existence see Polanyi, *The Great Transformation*.

17. On contemporary institutionalization of this see Shonfield, *Modern Capitalism*. See also Waterston, *Development Planning*.

18. Gallup Poll, reported in "Poll Finds Crime Exceeds Reports," *New York Times*, 14 January 1973.

19. For a suggestive fictional portrait of a future society dominated by racial violence see Brunner, *The Jagged Orbit*.

20. See Conner, *Deviance in Soviet Society*.

21. See Hanna, *The Revolt of the Body*; Kay Johnson, "Proximity," in Roszak, *Sources*; and Stanley Leleman, "The Body Is All," in *The Geocentric Experience*.

22. Wickler, *Sexual Code*.

23. Sykes, *The Cool Millennium*, p. 15.

24. Grant, *Technology and Empire*, p. 40.

25. See Comfort, *The Nature of Human Nature*, p. 206, circa. For a less disciplined but suggestive viewpoint see Brown, *Life Against Death* and *Love's Body*. For a critical viewpoint see Thompson, *Edge of History*, pp. 71–72.

26. Adler, *Superiority and Social Instinct*. Also Ansbacker, *The Individual Psychology of Alfred Adler*, especially p. 411; and Way, *Alfred Adler*, pp. 201–210.

27. Bennis and Slater, *The Temporary Society*. But see Packard, *A Nation of Strangers*; and Cox, *On Not Leaving It to the Snake*, pp. 112–123.

28. Becker, *Structure of Evil*, p. 276.

29. Love, Montagu argues, "is no mere creation of men but is grounded in the biological structure of man as a functioning organism" (*Being Human*, p. 98). Its role in the political community is the theme of De Grazia, *The Political Community*. On the destruction of community in America by liberalism see McWilliams, *The Idea of Fraternity in America*.

30. On social and political community see also Friedrich, *Nomos II: Community*; Minar and Greer, *The Concept of Community*; Simon, *Freedom and Community*; and Cassinelli, "The National Community."

31. On the isolating role of the family in modern American society see Slater, *The Pursuit of Loneliness*.

32. *The Social Contract*, 2:8.

33. A recent important treatment, unfortunately based on the premise of the social contract, is Rawls, *A Theory of Justice*. See also Pieper, *Justice*.

34. For Aristotle's views on justice and equality see *Ethics*, Bk. 5, and *Politics*, especially Bk. 3, ch. 9. Some modern approaches are: Pennock and Chapman, *Nomos IX: Equality*; Jouvenel, *The Ethics of Redistribution*; Knight, *The Ethics of Competition*; Jay, *The Socialist Case*. See also Moore, "But Some Are More Equal than Others"; Thurow, "Towards a Definition of Economic Justice"; and Vickers, "Changing Ethics of Distribution."

35. On the relationship of education and equality see Jencks et al., *Inequality*.

36. Lane, "The Fear of Equality."

37. "Biology, Darwinism, and Political Science."

38. Ibid., p. 21.

39. See Kahn, *The Sense of Injustice*.

40. For a radically opposed view see Jouvenel, *Sovereignty*.

41. Dubos, *So Human*, p. 155.

42. Much research on the effects of crowding has been done on animals; the extent to which it can be extrapolated to man is debated. See Bird, *The Crowding Syndrome*; and Calhoun, "Plight of the Ik and Kaiadelt . . ."

43. Searles, *The Nonhuman Environment*, pp. 5–6. See also Ilitis et al., "Criteria for an Optimum Human Environment."

44. *So Human*, p. 8.

45. One leading political scientist has even been concerned with what civil rights would have to be extended to "thinking" robots. See Lasswell, "The Political Science of Science," p. 976.

46. See Dunn, *Development*, pp. 21–23, circa.

47. On the need for work and its relation to leisure see Comfort, *Being Human*, pp. 6–7, 193, circa. Some studies indicate that rats prefer to work for food rather than obtain it free (Singh, "The Pied Piper vs. the Protestant Ethic").

48. *Freedom*, p. 128.

49. The classic study of the role of play in social history, however faulty, is still Huizinga, *Homo Ludens*. See also Caillois, *Man, Play, and Games*.

50. Kelly, *A Theory of Personality*. The economist Boulding expresses a similar position in economic terms by saying that well-being is both a "stock" and a "flow." *Beyond Economics*, p. 282.

13. The Common Good: Present and Future

1. *Politics* 3:12; see also *Ethics* 8:9.

2. *Summa Theologica*, I–II, Q. 90a. On Aquinas and the common good see Gilby, *Between Principality and Polity*; Koninck, *De la primauté du bien commun contre les personalistes*; and Maritain, *The Person and the Common Good*. Also Adler, *Common Sense*, pp. 21–22.

3. Manuscript, quoted in Baumgardt, *Bentham and the Ethics of Today*, p. 571.

4. For Green's concept of the common good see his *Works*, vol. 2.

5. The classic exposition is Pigou, *The Economics of Welfare*. See also Myint, *Theories of Welfare Economics*; and Timlin, "Theories of Welfare Economics."

6. Hobhouse, *Liberalism*.

7. See Smith, *Democracy and the Public Interest*. Henry B. Mayo says that "democracy is a political system . . . providing the machinery and opportunity for individuals to pursue their own private ends" (*An Introduction to Democratic Theory*, p. 249). Jouvenel locates the common good in the social bond itself rather than in the satisfaction of any discrete interests (*Sovereignty*, pp. 111, 123, 300). An intermediate position between considering the public interest as substantive or as purely procedural is to be found in Ranney and Kendall, *Democracy and the American Party System*.

8. On the relationship between liberalism and classical theory see Mansfield, "Disguised Liberalism."

9. *Social Contract* 2:1.

10. For differences between the Anglo-American approach to democracy and approaches based more on Rousseau's thought see Carr, *The Twenty Years Crisis*.

11. The seminal expression of group theory is Bentley's *The Process of Government*. For a description of Bentley's ideas see Taylor, "Arthur F. Bentley's Political Science." For later versions of group theory see Gross, *The Legislative Struggle*; Truman, *The Governmental Process*; and Latham, "The Group Basis of Politics."

The term "pluralism" as used by these contemporary theorists has a different meaning from that given it by earlier English and American political theorists. For a description of earlier pluralist theories see Hsiao, *Political Pluralism*. For an exposition of contemporary pluralism see Dahl, *Pluralist Democracy in the United States*. Closely allied ideologically to contemporary pluralism is the methodological and moral individualism represented in certain game theory and econometric approaches to politics, exemplified by such works as Buchanan and Tullock, *Calculus of Consent*; Buchanan, "An Individualistic Theory of the Political Process," in Easton, *Varieties of Political Theory*; and Olson, *The Logic of Collective Action*. For criticism of these approaches see Connolly, *The Bias of Pluralism* (especially Brian Barry, "The Public Interest"); McCoy and Playford, *Apolitical Politics*; and Morgan, *The Human Predicament*. Also Baskin, "American Pluralism"; Hale, "The Cosmology of Arthur F. Bentley"; Kress, "The Web and the Tree"; Odegard, "The Group Basis of Politics"; Partridge, "Politics, Philosophy, Ideology"; and Ricci, "Democracy Attenuated."

12. Some recent criticisms of the concept of the public interest which reflect this liberal attitude are Flathman, *The Public Interest*; Held, *The Public Interest and Individual Interests*; Schubert, *The Public Interest*; Sorauf, "The Public Interest Reconsidered"; Gross, *Legislative Struggle*; and Truman, *Governmental Process*. Other pertinent works are Friedrich, *Nomos V: The Public Interest*; Leys, *Philosophy and the Public Interest*; and Redford, *Democracy in the Administrative State*. See also Cassinelli, "The Concept of the Public Interest"; and Downs, "The Public Interest." For a more positive approach to the concept of the public interest see Cochran, "The Politics of Interest"; Conrad, "Rationality and Political Science"; and Lowi, "The Public Philosophy: Interest Group Liberalism."

13. *The New State*, p. 288; see also p. 295. Other examples of Follett's pioneering work in this area are *Creative Experience* and *Dynamic Administration*. See also Fox, "Mary Parker Follett: The Enduring Contribution."

14. Davies, *Human Nature in Politics*, p. 145. For views which reflect a liberal position see Arrow, *Social Choice and Individual Values* and "Public and Private Values" in Hook, *Human Values and Economic Policy*; and B. Brandt, "Personal Values and the Justification of Institutions" in ibid.

15. *The Public and Its Problems*, especially pp. 35, 67.

16. Ibid., p. 207.

17. *The Public Philosophy*, p. 40.

18. Quoted in Adler, *Common Sense*, p. 7.

19. Truman, *Governmental Process*. For comments on this point see Mackenzie, "Pressure Groups"; also Plamanetz, "Interests."

20. On the analogy between competitive exclusion and liberal society see Garrett Hardin, "The Cybernetics of Competition," in Shepard and McKinley, *Subversive Science*.

21. See Aberle, "The Functional Prerequisites of Society"; also Kluckhohn, "Ethical Relativity: Sic et Non"; and Wallace, "Perceptions of Order and Richness in Human Societies."

22. On the significance of energy see Cottrell, *Energy and Society*; Garvey, *Energy, Ecology, Economy*; and Odum, *Environment, Power, and Society*.

23. SCEP, *Man's Impact on the Global Environment*; also Broecker, "Enough Air."

24. Weinberg, "Social Institutions and Nuclear Energy."

25. See Sears, "The Inexorable Problem of Space."

26. For a reductio ad absurdum see Fremlin, "How Many People Can the Earth Support?", in Kostelanetz, *Human Alternatives*.

27. Nisbet, *The Social Bond*, pp. 244–245.

28. See Jeremy A. Sabloff, "The Collapse of Classic Maya Civilization," in Harte and Sokolow, *Patient Earth*.

29. On the relationship of political order to economic stability and development see Huntington, *Political Order in Changing Societies*.

30. See Wallace, *Culture and Personality*, pp. 29–41.

31. The existence of a plurality of cultures within a single political society is often the result of conquest or, in modern times, of colonialism. See the discussion of the concept of "plural society" in Ferkiss, "Race and Politics in Trinidad and Guyana."

32. Olson argues that "a society will . . . be more likely to cohere if people are socialized to have diverse wants with respect to private goods and similar wants with respect to collective goods" ("The Relationship between Economics and the Other Social Sciences," in Lipset, *Politics and the Social Sciences*, p. 151).

33. World history is interpreted in these terms in Toynbee, *A Study in History*.

34. Creation of a willed future as the purpose of planning is stressed in Jantsch, *Perspectives of Planning*. See especially the contribution of Ozbekhan, "Toward a General Theory of Planning."

35. See Fairlie, *The Kennedy Promise*.

36. See Polak, *The Image of the Future*.

37. Kariel, *Open System*, p. 142.

38. For opposed views see Callahan, "What Obligations Do We Have to Future Generations?"; and Golding, "Obligations to Future Generations" and "Ethical and Value Issues in Population Limitation and Distribution in the United States." See also Boulding, *Beyond Economics*, pp. 283–284; Dubos, *The Torch of Life*, pp. 122–123; Stearns, "Ecology and the Indefinite Unborn"; Taviss, "Futurology and the Problem of Values"; and Wagner, "Futurity Morality."

39. Marek, *Yestermorrow*, p. 43. Theodore Gordon warns of "the tyranny of the present" in "The Current Methods of Futures Research," in Toffler, *Futurists*, p. 189.

40. On creating the future see Gabor, *Inventing the Future*; Michael, *Future Responsive Societal Learning*; Roslansky, *Shaping the Future*; Moles, "The Future-Oriented Society"; and Platt, "How Men Can Shape their Future."

41. See Vickers, *The Art of Judgment*, pp. 31–35.

15. Necessary Utopia I: The Politics of Ecological Humanism

1. On recent developments in forms of opposition to established institutions see Goodman, *The Movement for a New America*.

2. See for example Wagar, *Building the City of Man*.

3. See Dunn, *Development*; Michael, *Learning*.

4. On dependency and social development see Bendix, "Tradition and Modernity Reconsidered."

5. Conceivably, of course, bombs set off by small nations or even criminal groups could start a general conflagration if one of the major powers were to attribute them to its rivals.

6. See Groth, *Comparative Politics: A Distributive Approach.* Britain is apparently doing better than the United States in dealing with environmental problems; how much this should be attributed to a more unified governmental structure and how much to public habits of cooperation for social goals is a matter for dispute. See Alfred Friendly, "Britain's Peter Walker: Power to Improve the Environment," Washington *Post*, 15 May 1971; Bayard Webster, "Briton Says His Nation Leads U.S. in Fight to Save Environment," *New York Times*, 27 May 1971.

7. The efforts of Rexford Guy Tugwell and his colleagues at the Center for the Study of Democratic Institutions tend to fall into this category. See Tugwell, "Model for a New Constitution"; Buchanan, "So Reason Can Rule"; Boyd, "Comments on the Tugwell Constitution."

8. On the problems involved in providing information to government see Roback, "Do We Need a Department of Science and Technology?"; Perl, "The Scientific Advisory System"; and Mosher, "Needs and Trends in Congressional Decision-Making."

9. See Junck, "Evolution and Revolution in the West," in Toffler, *Futurists*, p. 81–82. Also Cellarius and Platt, "Councils of Urgent Studies"; Platt, "What We Must Do" and "How Men Can Shape Their Future."

10. For an optimistic view of the computer as a means of social control see Calder, *Technopolis*, especially p. 208, circa; Hayashi, "The Information-Centered Society," in Toffler, *Futurists*; and Parker and Dunn, "Information Technology." For a less optimistic view see Weizenbaum, "On the Impact of the Computer on Society."

11. On the systems approach see Churchman, *The Systems Approach*, and Laszlo, *Systems View.*

12. On the divisive nature of the American political process see Burns, *The Deadlock of Democracy*; Drucker, "A Key to American Politics"; and Fisher, "Unwritten Rules of American Politics." The conflict between pluralism and the concept of the public interest in environmental matters is discussed in Wengert, *Natural Resources and the Political Struggle.*

13. Quoted in Chamberlain, *Beyond Malthus*, p. 135.

14. See generally Kendall, *The Conservative Affirmation*, and Ranney, *Democracy and the American Party System.*

15. For a recent criticism of earlier support for a responsible two-party system see Kirkpatrick, "Toward a More Responsible Two-Party System."

16. A classic example was the power of mining, timber, and grazing interests over national environmental policy during the time Representative Aspinall of Colorado was chairman of the Interior Committee. See Dennis Farney, "U.S. Presidents Come and Go, but the Power of Rep. Aspinall Persists," *Wall Street Journal*, 21 January 1972; and Philip Fradkin, "Rep. Aspinall, Boss of the West," Washington *Post*, 5 December 1971.

17. Bell, "The Commission on the Year 2000," p. 268.

18. Data indicates that key legislators are out of step with public opinion. See Tim O'Brien, "Business Poll Finds Legislators Don't Mirror Public Opinion," Washington *Post*, 23 August 1971; and Warren Weaver, "The Voters vs. House Leaders," *New York Times*, 23 August 1971.

19. On planning see especially Doob, *The Plans of Men*, and Jantsch, *Perspectives.*

20. On distinctions among types of planning see Braybroke and Lindblom, *A Strategy of Decision.*

21. Simon, *Sciences of the Artificial*, p. 112.

22. Ibid., p. 111.

23. On public participation in planning see Friedmann, *Transactive Planning* and *Retracking America*.

24. See Michael, *Learning*. This resembles the "mixed scanning" discussed in Etzioni, *Active Society*, pp. 282–309. See also Etzioni and Heidt, *Societal Guidance*, and Breed, *The Self-Guiding Society*.

25. Bachrach, *The Theory of Democratic Elitism*, p. 23.

26. Waterston, *Development Planning*.

27. Joint Economic Committee, U.S. Congress, *Restoration of Effective Sovereignty to Solve Social Problems*, pp. 9–11.

28. Lowi, *The Politics of Disorder*, p. viii. See also Wolff, *Liberalism*.

29. Follett, *The New State*, p. 309. See also White, "Social Change and Administrative Adaptation," in Marini, *New Public Administration*.

30. Remarks at Princeton (N.J.), 21 April 1965; quoted by Glenn D. Paige in Riggs, *Frontiers in Development Administration*, p. 167.

31. Buchanan and Tullock, *Calculus*, p. 96.

32. For an attempt to deal with this problem see Kendall and Carey, "The Intensity Problem and Democratic Theory."

33. On the problem of relating administrative discretion to governmental responsibility to the governed see Michael M. Harmon, "Normative Theory and Public Administration: Some Suggestions for a Redefinition of Administrative Responsibility," in Marini, *New Public Administration*.

34. Joint Economic Committee, *Restoration*, p. 7.

35. Ibid., p. 11.

16.   Necessary Utopia II: The Economics of Ecological Humanism

1. It can be argued that such terms as balance and order originated as social concepts and then became the basis of scientific concepts later on. But in recent intellectual history the movement of metaphor has been in the opposite direction.

2. On the limits of conventional economics see Schumacher, *Small Is Beautiful*; Weisskopf, *Alienation and Economics*; Berle, "What GNP Doesn't Tell Us"; Heilbroner, "On the Limited 'Relevance' of Economics"; Lekachman, "Humanizing GNP"; Silk, "Does Economics Ignore You?"; and Kapp, *The Social Costs of Business Enterprise*.

3. See Dorfman, *Economics of the Environment*; also the papers by Paul Davidon, John Halder, and Allan V. Kneese in Garnsey and Hibbs, *Social Sciences and the Environment*.

4. Recently economists such as James Tobin and William D. Nordhaus have been trying to develop such a standard. See Robert Reinhold, "MEW or NEW, How Does the Economy Grow," *New York Times*, 29 July 1973.

5. *Beyond Economics*, pp. 280–281.

6. Ibid., p. 284.

7. On the history of the concept of usury see Nelson, *The Idea of Usury*.

8. For some creative approaches to what I call ecophysics see Daly, *Toward a Steady-State Economy*; Odum, *Environment*; Watt, *Principles of Environmental Science*; Stanley A. Cain, "Can Ecology Provide the Basis for a Synthesis among the Social Sciences," in Garnsey and Hibbs, *Social Sciences*; Daly, "On Economics as a Life Science"; Garrett Hardin, "The Cybernetics of Competition: A Biologist's View of Society," in Shepard and McKinley, *Subversive Science*; Murray, "What the Ecologists Can Teach the Economists";

Odum, "The Strategy of Ecosystem Development"; and Spilhaus, "Ecolibrium."

On problems and techniques of environmental controls see Caldwell, *Environment: A Challenge for Modern Society* and "Environment: A New Focus for Public Policy." See also Murphy, *Governing Nature;* and Winthrop, "Environmental Dilemma" and "Total Environmental Management."

9. Hardin, "The Tragedy of the Commons"; also Crowe, "The Tragedy of the Commons Revisited."

10. Platt, "Lock-Ins and Multiple Lock-Ins in Collective Behavior" and "Social Traps." See also Schelling, "On the Ecology of Micromotives."

11. Mishan, "To Grow or Not to Grow?," p. 10.

12. See Herman E. Daly, "Toward a Steady-State Economy," in Harte and Sokolow, *Patient Earth*, p. 231.

13. On the concept of the steady-state as such see Sears, "The Steady State: Physical Law and Moral Choice," in Shepard and McKinley, *Subversive Science.* On the relationship between growth and welfare see Daly, "Economics as a Life Science" and *Steady State Economy;* Mishan, *The Costs of Economic Growth* and *Growth: The Price We Pay;* and Passell and Ross, *Affluence and Its Enemies.* On recent American debate on the subject of growth see White House National Goals Research Staff, *Toward Balanced Growth: Quantity with Quality;* and Stanford E. Slesser, "The Nation Debates an Issue: The Economy vs. the Environment," *Wall Street Journal*, 3 November 1971.

14. On the role of miniaturization in evolution see Odum, *Environment,* p. 158.

15. On land use problems and planning see Whyte, *The Last Landscape;* "Planning the Second America," Washington *Post*, 20 November 1971; Dennis Farley, "The Unsolved Problems of Land Use," *Wall Street Journal*, 2 February 1973; and Gladwin Hill, "Panel Foresees Difficulty in Limiting Use of Land," *New York Times*, 20 May 1973.

16. Calder, *The Environment Game;* See also Shepard, *The Tender Carnivore and the Sacred Game.*

17. Mackaye, *The New Exploration*, p. 147.

18. *Value Systems and Social Process*, p. 21.

19. See Reagan, *The Managed Economy;* and Morris, "Is the Corporate Economy a Corporate State?"

20. On the social impact of technology see Ferkiss, *Technological Man;* Mesthene, *Technological Change: Its Impact on Man and Society;* and Taviss, *Our Tool Making Society.* Also Hoelscher, "Technology and Social Change"; and Rosenberg, "Technology and the Environment."

21. McDermott, "Technology: The Opiate of the Intellectuals."

22. Habermas takes an essentially Marcusian position on this issue in *Rational Society;* see especially pp. 84–85.

23. Teich, *Technology and Man's Future*, p. ix.

24. On science policy and its problems see Brooks, *The Government of Science;* Baram, "Social Control of Science and Technology"; Benn, "On the Control of Science"; and Salomon, "Science Policy and Its Myths." See also the files of the journal *Minerva.*

25. Leiss, *Domination.*

26. *Science in the Modern World*, p. 98.

27. On the philosophical and ethical aspects of technology control see especially Barbour, *Science and Secularity;* Faramelli, *Technethics;* and Simon, *Philosophy of Democratic Government*, pp. 260–322. On the techniques and problems of technology control see especially Calder, *Technopolis*, and the National Academy of Sciences, *Technology: Processes of Assessment and*

Choice. On public attitudes toward technology control see La Porte and Metlay, *They Watch and Wonder*; and Taviss, "A Survey of Popular Attitudes toward Technology Control."

28. Report of National Academy of Sciences, *Technology*, p. 126.

29. Ibid., p. 32.

30. *The Dominion of Man*, p. 46.

31. Means, *The Ethical Imperative*, p. 255.

32. On population growth and its effect a useful book is Chamberlain, *Beyond Malthus*.

33. Calhoun, "Population Density and Social Pathology"; Galle and Mc-Pherson, "Population Density and Pathology." For a vivid fictional portrait of life in a crowded future world see Brunner, *Stand on Zanzibar*.

34. In some nations the pet population explosion is itself becoming a menace to social amenities and a drain on resources. See Barbara Eisenberg, "What's Next: How about Being Overrun by Cats and Dogs?," *Wall Street Journal*, 22 November 1972.

35. For an extreme example see Bozell, "Sin of Head Counts." For the position of the Vatican on birth control see Pope Paul VI's encyclical *Humanae Vitae*. This pronouncement excited much vocal dissent from both theologians and ordinary churchgoers, and has been virtually repudiated by several national hierarchies. It has been widely ignored by Catholics, especially in the United States. See Greeley, "The End of American Catholicism?," and Westhoff and Bumpass, "The Revolution in Birth Control Practices of U.S. Catholics."

36. Odum, *Environment*, p. 162.

37. Darlington, *The Evolution of Man and Society*, pp. 59, 672.

38. On the controversy over population policy in the United States see the Presidential Commission report, *Population and the American Future*. Black leaders have been especially hostile to any discussion of population control. See Gregory, "My Answer to Genocide," and Hare, "Black Ecology." But black women may have different ideas. See Ralph Z. Hallow, "The Blacks Cry Genocide," and Brian Sullivan, "Blacks Discount 'Genocide' in Birth Curbs," *New York Times*, 13 February 1972; Jack Rosenthal, "Birthrates Found in Sharp Decline among Poor Women," *New York Times*, 3 March 1972. Underdeveloped nations often take the same chauvinistic attitude toward population control that some leaders of minority groups do. See for example Novitski, "Brazil Shunning Population Curbs," *New York Times*, 13 February 1972.

39. See Riga, "Overpopulation and the American Catholic Conscience."

40. See Kolko, *Wealth and Power*.

41. See Bottomore, *Elites in Modern Society*; Parkin, *Class Inequality and Political Order*.

42. Many blue-collar workers refused to vote for McGovern in 1972 because of this attitude. See John Herbers, "Issues Fade—Party's Getting Rough," *New York Times*, 8 October 1972.

43. See Horowitz, "The Environmental Cleavage"; McCaull, "The Politics of Technology"; and Jon Margolis, "Land of Ecology," in Disch, *Conscience*.

44. As suggested by Bourrecauld, "The Paradoxes of Welfare."

45. The federal Social Security program is an example of a governmental welfare institution which has succeeded because of its wide constituency. The failure of President Johnson's "war on poverty" is certainly attributable in large part to the fact that the middle class participated only as providers of services and not as recipients with a direct interest in how effectively the services were provided.

46. See Kinkade, *A Walden Two Experiment*.

47. For Fourier's ideas see *Harmonian Man*.

17. Politics and Culture in a Humane Society

1. Just how far a society can and should go in enforcing its concept of psychological normality is increasingly under debate. See Kittrie, *The Right to Be Different*, and Szasz, *Ideology and Insanity*.

2. For developments in this field see the periodical *Radical Software*.

3. For a similar position see Berns, *Freedom, Virtue, and the First Amendment*.

4. See Wolin, *Politics*, p. 345.

5. "Repressive Tolerance" in Wolff, *A Critique of Pure Tolerance*, p. 90. See also the essay by Moore, "Tolerance and the Scientific Outlook," in ibid.

6. One of the few modern writers to be able to speak of patriotism seriously was the late Paul Goodman, as noted in Levine, "Paul Goodman, Outsider Looking In." See also Rock, "Alienation, Yes; Patriotism, Yes."

7. Quoted in Dubos, *So Human*, p. 235.

8. *Freedom*, p. 186.

9. "The Politics of Ecology," in Disch, *Conscience*, p. 159.

10. See William Ophuls, "Leviathan or Oblivion?" in Daly, *Steady-State Economy*, and Ophuls, "The Return of Leviathan."

11. *Neolithic Conservatism*, p. 191.

12. See Gross, "Friendly Fascism."

18. Getting There from Here: The Immanent Revolution

1. Shepard and McKinley, *Subversive Science*, p. 9.

2. Bennis, *The Planning of Change*, p. 3.

3. Hockett and Ascher, "The Human Revolution." See also Montagu, *The Human Revolution*.

4. For an interpretation that stresses the role of consciousness and violence in the process of modernization see Moore, *Social Origins of Dictatorship and Democracy*.

5. Nisbet, *The Social Bond*, p. 331–332.

6. Bendix, "Tradition and Modernity Reconsidered."

7. The role of population change in social change is stressed in Moore, *Social Change*. See also Chamberlain, *Beyond Malthus*.

8. Nisbet, *Social Bond*, p. 316.

9. Wallace, *Culture*, p. 144.

10. Platt in Carovillano and Skehan, *Science and the Future of Man*, pp. 86–87.

11. Nisbet, *Social Bond*, pp. 320–321.

12. *Technological Man*, p. 246.

13. Ibid., pp. 250–254.

14. Davies, "Toward a Theory of Revolution." See also Moore, *Social Change*, p. 82.

15. Lifton, *History and Human Survival*, especially pp. 184–186.

16. Weinstein, *The Political Experience*, p. 15.

17. On American class structure, especially on relevant social psychological aspects, see Dahrenwand, "Urban Leadership" in Duhl, *Urban Condition*;

also Kohn, "Social Class and Parental Values"; Bazelon, *Power in America*; Bensman and Vidich, *The New American Society*; Friedman, *Overcoming Middle Class Rage*; Howe, *The White Majority*; and Milner, *The Failure of Success*.

18. The terms are from Apter, *Choice and the Politics of Allocation*. On the radicalization of elites in all developed countries see Apter, in ibid., pp. 72–104; and Inglehart, "The Silent Revolution in Europe."

19. See Mead, *Culture and Commitment*; also Flacks, "Young Intelligentsia in Revolt."

20. Means, *Ethical Imperative*, p. 57. On the middle class as an agent of and participant in value change see Mitchener, "The Revolution in Middle Class Values"; and Nicholson, *The Radical Suburb*.

21. Quoted in Robert J. Donovan, "America's Middle Class vs. 'The Elitist's'," San Francisco *Chronicle*, 30 October 1972. See also Kahn, "The Squaring of America."

22. "About Equality," p. 43. See Podhoretz, "Laureate of the New Class," and Wildavsky, *The Revolt Against the Masses*, especially pp. 29–51, for other views hostile to the emerging value changes among the middle class.

23. *The Anatomy of Revolution*, pp. 39–49. The present upper-class and middle-class base of the movement for ecological reform is illustrated in Barkley and Weissman, "The Eco-Establishment"; Greene, "The Militant Malthusians"; and Weissman, "Why the Population Bomb Is a Rockefeller Baby."

24. On piecemeal and incremental evolutionary change see Moore, *Social Change*, pp. 33, 71; also Gotesky and Laszlo, "Evolution-Revolution."

25. Quoted in Flacks, "Strategies for Radical Social Change," p. 10.

26. See Wagar, *Building*, pp. 57–67.

27. On the commune as a more ecologically sound society see Corr and McLeod, "Getting It Together." On communes as a factor in the cultural revolution generally see Houriet, *Getting Back Together*; and Fairfield, *Communes U.S.A.* For a hostile view see Katz, *Armed Love*.

28. On the politics of ecology generally see Nelson, "Politics and Ecology", and Wheeler, "The Politics of Ecology." On the current failure of American political parties as tools for political mobilization see Ladd, *American Political Parties*, especially pp. 245–311. On political action related to environmental problems see de Bel, *The Voter's Guide to Environmental Politics*; Environmental Action, *Earth Tool Kit* and *Earth Day*; and Mitchell and Stallings, *Ecotactics*. See also Gladwin Hill, "Environment as Election Issue Grows at State and Local Level," *New York Times*, 1 October 1972, and Hill, "Environment Vote a Factor in 50 Congress Seats," in ibid., 12 November 1972.

29. Interview with Bob Hunter, Vancouver *Sun*, 8 June 1973.

30. See Cox, *Snake*.

31. Thus in 1973 the largest chapter of a leading environmental organization, the Sierra Club, sided with oil workers in a strike over refinery health hazards. See Deborah Shapley, "Shell Strike: Ecologists Refine Relations with Labor," *Science* 180 (1973): 166, and Stuart Auerback, "The First Strike over Potential Hazards to Health," San Francisco *Examiner and Chronicle*, 4 March 1973. To make such support most valuable, such organizations would also probably have to be prepared to support management against unions in appropriate situations. On the generally strained relations between environmental and labor groups see Woodcock, "Labor and the Politics of Environment"; Leonard Blaikie, "Labor-Liberal Alliance Expected," Oakland *Tribune*, 15 February 1973; and Adams, "The Quiet Confrontation."

32. On the relationship between cultural and social change see Martindale, *Social Life and Cultural Change*. On the cultural revolution and value

change in the United States see Braden, *The Age of Aquarius*; Kahn and Bruce-Briggs, *Things to Come*, pp. 83–113; Kelly, *Youth, Humanism, and Technology*; Kirkendall and Whitehurst, *The New Sexual Revolution*; Leonard, *The Transformation*; Packard, *The Sexual Wilderness;* Thompson, *At the Edge*; White, *Machina Ex Deo*; Winter, *Being Free*; Yankelovitch, *Changing Values*; Pitts, "The Counter Culture"; Reader, "Watching the Revolution Uncork"; *Time* essay, "Second Thoughts about Man"; Wilson, "How Our Values Are Changing"; Don Branning, "Four Ways of Life—Pick One," San Francisco *Chronicle*, 17 January 1973; and the editorial "A Year of Emerging Patterns," *Wall Street Journal*, 29 December 1972.

On the movement to do away with the dualism between the scientific and humanistic/mystical worldviews see Blackburn, "Sensuous-Intellectual Complementarity in Science"; Clarke and Watts, "Technology and Mysticism"; Maslow, "Toward a Humanisitc Biology"; and the collection *The Geocentric Experience*. For related material see also Brockman and Rosenfeld, *Real Time I*, and Tart, "States of Consciousness and State-Specific Sciences."

33. See Adler, *Common Sense*, p. 200. But, contrary to liberal assumptions, a strong belief in a value-position can be accompanied by a high level of political tolerance rather than a propensity toward violence and arbitrary action. See Putnam, "Studying Elite Political Culture."

34. On the limited use of violence as a tactic see Love, *Ecotage*. Also Ronald M. Legro, "Ecology Plus Sabotage Adds up to Bad News for Pollution Firms," *Wall Street Journal*, 27 July 1971. A limited amount of "discretionary violence" is not necessarily incompatible with social order, according to sociologist Daniel Bell. See Kahn and Bruce-Briggs, *Things to Come*, p. 68.

35. On the high degree of violence historically characteristic of western society see McNeill, *The Rise of the West*, especially p. 539ff.

19. Ecological Humanism and Planetary Society: The Restoration of Earth and Beyond

1. See B. Hubbard, "From Meaninglessness to New Worlds," and E. Hubbard, "The Need for New Worlds." On the space program see also Ehricke, "Terrestrial Imperative"; Lewis, "End of Apollo"; and Welford, "Last Apollo Wednesday: Scholars Assess Progress," *New York Times*, 3 December 1972.

2. Quoted in Dick Hallgren, "He Points Man's Way to the Stars," San Francisco *Chronicle*, 12 July 1972.

3. Clarke, *Profiles*, pp. 115–116.

4. See "The Greening of the Astronauts" and "God, Man, and Apollo."

5. The term "stargate" is borrowed from Braden, *Age of Aquarius*.

6. For necessarily unsatisfactory attempts to deal with problems of global concern from a liberal perspective see Ruggie, "Collective Goods and Future International Organization"; and Russett and Sullivan, "Collective Goods and International Organization." Some of the problems of the liberal approach are discussed in Cowhey, "The Theory of Collective Goods and the Future Regime of Ocean Space." For related theoretical discussion see Froelich, *Political Leadership and Collective Goods*.

7. For contrary views see Wagar, *Building*, pp. 142–146, and Adler, *Common Sense*, pp. 178–179.

8. See Morgenthau, *Politics among Nations*, pp. 315–317. This is the position adopted by Wagar, *Building*, p. 33. For views closer to my own see the essentially "functionalist" approaches of Haas, *Beyond the Nation State*;

Wheeler, "Making the World One"; and the systems-perspective of Laszlo, *The World System*.

9. Kahn and Bruce-Briggs, *Things to Come*, pp. 44–45.

10. Peace and war are discussed in an ecological context in Falk, *This Endangered Planet*, and Sprout, *Toward a Politics of the Planet Earth*.

11. On war and peace generally see the invaluable handbook by Pickus and Wioto, *To End War*.

12. On the relationship of national and international problems of ecology see North and Choucri, "Population, Technology, and Resources in the Future International System."

13. On the efforts toward global ecosystem management see Black and Falk, *The Future of the International Legal Order*, vol. 4; Caldwell, *In Defense of Earth*; Fleming, *The Issues of Survival*; Key and Skolnikoff, *World Ecocrisis*; Wilson, *International Environmental Action*; Atkeson, "Post-Stockholm"; Humpstone, "Pollution"; and Strong, "One Year after Stockholm." See also Kennan, "To Prevent a World Wasteland."

14. For useful attempts to raise this problem see Goulet, *The Cruel Choice*, and Guy Hunter, "The Search for a New Strategy," in Dinwiddy, *Aid Performance and Development Policies of Western Countries*.

15. Quoted in Adler, *Common Sense*, p. 184.

16. On relations between developed and underdeveloped nations see Shourie, "Growth, Poverty and Inequalities," and Woodhouse, "Revisioning the Future of the Third World."

17. The effect of technological change on the international system is discussed in Skolnikoff, "The International Functional Implications of Future Technology" and "Science and Technology."

18. Gunnar Myrdal notes that only a tiny stratum of the population of poorer nations benefits from development along current lines, and it is this unrepresentative ruling segment which raises the cry that attempts to control growth on a global basis hurt the interests of the poorer nations as a whole. See Jacobsoen, "Gunnar Myrdal Comments."

19. The phrase is from Polanyi, "The Republic of Science."

20. Junck argues that we are already breeding the first "planetarians." See "Evolution and Revolution in the West" in Toffler, *Futurists*, p. 74. On the emergence of world culture see Northrop, *Meeting of East and West*; Rudhyar, *The Planetization of Consciousness*; Wagar, *Building*, pp. 77–96; and Mazrui, "World Order through World Culture."

21. Wagar argues that a new world religion will arise directly out of the experience of struggle to rebuild the world (*Building*, p. 56). Mesthene see technology as contributing to syncretism in "Technological Change and Religious Unification."

22. Wagar, *Building*, p. 57.

23. See for example the Club of Rome, *Limits to Growth* and other publications; also Sakharov, *Progress*. On scientists and disarmament see Rotblatt, *Scientists in the Quest for Peace*.

24. See Thompson, *At the Edge*, pp. 100, circa.

25. Quite literally. See Platt, *Step to Man*, p. 151.

26. Quoted in Robert Reinhold, "Study Warns of Perils in World Growth," *New York Times*, 27 February 1972.

27. Stent, *The Coming of the Golden Age*. But see also Glass, "Science: Endless Horizons or Golden Age?"

28. Adler, *Common Sense*, p. 203.

29. Carrell, *Man the Unknown*.

30. See generally his "Biological Possibilities for the Human Species in the Next Ten Thousand Years," in Wolstenholme, *Man and His Future*. On human potentialities and the possibilities of future human development see Hanna, *Bodies in Revolt*; De Ropp, *Sex Energy*; Francoeur, *Eve's New Rib*; Jonas, *Visceral Learning*; Leonard, *Transformation*; Lilly, *The Center of the Cyclone* and *Programming*; Metzner, *Maps of Consciousness;* Naranjo and Ornstein, *On the Psychology of Meditation*; Stulman and Laszlo, *Emergent Man*; Teyler, *Altered States of Awareness*; Walker, *The Extra-Sensory Mind;* Dunn, "Evolution of the Mind and Human Potential"; Collier, "Brain Power"; and Stulman, "Beyond Crises."

31. The phrase is from Platt, *The Step to Man*.

Epilogue

1. On People's Park see Weisberg, *Beyond Repair*, pp. 165–166; Miller, "People's Park"; and Schultz, "People's Park."

2. For examples of antiecological backlash, intellectual and political, see Crosland, *Social Democratic Britain*; Grayson and Shepard, *The Disaster Lobby*; Maddox, *The Doomsday Syndrome*; Neuhaus, *In Defense of People*; Crenson, "Pollution Isn't THAT Serious"; Dwyer, "Ecology—Our New Religion"; Hare, "Black Ecology"; O'Brien, "The 'Ecology' of the Slums"; Stans, "Environment: Wait a Minute"; Kaysen, "The Computer that Printed Out W.O.L.F."; Lowry, "The New Religecology."

3. Elliot Richardson, then H.E.W. secretary, quoted in Robert Reinhold, "Warning on Growth Assayed," *New York Times*, 3 March 1972.

4. On the energy crisis see Ridgeway, *The Last Play*; Rocks and Runyon, *The Energy Crisis*; Roberts, "Is There an Energy Crisis?"

5. See Rowland, *The Plot to Save the World*.

BIBLIOGRAPHY*

1. Books

Adler, Alfred. *Superiority and Social Interest*. Evanston: Northwestern University Press, 1964.

Adler, Mortimer J. *The Common Sense of Politics*. New York: Holt, Rinehart and Winston, 1971.

————. *The Difference of Man and the Difference It Makes*. Cleveland: World Publishing Co., 1967.

Alexander, Samuel. *Space, Time, and Deity*. London: Macmillan, 1927.

Allport, Gordon. *Becoming: Basic Considerations for a Psychology of Personality*. New Haven: Yale University Press, 1955.

Allsopp, Bruce. *The Garden Earth*. New York: William Morrow, 1972.

Alves, Rubem A. *A Theology of Hope*. New York: Corpus Books, 1971.

Anderson, Walt, ed. *Politics and Environment*. Pacific Palisades, Calif.: Goodyear, 1970.

Ansbacher, Heinz L., and Ansbacher, Rowena R. *The Individual Psychology of Alfred Adler*. London: George Allen and Unwin, 1958.

Anshen, Ruth Nanda, ed. *Alfred North Whitehead: His Reflections on Man and Nature*. New York: Harper and Row, 1961.

————. *Freedom: Its Meaning*. New York: Harcourt, Brace, 1940.

Apter, David E. *Choice and the Politics of Allocation*. New Haven: Yale University Press, 1971.

Ardrey, Robert. *African Genesis*. New York: Atheneum, 1961.

————. *The Social Contract*. New York: Atheneum, 1970.

————. *The Territorial Imperative*. New York: Laurel Books, 1971.

Aristotle. *Ethics*. Baltimore: Penguin Books, 1953.

————. *Politics*. Oxford: Clarendon Press, 1948.

Arrow, Kenneth J. *Social Choice and Individual Values*. New Haven: Yale University Press, 1963.

Bachrach, Peter. *The Theory of Democratic Elitism*. Boston: Little, Brown, 1967.

Baier, Kurt, and Rescher, Nicholas, eds. *Values and the Future*. New York: The Free Press, 1969.

Baillie, John. *The Belief in Progress*. London: Oxford University Press, 1950.

Baltazar, Eulalio R. *God within Process*. New York: Newman Press, 1970.

Baran, Paul A., and Sweezy, Paul M. *Monopoly Capitalism*. New York: Monthly Review Press, 1966.

————. *The Political Economy of Growth*. New York: Monthly Review Press, 1957.

Barbour, Ian G. *Science and Secularity*. New York: Harper and Row, 1970.

Barnett, Lincoln. *The Universe and Dr. Einstein*. New York: William Sloane Associates, 1948.

Bates, Marston. *Gluttons and Libertines*. New York: Vintage Books, 1971.

---

* Books and articles with more than one author are listed only once, according to the name of the senior author. Journal articles are cited only by initial page. Editions of books cited are not necessarily those of original publication.

325

————. *The Prevalence of People*. New York: Scribner's, 1955.

Bateson, Gregory, and Ruesch, J. *Communication: The Social Matrix of Psychiatry*. New York: W. W. Norton, 1951.

Bauer, Raymond A. *Second-Order Consequences*. Cambridge: M.I.T. Press, 1969.

Baumgardt, David. *Bentham and the Ethics of Today*. Princeton: Princeton University Press, 1952.

Bay, Christian. *The Structure of Freedom*. New York: Atheneum, 1968.

Bazelon, David T. *Power in America*. New York: New American Library, 1967.

Beard, Charles A. *An Economic Interpretation of the Constitution of the United States*. New York: Macmillan, 1954.

Beck, William S. *Modern Science and the Nature of Life*. New York: Harcourt, Brace, 1957.

Becker, Ernest. *The Structure of Evil*. New York: George Braziller, 1968.

Ben-David, Joseph. *The Scientist's Role in Society*. Englewood Cliffs, N.J.: Prentice-Hall, 1971.

Benedict, Ruth. *Patterns of Culture*. Boston: Houghton Mifflin, 1961.

Benello, C. George, and Roussopoulos, Dimitri, eds. *The Case for Participatory Democracy*. New York: Grossman, 1971.

Bennis, Warren G.; Benne, Kenneth D.; and Chin, Robert, eds. *The Planning of Change*. 2d ed. New York: Holt, Rinehart and Winston, 1969.

Bennis, Warren G., and Slater, Philip. *The Temporary Society*. New York: Harper and Row, 1968.

Bensman, Joseph, and Vidich, Arthur J. *The New American Society*. Chicago: Quadrangle Books, 1971.

Bentley, Arthur F. *The Process of Government*. Evanston: The Principia Press, 1949.

Berger, Peter, and Luckmann, Thomas. *The Social Construction of Reality*. New York: Anchor Books, 1966.

Berkowitz, Leonard. *Aggression: A Social Psychological Analysis*. New York: McGraw-Hill, 1962.

————, ed. *Roots of Aggression*. Chicago: Aldine-Atherton, 1969.

Berns, Walter. *Freedom, Virtue, and the First Amendment*. Chicago: Henry Regnery, 1967.

Bertalanffy, Ludwig von. *General System Theory*. New York: George Braziller, 1972.

————. *Problems of Life*. New York: Harper and Bros., 1960.

————. *Robots, Men, Minds*. New York: George Braziller, 1967.

Bidney, David, ed. *The Concept of Freedom in Anthropology*. The Hague: Mouton, 1963.

————. *Theoretical Anthropology*. New York: Columbia University Press, 1953.

Bigelow, Robert. *The Dawn Warriors*. Boston: Little, Brown, 1969.

Bird, Caroline. *The Crowding Syndrome*. New York: David McKay, 1972.

Black, Cyril E., and Falk, Richard A., eds. *The Future of the International Legal Order. Vol. 4: The Structure of the International Environment*. Princeton: Princeton University Press, 1973.

Black, John. *The Dominion of Man*. Edinburgh: Edinburgh University Press, 1970.

Blandino, Giovanni, S.J. *Theories of the Nature of Life*. New York: Philosophical Library, 1969.

Bleibtreu, John N. *The Parable of the Beast*. New York: Collier Books, 1969.

Blissett, Marlan. *Politics in Science*. Boston: Little, Brown, 1972.

Bodard, Lucian. *Green Hell: Massacre of the Brazilian Indians*. New York: Outerbridge and Dienstfrey, 1972.

Boden, Margaret A. *Purposive Explanation in Psychology.* Cambridge: Harvard University Press, 1972.
Bonifazi, Conrad. *A Theology of Things.* Philadelphia: J. B. Lippincott, 1967.
Bookchin, Murray. *Post-Scarcity Anarchism.* Berkeley: Ramparts Press, 1971.
Borghese, Elizabeth Mann, ed. *Pacem In Maribus.* New York: Dodd, Mead, 1972.
Bottomore, Tom. *Elites in Modern Society.* Baltimore: Penguin Books, 1968.
Boughey, Arthur S. *Man and the Environment.* New York: Macmillan, 1971.
Boulding, Kenneth. *Beyond Economics.* Ann Arbor: University of Michigan Press, 1968.
Braden, William. *The Age of Aquarius.* Chicago: Quadrangle Books, 1970.
———. *The Private Sea.* Chicago: Quadrangle Books, 1967.
Braybrooke, David, and Lindblom, Charles A. *A Strategy of Decision.* New York: The Free Press, 1963.
Breed, Warren. *The Self-Guiding Society.* New York: The Free Press, 1971.
Brinton, Crane. *The Anatomy of Revolution.* New York: Vintage Books, 1965.
Brockman, John, and Rosenfeld, Edward, eds. *Real Time I.* New York: Anchor Books, 1973.
Brooks, Harvey. *The Government of Science.* Cambridge: M.I.T. Press, 1968.
Brown, James A. C. *Freud and the Post-Freudians.* Baltimore: Penguin Books, 1964.
Brown, Lester. *Seeds of Change.* New York: Praeger, 1970.
———. *World Without Borders.* New York: Random House, 1972.
Brown, Norman O. *Life Against Death.* Middletown, Conn.: Wesleyan University Press, 1959.
———. *Love's Body.* New York: Random House, 1966.
Brown, Robert E. *Charles Beard and the Constitution.* Princeton: Princeton University Press, 1956.
Brownell, Baker. *The Human Community.* New York: Harper and Bros., 1950.
Brubaker, Stuart. *To Live On Earth.* Baltimore: Johns Hopkins University Press, 1972.
Brunner, John. *The Jagged Orbit.* (A novel.) New York: Ace Books, 1969.
———. *Stand On Zanzibar.* (A novel.) New York: Ballantine Books, 1970.
Buchanan, James M., and Tullock, Gordon. *The Calculus of Consent.* Ann Arbor: University of Michigan Press, 1962.
Buckley, Walter. *Sociology and Modern Systems Theory.* Englewood Cliffs, N.J.: Prentice-Hall, 1967.
Buckley, William F., Jr., ed. *Did You Ever See a Dream Walking?* Indianapolis: Bobbs-Merrill, 1970.
Buettner-Janusch, John. *Origins of Man: Physical Anthropology.* New York: John Wiley, 1966.
Burch, William R., Jr. *Daydreams and Nightmares.* New York: Harper and Row, 1971.
Burdick, Eugene. *The 480.* (A novel.) New York: McGraw-Hill, 1964.
Burns, James MacGregor. *The Deadlock of Democracy.* Englewood Cliffs, N.J.: Prentice-Hall, 1963.
Caillois, Roger. *Man, Play, and Games.* New York: The Free Press, 1961.
Calder, Nigel. *The Environment Game.* London: Panther Science Books, 1969.
———. *The Mind of Man.* New York: Viking Press, 1971.
———. *Technopolis: Social Control of the Uses of Science.* New York: Simon and Schuster, 1970.
———. *The Violent Universe.* New York: Viking Press, 1970.
Caldwell, Lynton Keith. *Environment: A Challenge for Modern Society.* Garden City, N.Y.: Natural History Press, 1970.

————. *In Defense of Earth*. Bloomington: Indiana University Press, 1972.

Calhoun, John C. *A Disquisition on Government*. New York: Peter Smith, 1943.

Callan, Hilary. *Ethology and Society*. Oxford: Clarendon Press, 1970.

Campbell, Keith. *Body and Mind*. New York: Anchor Books, 1970.

Carothers, J. Edward. *The Pusher and the Pulled*. Nashville: Abingdon Press, 1968.

————; Mead, Margaret; McCracken, Daniel D.; and Shinn, Roger L., eds. *To Love or to Perish*. New York: Friendship Press, 1972.

Carovillano, Robert L., and Skehan, James W., S.J. *Science and the Future of Man*. Cambridge: M.I.T. Press, 1970.

Carr, Edward H. *The New Society*. New York: St. Martin's Press, 1960.

————. *The Twenty Years Crisis, 1919–1939*. London: Macmillan, 1954.

Carrell, Alexis. *Man the Unknown*. New York: Harper and Bros., 1939.

Carthy, J. D. and Ebling, F. J., eds. The Natural History of Aggression. London: Academic Press, 1964.

Cauthen, Kenneth. *Christian Biopolitics*. Nashville: Abingdon Press, 1971.

————. *Science, Secularization and God*. Nashville: Abingdon Press, 1969.

Caws, Peter. *Science and the Theory of Value*. New York: Random House, 1967.

Center for the Study of Democratic Institutions. *Natural Law and Modern Society*. Cleveland: World Publishing Co., 1966.

Chamberlain, Neil. *Beyond Malthus*. New York: Basic Books, 1970.

Charter, S. P. R. *Man On Earth*. New York: Grove Press, 1970.

Chein, Isador. *The Science of Behavior and the Image of Man*. New York: Basic Books, 1972.

Churchman, C. West. *The Systems Approach*. New York: Delacorte Press, 1966.

Cippola, Carlo. *The Economic History of World Population*. Baltimore: Penguin Books, 1962.

Clark, George Norman. *The Seventeenth Century*. Oxford: Oxford University Press, 1947.

Clarke, Arthur C. *Profiles of the Future*. New York: Bantam Books, 1964.

Collingwood, R. G. *The Idea of Nature*. New York: Oxford University Press, 1945.

Comfort, Alex. *The Nature of Human Nature*. New York: Harper and Row, 1967.

Commission on Population Growth and the American Future. *Population and the American Future*. New York: New American Library, 1972.

Committee on Resources and Man—National Academy of Science—National Resource Council. *Resources and Man*. San Francisco: W. H. Freeman, 1969.

Commoner, Barry. *The Closing Circle*. New York: Knopf, 1971.

Conner, Walter D. *Deviance in Soviet Society*. New York: Columbia University Press, 1972.

Connolly, William E., ed. *The Bias of Pluralism*. New York: Atherton Press, 1969.

Cook, Fred J. *The Corrupted Land*. New York: Macmillan, 1966.

Coon, Carleton. *Living Races of Man*. New York: Knopf, 1965.

Cottrell, Fred. *Energy and Society*. New York: McGraw-Hill, 1955.

Cowie, Leonard W. *Seventeenth-Century Europe*. New York: Frederick Ungar, 1960.

Cox, Harvey. *On Not Leaving It to the Snake*. New York: Macmillan, 1967.

————. *The Secular City*. New York: Macmillan, 1965.

Cox, Peter R., and Peel, John, eds. *Population and Pollution*. New York: Academic Press, 1972.

Crick, Francis. *Of Molecules and Men*. Seattle: University of Washington Press, 1966.

Crosland, Anthony, M. P. *Social Democratic Britain*. London: Fabian Tracts, 1971.

Cutler, Donald R., ed. *The Religious Situation, 1969.* Boston: Beacon Press, 1969.
Dahl, Robert A. *Pluralist Democracy in the United States.* Chicago: Rand McNally, 1967.
Daly, H. E., ed. *Toward a Steady-State Economy.* San Francisco: W. H. Freeman, 1973.
Darling, F. Fraser. *Wilderness and Plenty.* New York: Ballantine Books, 1971.
Darlington, C. D. *The Evolution of Man and Society.* New York: Simon and Schuster, 1970.
Dasmann, Raymond F. *The Last Horizon.* New York: Macmillan, 1963.
Davies, James C. *Human Nature in Politics.* New York: John Wiley, 1963.
de Bel, Garrett, ed. *The Environmental Handbook.* New York: Ballantine Books, 1970.
———. *The Voter's Guide to Environmental Politics.* New York: Ballantine Books, 1970.
De Grazia, Sebastian. *The Political Community.* Chicago: University of Chicago Press, 1948.
Delgado, Jose M., M.D. *Physical Control of the Mind.* New York: Harper and Row, 1969.
De Ropp, Robert S. *Sex Energy.* New York: Delta Paperbacks, 1972.
Dewey, John. *The Public and Its Problems.* Chicago: Gateway Books, 1946.
———. *The Quest for Certainty.* New York: Milton, Balch, 1929.
———. *The Theory of Valuation.* Chicago: University of Chicago Press, 1939.
Dice, Lee R. *Man's Nature and Nature's Man.* Ann Arbor: University of Michigan Press, 1955.
Didion, Joan. *Slouching toward Bethlehem.* New York: Delta Books, 1969.
Dinwiddy, Bruce, ed. *Aid Performance and Development Policies of Western Countries.* New York: Praeger, 1973.
Disch, Robert, ed. *The Ecological Conscience.* Englewood Cliffs, N.J.: Prentice-Hall, 1970.
Dobzhansky, Theodosius. *The Biological Basis of Human Freedom.* New York: Columbia University Press, 1956.
———. *The Biology of Ultimate Concern.* New York: New American Library, 1967.
———. *Mankind Evolving.* New Haven: Yale University Press, 1962.
Dolbeare, Kenneth M. *Directions in American Political Thought.* New York: John Wiley, 1969.
———, and Dolbeare, Patricia. *American Ideologies.* Chicago: Markham, 1971.
Domhoff, G. William. *The Higher Circles.* New York: Vintage Books, 1971.
———. *Who Rules America?* Englewood Cliffs, N.J.: Prentice-Hall, 1967.
Doob, Leonard W. *The Plans of Men.* Hamden, Conn.: Archon Books, 1968.
Dorfman, Robert, and Dorfman, Nancy S., eds. *Economics of the Environment.* New York: W. W. Norton, 1972.
Driekurs, Rudolf, M.D. *Fundamentals of Adlerian Psychology.* New York: Greenberg, 1950.
Dreyfus, Hubert L. *Alchemy and Artificial Intelligence.* Santa Monica: Rand Corporation, 1965.
———. *What Computers Can't Do.* New York: Harper and Row, 1971.
Droscher, Vitus B. *The Friendly Beast.* New York: Harper and Row, 1972.
Dubos, Rene. *A God Within.* New York: Scribners, 1972.
———. *Man Adapting.* New Haven: Yale University Press, 1965.
———. *Reason Awake.* New York: Columbia University Press, 1970.
———. *So Human An Animal.* New York: Scribners, 1968.
———. *The Torch of Life.* New York: Simon and Schuster, 1962.

Duhl, Leonard J., M.D., ed. *The Urban Condition*. New York: Basic Books, 1963.

Dunn, Edgar S., Jr. *Economic and Social Development*. Baltimore: Johns Hopkins University Press, 1971.

Dye, Thomas R., and Zeigler, L. Harmon. *The Irony of Democracy*. Belmont, Calif.: Wadsworth, 1971.

Easton, David. *The Political System*. New York: Knopf, 1953.

——, ed. *Varieties of Political Theory*. Englewood Cliffs, N.J.: Prentice-Hall, 1966.

Edel, Abraham. *Method in Ethical Theory*. London: Routledge and Kegan Paul, 1963.

Edel, Mary, and Edel, Abraham. *Anthropology and Ethics*. Rev. Ed. Cleveland: The Press of Case Western Reserve University, 1968.

Editors of *The Ecologist*. *Blueprint for Survival*. Boston: Houghton Mifflin, 1972.

Ehrenfeld, David. *Conserving Life on Earth*. New York: Oxford University Press, 1972.

Ehrlich, Paul. *The Population Bomb*. New York: Ballantine Books, 1971.

——, and Ehrlich, Anne. *Population: Resources: Environment*. San Francisco: W. H. Freeman, 1970.

——, and Harriman, Richard L. *How to Be a Survivor*. New York: Ballantine Books, 1971.

Eisenberg, J. F., and Dillon, Wilton S., eds. *Man and Beast: Comparative Social Behavior*. Washington: Smithsonian Institution Press, 1971.

Ekrich, Arthur A., Jr. *Man and Nature in America*. New York: Columbia University Press, 1963.

Elder, Frederick. *Crisis in Eden*. Nashville: Abingdon Press, 1970.

Eliade. Mercea. *The Myth of the Eternal Return*. Princeton: Princeton University Press, 1954.

Ellul, Jacques. *False Presence in the Kingdom*. New York: Seabury Press, 1972.

——. *The Political Illusion*. New York: Knopf, 1967.

——. *The Technological Society*. New York: Knopf, 1964.

Engels, Frederick. *Dialectics of Nature*. New York: International Publishers, 1940.

Environmental Action. *Earth Day—The Beginning*. New York: Ballantine Books, 1970.

——. *Earth Tool Kit*. New York: Pocket Books, 1971.

Etzioni, Amitai. *The Active Society*. New York: The Free Press, 1968.

——, and Heidt, Sarajane, eds. *Societal Guidance*. New York: Thomas Y. Crowell, 1969.

Fairfield, Richard. *Communes U.S.A.* Baltimore: Penguin Books, 1972.

Fairlie, Henry. *The Kennedy Promise*. New York: Doubleday, 1973.

Falk, Richard A. *This Endangered Planet*. New York: Random House, 1971.

Faramelli, Norman. *Technethics*. New York: Friendship Press, 1971.

Ferkiss, Victor C. *Technological Man*. New York: George Braziller, 1969.

Flathman, Richard E. *The Public Interest*. New York: John Wiley, 1966.

Fleming, D. F. *The Issues of Survival*. New York: Doubleday, 1972.

Follett, Mary Parker. *Creative Experience*. New York: Longmans, Green, 1924.

——. *Dynamic Administration*. New York: Harpers, 1942.

——. *The New State*. New York: Longmans, Green, 1918.

Foss, Philip O. *Politics and Grass*. Seattle: University of Washington Press, 1960.

Fourier, Charles. *Harmonian Man*. New York: Anchor Books, 1971.

Francoeur, Robert T. *Eve's New Rib*. New York: Harcourt Brace Jovanovich, 1972.

———. *Evolving World, Converging Man*. New York: Holt, Rinehart and Winston, 1970.

———. *Perspectives in Evolution*. Baltimore: Helicon Press, 1965.

———. *Utopian Motherhood*. New York: Doubleday, 1970.

Frank, Lawrence K. *Nature and Human Nature*. New Brunswick: Rutgers University Press, 1951.

Freeman, A. Myrick III; Haveman, Robert H.; and Kneese, Alan V. *The Economics of Environmental Policy*. New York: John Wiley, 1973.

Friedman, Maurice. *To Deny Our Nothingness*. New York: Delacorte Press, 1967.

Friedman, Murray, ed. *Overcoming Middle Class Rage*. Philadelphia: Westminster Press, 1971.

Friedmann, John. *Transactive Planning*. New York: Doubleday, 1972.

Friedrich, Carl J. *The Age of the Baroque: 1610–1660*. New York: Harper and Row, 1952.

———, ed. *Nomos II: Community*. New York: Liberal Arts Press, 1959.

———, ed. *Nomos IV: Liberty*. New York: Atherton Press, 1962.

———, ed. *Nomos V. The Public Interest*. New York: Atherton Press, 1966.

Froelich, Norman; Oppenheimer, Joe A.; and Young, Oran F. *Political Leadership and Collective Goods*. Princeton: Princeton University Press, 1971.

Fromm, Erich. *Beyond the Chains of Illusion*. New York: Simon and Schuster, 1962.

———. *The Heart of Man*. New York: Harper's, 1964.

———. *The Revolution of Hope*. New York: Bantam Books, 1968.

———, ed. *Socialist Humanism*. New York: Anchor Books, 1966.

Fuller, Buckminister; Walker, Eric A.; and Killian, James R., Jr. *Approaching the Benign Environment*. New York: Collier Books, 1970.

———. *Earth, Inc*. New York: Anchor Books, 1973.

———. *Operating Manual for Spaceship Earth*. Carbondale: Southern Illinois University Press, 1969.

———. *Utopia or Oblivion*. New York: Bantam Books, 1969.

Fustel de Coulanges, Nuna. *The Ancient City*. New York: Anchor Books, 1956.

Gabor, Dennis. *Innovations: Scientific, Technological, and Social*. New York: Oxford University Press, 1972.

———. *Inventing the Future*. Hammondsworth, Eng.: Penguin Books, 1969.

———. *The Mature Society*. New York: Praeger, 1972.

Galbraith, John Kenneth. *The New Industrial State*. 2d rev. ed. Boston: Houghton Mifflin, 1972.

Garnsey, Morris E., and Hibbs, James R., eds. *Social Sciences and the Environment*. Boulder: University of Colorado Press, 1967.

Garvey, Gerald. *Energy, Ecology, Economy*. New York: W. W. Norton, 1972.

The Geocentric Experience. Los Gatos, Calif.: Maplighters Roadway Press, 1972.

Gilby, Thomas, O.P. *Between Principality and Polity*. London: Longmans, Green, 1953.

Gill, David M. *From Here to Where?* Geneva: World Council of Churches, 1970.

Ginsburg, Morris. *Essays in Sociology and Social Philosophy, Vol. I: On the Diversity of Morals*. London: Heinemann, 1956.

———. *Reason and Experience in Ethics*. London: Oxford University Press, 1956.

Girvetz, Harry K. *From Wealth to Welfare*. Stanford: Stanford University Press, 1950.

Glacken, Clarence J. *Traces on the Rhodian Shore*. Berkeley: University of California Press, 1967.

Goldman, Marshall L. *Ecology and Economics*. Englewood Cliffs, N.J.: Prentice-Hall, 1972.

————. *The Spoils of Progress*. Cambridge: M.I.T. Press, 1972.

Goodman, Mitchell. *The Movement Toward a New America*. Philadelphia: Pilgrim Press, 1970.

Goodman, Paul. *The New Reformation and Notes of a Neolithic Conservative*. New York: Random House, 1970.

————, ed. *Seeds of Liberation*. New York: George Braziller, 1965.

Gorney, Roderic, M.D. *The Human Agenda*. New York: Bantam Books, 1973.

Gotesky, Rubin, and Laszlo, Ervin, eds. *Evolution—Revolution*. New York: Gordon and Breach, 1971.

Goulet, Denis. *The Cruel Choice*. New York: Atheneum, 1971.

Graham, George A., and Carey, George W., eds. *The Post Behavioral Era*. New York: David McKay, 1972.

Grant, George. *Technology and Empire*. Toronto: House of Anansi, 1969.

Grayson, Melvin, and Shepard, Thomas. *The Disaster Lobby*. New York: McGraw-Hill, 1973.

Greeley, Andrew. *Unsecular Man*. New York: Schocken Books, 1972.

Green, T. H. *Works. Vol. II*. London: Longmans, Green, 1886.

Gross, Bertram M., ed. *Action Under Planning*. New York: McGraw-Hill, 1967.

————, ed. *A Great Society?* New York: Basic Books, 1968.

————. *The Legislative Struggle*. New York: McGraw-Hill, 1953.

Guardini, Romano. *The End of the Modern World*. New York: Sheed and Ward, 1956.

Haas, Ernest B. *Beyond the Nation State*. Stanford: Stanford University Press, 1968.

Habermas, Jurgen. *Toward a Rational Society*. Boston: Beacon Press, 1970.

Hacker, Andrew. *The End of the American Era*. New York: Atheneum, 1970.

Haines, Charles Grove. *The Revival of National Law Concepts*. Cambridge: Harvard University Press, 1950.

Hall, Gus. *Ecology: Can We Survive Under Capitalism?* New York: International Publishers, 1972.

Hamilton, Michael, ed. *The New Genetics and the Future of Man*. Grand Rapids, Mich.: William B. Eerdmans, 1972.

————, ed. *This Little Planet*. New York: Scribner's, 1970.

Hanna, Thomas. *Bodies in Revolt*. New York: Holt, Rinehart and Winston, 1970.

Hardin, Garrett. *Exploring New Ethics for Survival*. New York: Viking Press, 1972.

————. *Nature and Man's Fate*. New York: Rinehart and Co., 1958.

Hardy, Sir Alister. *The Living Stream*. New York: Harper and Row, 1967.

Harrington, Michael. *Socialism*. New York: Saturday Review Press, 1972.

Hart, Jeffrey. *The American Dissent*. New York: Doubleday, 1966.

Harte, John, and Socolow, Robert H., eds. *Patient Earth*. New York: Holt, Rinehart and Winston, 1971.

Hartshorne, Charles. *Beyond Humanism*. Chicago: Willett, Clark, 1937.

————. *A Natural Theology for Our Time*. LaSalle, Ill.: Open Court Publishing Co., 1967.

Hartz, Louis. *The Liberal Tradition in America*. New York: Harcourt, Brace and World, 1955.

Haselden, Kyle, and Hefner, Philip. *Changing Man*. New York: Anchor Books, 1969.

Haspers, John. *Libertarianism: A Political Philosophy for Tomorrow*. Santa Barbara: Reason Press, 1972.

Hayek, Friedrich A. *The Constitution of Liberty*. Chicago: University of Chicago Press, 1960.

Hays, Samuel P. *Conservation and the Gospel of Efficiency.* Cambridge: Harvard University Press, 1959.

Held, Virginia. *The Public Interest and Individual Interests.* New York: Basic Books, 1970.

Henderson, Lawrence J. *The Fitness of the Environment.* Boston: Beacon Press, 1958.

Hill, Christopher. *The Century of Revolution 1603–1714.* London: Thomas Nelson, 1961.

Hobhouse, L. T. *Liberalism.* London: Oxford University Press, 1811.

Hoffer, Eric. *First Things, Last Things.* New York: Harper and Row, 1971.

Hook, Sidney, ed. *Human Values and Economic Policy.* New York: New York University Press, 1967.

Holmberg, Allan R. *Nomads of the Long Bow.* Garden City, N.Y.: Natural History Press, 1969.

Hoselitz, Bert F., ed. *Economics and the Idea of Mankind.* New York: Columbia University Press, 1965.

——, ed. *Theories of Economic Growth.* New York: The Free Press, 1960.

Houriet, Robert. *Getting Back Together.* New York: Coward, McCann, and Geoghegan, 1971.

Howe, Irving, ed. *The Radical Papers.* New York: Anchor Books, 1966.

——. *Steady Work.* New York: Harcourt, Brace and World, 1966.

Howe, Louise Kapp, ed. *The White Majority.* New York: Random House, 1971.

Hsiao, Chuan Kung. *Political Pluralism.* New York: Harcourt, Brace, 1927.

Hsu, Francis. *Psychological Anthropology.* Homewood, Ill.: Dorsey, 1961.

Hudson, Michael. *Super Imperialism.* New York: Holt, Rinehart and Winston, 1973.

Huizinga, Johan. *Homo Ludens.* London: Temple Smith, 1970.

Huntington, Samuel. *Political Order in Changing Societies.* New Haven: Yale University Press, 1968.

Huth, Hans. *Nature and the American.* Berkeley: University of California Press, 1957.

Huxley, T. H., and Huxley, Julian. *Evolution and Ethics 1893–1943.* London: The Pilot Press, 1947.

Jacobs, Paul, and Landau, Saul. *The New Radicals.* New York: Vintage Books, 1966.

Jaeger, Werner. *Paideia: The Ideals of Greek Culture. Vol. III.* New York: Oxford University Press, 1944.

Jalee, Pierre. *The Pillage of the Third World.* New York: Monthly Review Press, 1968.

——. *The Third World in World Economy.* New York: Monthly Review Press, 1969.

Jantsch, Erich, ed. *Perspectives of Planning.* Paris: O.E.C.D., 1968.

Jay, Douglas. *The Socialist Case.* London: Faber and Faber, 1947.

Jencks, Christopher; Smith, Marshall; Acland, Henry; Bane, Mary Jo; Cohen, David; Gintis, Herbert; Heyns, Barbara; and Mickelson, Stephen. *Inequality.* New York: Basic Books, 1972.

Joint Economic Committee, Congress of the United States. *Restoration of Effective Sovereignty to Solve Social Problems.* Report of the Subcommittee on Urban Affairs. Washington: U.S. Government Printing Office, 1971.

Joll, James. *The Anarchists.* Boston: Little, Brown, 1964.

Jonas, Gerald. *Visceral Learning.* New York: Viking Press, 1973.

Jones, Howard Mumford. *O Strange New World.* New York: Viking Press, 1964.

Jouvenel, Bertrand de. *The Art of Conjecture.* New York: Basic Books, 1967.

———. *The Ethics of Redistribution*. Cambridge: Cambridge University Press, 1951.

———. *Sovereignty*. Chicago: University of Chicago Press, 1957.

Julian, Claude. *America's Empire*. New York: Pantheon, 1971.

Kahn, Edmund H. *The Sense of Injustice*. Bloomington: Indiana University Press, 1964.

Kahn, Herman, and Bruce-Briggs, B. *Things to Come*. New York: Macmillan, 1972.

———, and Weiner, Anthony J. *The Year 2000*. New York: Macmillan, 1967.

Kapp, Karl William. *The Social Costs of Business Enterprise*. New York: Schocken Books, 1971.

Kariel, Henry. *Open Systems*. Itaska, Ill.: F. A. Peacock, 1968.

Katz, Elia. *Armed Love*. New York: Holt, Rinehart and Winston, 1971.

Kaufman, Arnold. *The Radical Liberal*. New York: Atherton Press, 1968.

Kay, David A., and Skolnikoff, Eugene B., eds. *World Eco-Crisis: International Organizations in Response*. Madison: University of Wisconsin, 1972.

Kegley, Charles W., and Bietall, Robert W., eds. *Reinhold Niebuhr*. New York: Macmillan, 1956.

Keith, Sir Arthur. *Evolution and Ethics*. New York: G. P. Putnam's Sons, 1947.

Kelly, George A. *A Theory of Personality*. New York: W. W. Norton, 1963.

Kelly, Kevin. *Youth, Humanism, and Technology*. New York: Basic Books, 1972.

Kendall, Willmoore. *The Conservative Affirmation*. Chicago: Henry Regnery, 1963.

Kindleberger, Charles P., ed. *The International Corporation*. Cambridge: M.I.T. Press, 1971.

King, Richard. *The Party of Eros*. Chapel Hill: University of North Carolina Press, 1972.

Kinkade, Kathleen. *A Walden Two Experiment*. New York: William Morrow, 1973.

Kirk, Russell. *The Conservative Mind*. Chicago: Henry Regnery, 1953.

———. *A Program for Conservatives*. Chicago: Henry Regnery, 1954.

Kirkendall, Lester A., and Whitehurst, Robert N., eds. *The New Sexual Revolution*. Buffalo: Prometheus Books, 1971.

Kittrie, Nicholas N. *The Right to Be Different*. Baltimore: Johns Hopkins University Press, 1971.

Kneese, Allen V.; Rolfe, Sidney E.; and Harned, Joseph W. *Managing the Environment*. New York: Praeger, 1971.

Knelman, Fred H., ed. *1984 and All That*. Belmont, Calif.: Wadsworth, 1971.

Knight, Frank H. *The Ethics of Competition*. New York: Harper's, 1935.

Koestler, Arthur, and Smythies, J. R., eds. *Beyond Reductionism*. Boston: Beacon Press, 1971.

———. *The Ghost in the Machine*. New York: Macmillan, 1967.

Kolko, Gabriel. *The Triumph of Conservatism*. New York: The Free Press, 1963.

———. *Wealth and Power in America*. New York: Praeger, 1962.

Koninck, Charles de. *De la primauté du bien commun contre les personalistes*. Quebec City: Editions Université Laval, 1943.

Kormondy, Edward J. *Concepts of Ecology*. Englewood Cliffs, N.J.: Prentice-Hall, 1969.

Kostelanetz, Richard, ed. *Human Alternatives*. New York: William Morrow, 1971.

———, ed. *Social Speculations*. New York: William Morrow, 1971.

Kramnick, Isaac, ed. *Essays in the History of Political Thought*. Englewood Cliffs, N.J.: Prentice-Hall, 1969.

Krimerman, Leonard I., and Perry, Lewis, eds. *Patterns of Anarchy*. New York: Anchor Books, 1966.

Krutch, Joseph Wood. *The Great Chain of Life*. Boston: Houghton Mifflin, 1957.

Kuhns, William. *The Post-Industrial Prophets*. New York: Weybright and Talley, 1971.

Ladd, Everett Carll, Jr. *American Political Parties*. New York: W. W. Norton, 1970.

La Porte, Todd, and Metlay, Daniel. *They Watch and Wonder*. Berkeley: Institute of Governmental Studies, University of California, 1973.

Laszlo, Ervin. *The Systems View of the World*. New York: George Braziller, 1972.

———, ed. *The World System*. New York: George Braziller, 1973.

Leach, Gerald. *The Biocrats*. Hammondsworth, Eng.: Penguin Books, 1972.

Leiss, William. *The Domination of Nature*. New York: George Braziller, 1972.

Leonard, George B. *The Transformation*. New York: Delacorte Press, 1972.

Leopold, Aldo. *A Sand County Almanac*. New York: Oxford University Press, 1949.

Leys, Wayne A. R., and Perry, Charner Marquis. *Philosophy and the Public Interest*. Chicago: Committee to Advance Original Work in Philosophy, 1959.

Lieb, Irwin C. *Experience, Existence, and the Good*. Carbondale: University of Southern Illinois, 1961.

Lifton, Robert Jay. *History and Human Survival*. New York: Random House, 1970.

Lilly, John C., M.D. *The Center of the Cyclone*. New York: Julian Press, 1972.

———. *Programming and Metaprogramming in the Human Biocomputer*. New York: Julian Press, 1972.

Linton, Ralph. *The Cultural Background of Personality*. New York: Appleton-Century-Crofts, 1945.

Lippmann, Walter. *The Public Philosophy*. Boston: Little, Brown, 1955.

Lipset, Seymour, ed. *Politics and the Social Sciences*. New York: Oxford University Press, 1969.

———, and Bendix, Reinhard. *Social Mobility in Industrial Society*. Berkeley: University of California Press, 1964.

Livingston, John A. *One Cosmic Instant*. Boston: Houghton Mifflin, 1973.

Lockhard, Duane. *The Perverted Priorities of American Politics*. New York: Macmillan, 1971.

Loeb, Leo. *The Biological Basis of Individuality*. Springfield, Ill.: Charles C Thomas, 1945.

Loth, David, and Ernst, Morris L. *The Taming of Technology*. New York: Simon and Schuster, 1972.

Love, Sam. *Ecotage*. New York: Pocket Books, 1972.

Lovejoy, Arthur O. *The Great Chain of Being*. Cambridge: Harvard University Press, 1957.

Lowi, Theodore J. *The End of Liberalism*. New York: W. W. Norton, 1969.

———. *The Politics of Disorder*. New York: Basic Books, 1971.

Luce, Gay Gaer. *Body Time*. New York: Pantheon, 1971.

McCloskey, Maxine, and Gilligan, James P. *Wilderness and the Quality of Life*. San Francisco: Sierra Club, 1969.

McConnell, Brian. *Britain in the Year 2000*. London: New English Library, 1970.

McCoy, Charles A., and Playford, John, eds. *Apolitical Politics*. New York: Thomas Y. Crowell, 1967.

McDonald, Forrest. *We the People*. Chicago: University of Chicago Press, 1958.

McHale, John. *The Ecological Context*. New York: George Braziller, 1970.

——. *The Future of the Future*. New York: George Braziller, 1969.

McHarg, Ian. *Design With Nature*. Garden City, N.Y.: Natural History Press, 1969.

MacKay, Donald M. *Freedom of Action in a Mechanistic Universe*. London: Cambridge University Press, 1967.

——. *Information, Mechanism, and Meaning*. Cambridge: M.I.T. Press, 1969.

Mackaye, Benton. *The New Exploration*. Urbana: University of Illinois Press, 1962.

McLuhan, Marshall, and Fiore, Quentin. *War and Peace in the Global Village*. New York: Bantam Books, 1971.

McNeill, William. *The Rise of the West*. Chicago: University of Chicago Press, 1963.

McNeilly, F. S. *The Anatomy of Liberalism*. London: Macmillan, 1968.

McPhee, John. *Encounters With the Archdruid*. New York: Farrar, Straus and Giroux, 1971.

Macpherson, C. B. *The Political Theory of Possessive Individualism: Hobbes to Locke*. Oxford: Clarendon Press, 1962.

McWilliams, Wilson Carey. *The Idea of Fraternity in America*. Berkeley: University of California Press, 1973.

Maddox, John. *The Doomsday Syndrome*. New York: McGraw-Hill, 1972.

Magdoff, Harry. *The Age of Imperialism*. New York: Monthly Review Press, 1969.

Malthus, T. P. *Population: The First Essay*. Ann Arbor: University of Michigan Press, 1959.

Marcuse, Herbert. *An Essay on Liberation*. Boston: Beacon Press, 1968.

——. *Five Lectures*. Boston: Beacon Press, 1970.

——. *One Dimensional Man*. Boston: Beacon Press, 1964.

Marek, Kurt W. *Yestermorrow*. New York: Knopf, 1961.

Margenau, Henry. *Open Vistas*. New Haven: Yale University Press, 1961.

Marine, Gene. *America the Raped*. New York: Discus Books, 1969.

——, and Van Allen, Judith. *Food Pollution*. New York: Holt, Rinehart and Winston, 1972.

Marini, Frank, ed. *Toward a New Public Administration*. Scranton: Chandler, 1971.

Maritain, Jacques. *The Person and the Common Good*. New York: Scribner's, 1947.

Martin, P. S., and Wright, H. H., Jr., eds. *Pleistocene Extinctions*. New Haven: Yale University Press, 1967.

Martindale, Don. *Social Life and Cultural Change*. Princeton: Van Nostrand, 1962.

Marx, Leo. *The Machine in the Garden*. New York: Oxford University Press, 1964.

Marx, Wesley. *The Frail Ocean*. New York: Ballantine Books, 1969.

Maslow, Abraham H. *Religion, Values, and Peak-Experiences*. Columbus: Ohio State University Press, 1964.

——. *Toward a Psychology of Being*. 2d ed. New York: Van Nostrand Reinhold, 1968.

Matson, Floyd. *The Broken Image*. New York: George Braziller, 1965.

Mayo, Henry B. *An Introduction to Democratic Theory*. New York: Oxford University Press, 1960.

Mayr, Ernest. *Population, Species, and Evolution*. Cambridge: Belknap Press, 1970.

Mead, George Herbert. *Mind, Self, and Society*. Chicago: University of Chicago Press, 1934.

——. *The Philosophy of the Act*. Chicago: University of Chicago Press, 1932.

Mead, Margaret. *Culture and Commitment*. Garden City, N.Y.: Natural History Press, 1971.

——. *Twentieth Century Faith*. New York: Harpers, 1972.

Meadows, Donella H.; Meadows, Dennis L.; Randers, Jorgen; and Behrens, William W., III. *The Limits to Growth*. New York: Universe Books, 1972.

Means, Richard L. *The Ethical Imperative*. New York: Doubleday, 1969.

Meek, Ronald L., ed. *Marx and Engels on Malthus*. New York: International Publishers, 1954.

Meggysey, David. *Out of Their League*. Berkeley: Ramparts Press, 1970.

Megill, Kenneth A. *The New Democratic Theory*. New York: The Free Press, 1970.

Mesthene, Emmanuel G. *Technological Change*. Cambridge: Harvard University Press, 1970.

Metzner, Ralph. *Maps of Consciousness*. New York: Macmillan, 1972.

Michael, Donald N. *Future Responsive Societal Learning*. San Francisco: Jossey-Bass, 1973.

Miller, David L. *Individualism*. Austin: University of Texas Press, 1967.

Miller, Herman P. *Rich Man, Poor Man*. New York: Thomas Y. Crowell, 1971.

Mills, C. Wright. *The Causes of World War III*. New York: Simon and Schuster, 1958.

Milner, Esther. *The Failure of Success*. St. Louis, Warren H. Green, 1968.

Minar, David W., and Greer, Scott, eds. *The Concept of Community*. Chicago: Aldine, 1969.

——. *Ideas and Politics*. Homewood, Ill.: Dorsey, 1964.

Mintz, Morton, and Cohen, Jerry S. *America, Inc.* New York: Dial Press, 1971.

Mishan, E. J. *The Costs of Economic Growth*. Hammondsworth, Eng.: Penguin Books, 1969.

——. *Growth: The Price We Pay*. London: Staples Press, 1969.

——. *Welfare Economics*. New York: Random House, 1964.

Mitchell, John G., and Stallings, Constance L., eds. *Ecotactics*. New York: Pocket Books, 1970.

Monod, Jacques. *Chance and Necessity*. New York: Knopf, 1971.

Montagu, Ashley. *The Human Revolution*. Cleveland: World Publishing Co., 1965.

——, ed. *Man and Aggression*. New York: Oxford University Press, 1968.

——. *On Being Human*. New York: Hawthorn Books, 1966.

Moore, Barrington, Jr. *Social Origins of Dictatorship and Democracy*. Boston: Beacon Press, 1966.

Moore, Charles A., ed. *Essays in East-West Philosophy*. Honolulu: University of Hawaii Press, 1951.

Moore, Wilbert E. *Social Change*. Englewood Cliffs, N.J. Prentice-Hall, 1963.

——, ed. *Technology and Social Change*. Chicago: Quadrangle Books, 1972.

Morgan, George W. *The Human Predicament*. Providence: Brown University Press, 1968.

Morgenthau, Hans J. *Politics Among Nations*. 4th ed. New York: Knopf, 1967.

——. *Science: Servant or Master?* New York: New American Library, 1972.

Morris, Desmond. *The Human Zoo*. New York: McGraw-Hill, 1969.

——. *The Naked Ape*. New York: Dell Books, 1969.

Mumford, Lewis. *The Myth of the Machine*. 2 vols. New York: Harcourt Brace Jovanovich, 1967–1970.

Murphy, Earl Finbar. *Governing Nature*. Chicago: Quadrangle Books, 1967.

Myint, Hla. *Theories of Welfare Economics*. Cambridge: Harvard University Press, 1948.

Myrdal, Gunnar. *An American Dilemna*. New York: Harper's, 1944.

Nader, Ralph. *Unsafe at Any Speed*. New York: Grossman, 1972.

Nagel, Ernest. *The Structure of Science*. New York: Harcourt, Brace and World, 1961.

Naipaul, V. S. *The Loss of El Dorado*. New York: Knopf, 1970.

Naranjo, Claudio, and Ornstein, Robert E. *On the Psychology of Meditation*. New York: Viking Press, 1971.

Nasr, S. H. *The Encounter of Man and Nature*. London: George Allen and Unwin, 1968.

Nearing, Scott. *Freedom: Promise and Menace*. Harborside, Me.: Social Science Institute, 1961.

Needham, Joseph, and Green, David, eds. *Perspectives in Biochemistry*. London: Cambridge University Press, 1937.

————. *Science and Civilization in China. Vol. II*. Cambridge: Cambridge University Press, 1956.

————. *Time, the Refreshing River*. New York: Macmillan, 1943.

Nelson, Benjamin N. *The Idea of Usury*. 2d ed. Chicago: University of Chicago Press, 1969.

Neuhaus, Richard J. *In Defense of People*. New York: Macmillan, 1971.

Nicholson, Max. *The Environmental Revolution*. New York: McGraw-Hill, 1970.

Niebuhr, Reinhold. *Moral Man and Immoral Society*. New York: Scribner's, 1952.

Nisbet, Robert A. *The Social Bond*. New York: Knopf, 1970.

Nobile, Philip, ed. *The Con III Controversy*. New York: Pocket Books, 1971.

Nogar, Raymond J., O.P. *The Lord of the Absurd*. New York: Herder and Herder, 1966.

Northrop, Filmer S. *The Meeting of East and West*. New York: Macmillan, 1947.

————. *Philosophical Anthropology and Practical Politics*. New York: Macmillan, 1960.

Oakeshott, Michael J. *Rationalism in Politics and Other Essays*. New York: Basic Books, 1962.

Odum, Howard T. *Environment, Power, and Society*. New York: Wiley-Interscience, 1971.

Oglesby, Carl, ed. *The New Left Reader*. New York: Grove Press, 1969.

Olson, Mancur. *The Logic of Collective Action*. Cambridge: Harvard University Press, 1965.

Oppenheim, Felix E. *Dimensions of Freedom*. New York: St. Martin's Press, 1961.

Oppenheimer, J. Robert. *Science and the Common Understanding*. New York: Simon and Schuster, 1954.

Orr, John B., and Nichelson, F. Patrick. *The Radical Suburb*. Philadelphia: Westminster Press, 1970.

Orton, William A. *The Liberal Tradition*. New Haven: Yale University Press, 1945.

Ostrander, Sheila, and Schroeder, Lynn. *Psychic Discoveries Behind the Iron Curtain*. New York: Bantam Books, 1970.

Overman, Richard H. *Evolution and the Christian Doctrine of Creation*. Philadelphia: Westminster Press, 1967.

Packard, Vance. *A Nation of Strangers*. Philadelphia: David McKay, 1972.

————. *The Sexual Wilderness*. New York: Pocket Books, 1970.

Parker, Richard. *The Myth of the Middle Class*. New York: Liveright, 1972.

Parkin, Frank. *Class Inequality and Political Order*. New York: Praeger, 1971.
Parrington, Vernon L. *Main Currents in American Thought*. 3 vols. New York: Harcourt, Brace, 1927.
Parsons, Talcott. *The Structure of Social Action*. Glencoe, Ill.: The Free Press, 1949.
Passell, Peter, and Ross, Leonard. *Affluence and Its Enemies*. New York: Viking Press, 1973.
Passerin D' Entreves, A. P. *Natural Law*. London: Hutchinson's Universal Library, 1951.
Pearce, Joseph Chilton. *The Crack in the Cosmic Egg*. New York: Julian Press, 1971.
Pearl, Raymond. *Man the Animal*. Bloomington, Ind.: Principia Press, 1946.
Peccei, Aurelio. *The Chasm Ahead*. New York: Macmillan, 1969.
Pennock, J. Roland, and Chapman, John W., eds. *Nomos IX: Equality*. Chicago: Atherton Press, 1967.
Pfaff, William. *Condemned to Freedom*. New York: Random House, 1971.
Pickus, Robert, and Wioto, Richard. *To End War*. New York: Harper and Row, 1970.
Pieper, Joseph. *Justice*. New York: Pantheon Books, 1955.
Pigou, A. C. *The Economics of Welfare*. 4th ed. London: Macmillan, 1948.
Platt, John R., ed. *New Views of the Nature of Man*. Chicago: University of Chicago Press, 1965.
———. *The Step to Man*. New York: John Wiley, 1966.
Polak, Fred L. *The Image of the Future*. 2 vols. The Hague: A. J. Sijhoff, 1961.
Polanyi, Karl. *The Great Transformation*. Boston: Beacon Press, 1967.
Polanyi, Michael. *The Tacit Dimension*. New York: Anchor Books, 1966.
Potter, David M. *People of Plenty*. Chicago: Phoenix Books, 1958.
Potter, Van Rensselaer. *Bioethics: Bridge to the Future*. Englewood Cliffs, N.J.: Prentice-Hall, 1971.
President's Task Force on Science Policy. *Science and Technology: Tools for Progress*. Washington: U.S. Government Printing Office, 1970.
Price, Don K. *The Scientific Estate*. Cambridge: Belknap Press, 1965.
Pryde, Philip R. *Conservation in the Soviet Union*. New York: Cambridge University Press, 1972.
Quillian, W. F., Jr. *The Moral Theory of Evolutionary Naturalism*. New Haven: Yale University Press, 1945.
Ramsey, Paul. *Fabricated Man*. New Haven: Yale University Press, 1970.
Rand, Ayn. *The Anti Industrial Revolution*. New York: New American Library, 1971.
———. *For the New Intellectual*. New York: Random House, 1961.
Ranney, Austin, and Kendall, Willmoore. *Democracy and the American Party System*. New York: Harcourt Brace, 1956.
Raskin, Marcus. *Being and Doing*. New York: Random House, 1971.
Rathlesberger, James, ed. *Nixon and the Environment*. New York: Village Voice Books, 1972.
Rawls, John. *A Theory of Justice*. Cambridge: Harvard University Press, 1971.
Reagan, Michael. *The Managed Economy*. New York: Oxford University Press, 1962.
Redford, Emmette S. *Democracy in the Administrative State*. New York: Oxford University Press, 1968.
Reich, Charles A. *The Greening of America*. New York: Random House, 1970.
Reid, Leslie. *The Sociology of Nature*. Baltimore: Penguin Books, 1962.
Rensch, Bernard. *Biophilosophy*. New York: Columbia University Press, 1971.

————. *Homo Sapiens*. New York: Columbia University Press, 1972.

Report of the National Academy of Sciences. *Technology: Processes of Assessment and Choice*. Washington: U.S. Government Printing Office, 1969.

Revel, Jean-François. *Without Marx or Jesus*. New York: Doubleday, 1971.

Ridgeway, James. *The Last Play*. New York: Dutton, 1973.

Riggs, Fred W., ed. *Frontiers of Development Administration*. Durham: Duke University Press, 1970.

Robinson, Paul. *The Freudian Left*. New York: Harper and Row, 1969.

Rocks, Lawrence, and Runyon, Richard P. *The Energy Crisis*. New York: Crown Publishers, 1973.

Roe, Anne, and Simpson, George Gaylord, eds. *Behavior and Evolution*. New Haven: Yale University Press, 1958.

Rogow, Arnold, and Lasswell, Harold. *Power, Corruption and Rectitude*. Englewood Cliffs, N.J.: Prentice-Hall, 1963.

Rommen, Heinrich. *The Natural Law*. St. Louis: B. Herder, 1947.

Ropke, Wilhelm. *A Humane Economy*. Chicago: Henry Regnery, 1960.

Roslansky, John D., ed. *Shaping the Future*. New York: Fleet Academic Press, 1972.

————, ed. *The Uniqueness of Man*. Amsterdam: North Holland Publishing Company, 1969.

Rossiter, Clinton. *Conservatism in America*. 2d. ed. New York: Knopf, 1962.

Roszak, Theodore. *The Making of a Counter Culture*. New York: Anchor Books, 1969.

————, ed. *Sources*. New York: Harper and Row, 1972.

————. *Where the Wasteland Ends*. New York: Doubleday, 1972.

Rotblat, Joseph. *Scientists in the Quest for Peace*. Cambridge: M.I.T. Press, 1972.

Rothblatt, Ben, ed. *Changing Perspectives on Man*. Chicago: University of Chicago Press, 1968.

Rougement, Denis de. *Love in the Western World*. New York: Anchor Books, 1957.

Rowland, Wade. *The Plot to Save the World*. Toronto: Clarke, Irwin, 1973.

Rudhyar, Dane. *The Planetization of Consciousness*. New York: Harper and Row, 1971.

Ruggiero, Guido de. *The History of European Liberalism*. Boston: Beacon Press, 1961.

Runciman, W. G. *Relative Deprivation and Social Justice*. Berkeley: University of California Press, 1966.

Russell, Bertrand. *Power: A New Social Analysis*. New York: W. W. Norton, 1938.

Rust, Eric Charles. *Nature and Man in Biblical Thought*. London: Lutterworth Press, 1953.

Sakharov, Andrei P. *Progress, Coexistence, and Intellectual Freedom*. New York: W. W. Norton, 1968.

Sale, Kirkpatrick. *SDA*. New York: Random House, 1973.

Santmire, H. Paul. *Brother Earth*. New York: Thomas Nelson, 1970.

SCEP. *Man's Impact on the Global Environment*. Cambridge: M.I.T. Press, 1970.

Schmidt, Alfred. *The Concept of Nature in Marx*. London: NLB, 1971.

Schrodinger, Erwin. *Mind and Matter*. Cambridge: The University Press, 1958.

————. *What Is Life?* Cambridge: The University Press, 1955.

Schubert, Glendon A. *The Public Interest*. New York: The Free Press, 1961.

Schumacher, E. F. *Small Is Beautiful*. New York: Harper and Row, 1973.

Schwartz, Eugene S. *Overskill*. Chicago: Quadrangle Books, 1971.

Scientific American. Man and the Ecosphere. San Francisco: W. H. Freeman, 1971.

Scitovsky, Tibor. *Papers on Welfare and Growth.* Stanford: Stanford University Press, 1964.

Searles, Howard F., M.D. *The Nonhuman Environment.* New York: International Universities Press, 1960.

Seliger, Martin. *The Liberal Politics of John Locke.* New York: Praeger, 1969.

Sennett, Richard, and Cobb, Jonathan. *The Hidden Injuries of Class.* New York: Knopf, 1972.

Shepard, Paul. *Man in the Landscape.* New York: Knopf, 1967.

———. *The Tender Carnivore and the Sacred Game.* New York: Scribner's, 1973.

———, and McKinley, Daniel, eds. *The Subversive Science.* Boston: Houghton Mifflin, 1969.

Sherman, Franklin, ed. *Christian Hope and the Future of Humanity.* Minneapolis: Augsburg Publishing House, 1969.

Sherrington, Charles. *Man on His Nature.* New York: Macmillan, 1941.

Shonfield, Andrew. *Modern Capitalism.* New York: Oxford University Press, 1965.

Sigmund, Paul. *Natural Law in Political Thought.* Cambridge: Winthrop, 1971.

Simon, Herbert. *The Sciences of the Artificial.* Cambridge: M.I.T. Press, 1969.

Simon, Michael. *The Matter of Life.* New Haven: Yale University Press, 1971.

Simon, Yves R. *Freedom and Community.* New York: Fordham University Press, 1968.

———. *Philosophy of Democratic Government.* Chicago: University of Chicago Press, 1951.

Simons, Henry C. *Economic Policy for a Free Society.* Chicago: University of Chicago Press, 1968.

Simpson, George Gaylord. *Biology and Man.* New York: Harcourt, Brace and World, 1969.

———. *This View of Life.* New York: Harcourt, Brace and World, 1964.

Sittler, Joseph A. *The Care of the Earth and Other University Sermons.* Philadelphia: Fortress Press, 1964.

———. *Nature and Grace.* Philadelphia: Fortress Press, 1972.

Siu, R. G. H. *The Tao of Science.* Cambridge: M.I.T. Press, 1957.

Skinner, B. F. *Beyond Freedom and Dignity.* New York: Bantam Books, 1972.

Slater, Philip. *The Pursuit of Loneliness.* Boston: Beacon Press, 1971.

Slater, R. Giuseppe et al. *The Earth Belongs to the People: Ecology and Power.* San Francisco: People's Press, 1970.

Smith, Adam. *The Wealth of Nations.* New York: Appleton-Century-Crofts, 1957.

Smith, Henry Nash. *Virgin Land.* Cambridge: Harvard University Press, 1950.

Smith, Howard R. *Democracy and the Public Interest.* Athens: University of Georgia Press, 1960.

Smuts, Jan C. *Holism and Evolution.* New York: Macmillan, 1926.

Sorokin, Pitirim A. *The Ways and Power of Love.* Chicago: Henry Regnery, 1967.

Spitz, David, ed. *Political Theory and Social Change.* New York: Atherton Press, 1967.

Sprout, Harold, and Sprout, Margaret. *Toward a Politics of the Planet Earth.* New York: Van Nostrand Reinhold, 1971.

Stent, Gunther S. *The Coming of the Golden Age.* Garden City, N.Y.: Natural History Press, 1969.

Stevens, L. Clark. *EST, The Steersman's Handbook.* Santa Barbara: Capricorn Press, 1970.

Stone, Glenn C., ed. *A New Ethic for a New Earth*. Philadelphia: Friendship Press, 1971.

Storrs, Anthony. *Human Aggression*. New York: Atheneum, 1963.

Strauss, Anselm, ed. *The Social Psychology of George Herbert Mead*. Chicago: University of Chicago Press, 1956.

Strauss, Leo. *Natural Right and History*. Chicago: University of Chicago Press, 1953.

————. *The Political Philosophy of Hobbes*. Oxford: Clarendon Press, 1936.

————. *Thoughts on Machiavelli*. Seattle: University of Washington Press, 1969.

————. *What Is Political Philosophy?* Glencoe, Ill.: The Free Press, 1959.

————, and Cropsey, Joseph, eds. *History of Political Philosophy*. Chicago: Rand McNally, 1963.

Stulman, Julian, and Laszlo, Ervin, eds. *Emergent Man: His Chances, Problems, and Potentials*. New York: Gordon and Breach, 1973.

Subcommittee on Science, Research and Development, Committee on Science and Astronautics, U.S. House of Representatives, 91st Congress. *Technology Assessment*. Washington: U.S. Government Printing Office, 1970.

Sundquist, James L. *Politics and Policy*. Washington: Brookings Institution, 1968.

Sykes, Gerald. *The Cool Millennium*. Englewood Cliffs, N.J.: Prentice-Hall, 1967.

Szasz, Thomas S. *Ideology and Insanity*. New York: Anchor Books, 1970.

Taviss, Irene. *Our Tool Making Society*. Englewood Cliffs, N.J.: Prentice-Hall, 1972.

Tax, Sol. *Evolution After Darwin*. 3 vols. Chicago: University of Chicago Press, 1960.

Taylor, Gordon Rattray. *The Biological Time Bomb*. New York: Signet Books, 1969.

————. *The Doomsday Book*. Cleveland: World Publishing Co., 1970.

————. *Rethink: A Paraprimitive Solution*. London: Secker and Warburg, 1972.

Taylor, John C. *The New Physics*. New York: Basic Books, 1972.

————. *The Shape of Minds to Come*. New York: Weybright and Talley, 1971.

Teich, Albert H., ed. *Technology and Man's Future*. New York: St. Martin's Press, 1972.

Teilhard de Chardin, Pierre. *The Future of Man*. London: Fontana Books, 1964.

————. *Human Energy*. New York: Harcourt, Brace, 1971.

————. *Man's Place in Nature*. New York: Harper and Row, 1959.

————. *The Phenomenon of Man*. New York: Harper and Row, 1959.

Teodori, Massimo, ed. *The New Left: A Documentary History*. Indianapolis: Bobbs-Merrill, 1969.

Terry, Mark. *Teaching for Survival*. New York: Ballantine Books, 1971.

Teyler, Timothy J., ed. *Altered States of Awareness*. San Francisco: W. H. Freeman, 1972.

Theobald, Robert. *An Alternative Future for America II*. Chicago: Swallow Press, 1970.

Thomas, Franklin. *The Environmental Basis of Society*. New York: Century, 1925.

Thompson, William Irwin. *At the Edge of History*. New York: Harper and Row, 1971.

Thorpe, W. H. *Biology and the Nature of Man*. London: Oxford University Press, 1962.

Thorson, Thomas L. *Biopolitics*. New York: Holt, Rinehart and Winston, 1970.

Tiger, Lionel. *Men in Groups*. New York: Random House, 1969.

————, and Fox, Robin. *The Imperial Animal*. New York: Holt, Rinehart and Winston, 1971.

Tiselius, Arne, and Nilsson, Sam, eds. *The Place of Values in a World of Facts*. New York: Wiley Interscience, 1969.

Truman, David B. *The Governmental Process*. New York: Knopf, 1964.

Toffler, Alvin. *Future Shock*. New York: Bantam Books, 1971.

————, ed. *The Futurists*. New York: Random House, 1972.

Toynbee, Arnold. *A Study of History*. 12 vols. London: Oxford University Press, 1934–1961.

Trotsky, Leon. *Literature and Revolution*. New York: International Publishers, 1925.

Turnbull, Colin. *The Mountain People*. New York: Simon and Schuster, 1972.

Velikovsky, Immanuel. *Earth in Upheaval*, New York: Dell Books, 1969.

————. *Worlds in Collision*. New York: Dell Books, 1967.

Vickers, Geoffrey. *The Art of Judgment*. New York: Basic Books, 1965.

————. *Freedom in a Rocking Boat*. London: Allen Lane, 1970.

————. *Value Systems in Social Process*. New York: Basic Books, 1968.

Viereck, Peter. *Conservatism Revisited*. New York: Scribner's, 1950.

Volkov, G. *Era of Man or Robot?* Moscow: Progress Publishers, 1967.

Waddington, C. W. *The Ethical Animal*. New York: Atheneum, 1961.

————, ed. *Toward a Theoretical Biology. Vol. II*. Chicago: Aldine, 1968.

Wagar, W. Warren. *Building the City of Man*. New York: Grossman, 1971.

Walker, Kenneth. *The Extra-Sensory Mind*. New York: Harper's, 1972.

Walla, C. S., ed. *Toward Century 21*. New York: Basic Books, 1970.

Ward, Barbara, and Dubos, Rene. *Only One Earth*. New York: W. W. Norton, 1972.

Washburn, Sherwood L., ed. *Social Life of Early Man*. Chicago: Aldine, 1961.

Waskow, Arthur J. *Running Riot*. New York: Herder and Herder, 1970.

Waterston, Albert. *Development Planning*. Baltimore: Johns Hopkins University Press, 1965.

Watkins, Frederick W. *The Political Tradition of the West*. Cambridge: Harvard University Press, 1948.

Watt, Kenneth E. F. *Principles of Environmental Science*. New York: McGraw-Hill, 1973.

Watts, Alan. *The Book*. New York: Pantheon Books, 1966.

————. *Does It Matter?* New York: Vintage Books, 1971.

————. *Nature, Man, and Woman*. New York: Vintage Books, 1970.

————. *The Way of Zen*. New York: Pantheon, 1957.

Way, Louis. *Alfred Adler*. London: Penguin Books, 1956.

Weber, Max. *The Theory of Social and Economic Organization*. Glencoe, Ill.: The Free Press, 1947.

Weinstein, James. *The Corporate Ideal in the Liberal State*. Boston: Beacon Press, 1968.

Weinstein, Michael, ed. *The Political Experience*. St. Martin's Press, 1972.

Weisberg, Barry. *Beyond Repair*. Boston: Beacon Press, 1971.

Weiss, Paul. *Nature and Man*. New York: Henry Holt, 1947.

Weisskopf, Walter A. *Alienation and Economics*. New York: Dutton, 1971.

Weldon, Thomas D. *The Vocabulary of Politics*. London: Penguin Books, 1953.

Wellford, Harrison. *Sowing the Wind*. New York: Grossman, 1973.

Wengert, Norman. *Natural Resources and the Political Struggle*. New York: Doubleday, 1955.

Westin, Alan F., ed. *Information Technology in a Democracy*. Cambridge: Harvard University Press, 1971.

Wheeler, Harvey. *The Politics of Revolution*. Berkeley: Glendessary Press, 1971.

White, Lynn, Jr. *Machina ex Deo*. Cambridge: M.I.T. Press, 1968.

Whitehead, Alfred North. *The Concept of Nature*. Cambridge: Cambridge University Press, 1964.

————. *Nature and Life*. New York: Greenwood Publishers, 1968.

————. *Science and the Modern World*. New York: Mentor Books, 1948.

White House National Goals Research Staff. *Toward Balanced Growth: Quantity With Quality*. Washington: U.S. Government Printing Office, 1970.

Whyte, Lancelot Law. *Accent on Form*. New York: Harper and Bros., 1954.

Whyte, William H. *The Last Landscape*. New York: Anchor Books, 1968.

Wickler, Wolfgang. *The Biology of the Ten Commandments*. New York: McGraw-Hill, 1972.

————. *The Sexual Code*. New York: Doubleday, 1972.

Wildavsky, Aaron. *The Revolt Against the Masses*. New York: Basic Books, 1971.

Williams, William Appleton. *The Great Evasion*. Chicago: Quadrangle Books, 1968.

Wills, Garry. *Nixon Agonistes*. New York: Signet Books, 1971.

Wilson, Thomas W., Jr. *International Environmental Action*. New York: Dunellen, 1971.

Winter, Gibson. *Being Free*. New York: Macmillan, 1970.

————. *Elements For a Social Ethic*. New York: Macmillan, 1966.

Wolfe, Alan, and Surkin, Marvin, eds. *An End to Political Science*. New York: Basic Books, 1970.

Wolff, Robert Paul; Moore, Barrington, Jr.; and Marcuse, Herbert. *A Critique of Pure Tolerance*. Boston: Beacon Press, 1965.

————. *The Poverty of Liberalism*. Boston: Beacon Press, 1968.

Wolin, Sheldon S. *Politics and Vision*. Boston: Little, Brown, 1960.

Wolstenholme, Gordon, ed. *Man and His Future*. Boston: Little, Brown, 1963.

Woodcock, George. *Anarchism*. Cleveland: Meridian Books, 1961.

Wooldridge, Dean E. *Mechanical Man*. New York: McGraw-Hill, 1968.

Yankelovitch, Daniel, Inc. *The Changing Values on Campus*. New York: Washington Square Press, 1972.

Young, David P. *A New World in the Making*. Philadelphia: Westminster Press, 1972.

Zaehner, R. C. *Matter and Spirit: Their Convergence in Eastern Religions, Marx, and Teilhard de Chardin*. New York: Harper and Row, 1963.

Ziegler, Edward. *The Vested Interests*. New York: Macmillan, 1964.

Zijderfeld, Anton C. *The Abstract Society*. New York: Doubleday, 1970.

2.   Periodicals

Anderson, Terri. "World Priorities." *Environment* 14 (July–August 1972): 4.

Aberle, D. F.; Cohen, A. K.; Davis, A. K.: Levy, M. J., Jr.; and Sutton, F. X. "The Functional Prerequisites of a Society." *Ethics* 60 (1950): 110.

Adams, Gerald. "The Quiet Confrontation." *California Living* (14 January, 1973): 20.

Allee, W. C. "Where Angels Fear to Tread: A Contribution from General Sociology to Human Ethics." *Science* 97 (1943): 518.

Allen, Glen C. "The Is-Ought Question Reformulated and Answered." *Ethics* 82 (1972): 181.

Amara, Roy C. "Forecasting: From Conjectural Art Toward Science." *The Futurist* 6 (1972): 112.

Anderson, Paul. "More Futures Than One." *Playboy* 18 (October 1971): 97.

Atkeson, Timothy et al. "Post Stockholm: Influencing National Environmental Law and Practice Through International Law and Policy: Remarks." *American Journal of International Law* 66 (1972): 1.

Auden, W. H. "The Real World." *New Republic* 157 (9 December 1967): 25.

Aultman, Mark. "Technology and the End of Law." *American Journal of Jurisprudence* 17 (1972): 46.

Ayala, Francisco J. "Teleological Explanations in Evolutionary Biology." *Philosophy of Science* 37 (1970): 1.

Ayres, Robert D., and Kneese, A. V. "Production, Consumption, and Externalities." *American Economic Review* 59 (1969): 282.

Baran, Michael S. "Social Control of Science and Technology." *Science* 172 (1971): 535.

Barkley, Katherine, and Weissman, Steve. "The Eco-Establishment." *Ramparts* 8 (May 1970): 48.

Barnette, Henlee C. "Toward an Ecological Ethics." *The Review and Expositor* 69 (1972): 23.

Baskin, Darryl. "American Pluralism: Theory, Practice, Ideology." *Journal of Politics* 32 (1970): 71.

Bay, Christian. "Needs, Wants, and Political Legitimacy." *Canadian Journal of Political Science* 1 (1968): 241.

Beer, Stafford. "The Liberty Machine." *Futures* 3 (1972): 338.

Bell, Daniel. "The Commission on the Year 2000." *Futures* 3 (1972): 338.

———. "The Corporation and Society in the 1970's." *The Public Interest* no. 24 (Summer 1971): 5.

———. "Technocracy and Politics." *Survey* 16 (1971): 1.

———. "Technology, Nature, and Society." *American Scholar* 42 (1973): 385.

———. "Twelve Modes of Prediction . . . A Preliminary Sorting of Approaches in the Social Sciences." *Daedalus* 93 (Summer 1969): 845.

Bendix, Reinhard. "Tradition and Modernity Reconsidered." *Comparative Studies in Society and History* 9 (1967): 292.

Benn, Anthony Wedgwood; Tribus, Myron; Drucker, Daniel C.; and Mencher, Alan C. "On the Control of Science: Four Views." *Bulletin of the Atomic Scientists* 27 (December 1971): 23.

Berger, Henry W. "Organized Labor and Imperial Policy." *Society* 10 (November–December 1972): 94.

Berger, Peter L., and Berger, Betty. "The Blueing of America." *New Republic* 164 (1971): 20.

Berle, Adolf A. "What GNP Doesn't Tell Us." *Saturday Review* 51 (31 August 1968): 10.

Berry, R. Stephen. "Recycling, Thermodynamics, and Environmental Thrift." *Bulletin of the Atomic Scientists* 28 (May 1972): 8.

"The Biosphere." Special issue of *Scientific American.* 223 (September 1970).

Birch, Charles. "Purpose in the Universe: A Search for Wholeness." *Zygon* 6 (1971): 4.

Blackburn, Thomas R. "Sensuous-Intellectual Complementarity in Science." *Science* 172 (1971): 1003.

Bookchin, Murray. "Toward an Ecological Solution." *Ramparts* 8 (May 1970): 7.

Borgstrom, Georg A. "The World Food Crisis." *Futures* I (1970): 339.

Boulding, Kenneth. "Ecology and Environment." *Trans-Action* 7 (March 1970): 38.

———. "Is Scarcity Dead?" *The Public Interest,* no. 5 (Fall 1966): 36.

———. "Philosophy, Behavioral Science, and the Nature of Man." *World Politics* 12 (1960): 272.

Bourricaud, François. "The Paradoxes of Welfare." *Survey* 16 (1971): 43.

Boyd, William C. et al. "Comments on the Tugwell Constitution." *The Center Magazine* 4 (March–April 1971): 45.

Bozell, L. Brent. "Sin of Head Counts." *New York Times*, 13 October, 1971.

Brady, E. L., and Branscomb, L. M. "Information for a Changing Society." *Science* 175 (1972): 961.

Brandy, Karl. "Moral Presuppositions of a Free Enterprise Economy." *The Intercollegiate Review* 11 (1965): 111.

Branscomb, Lewis M. "Taming Technology." *Science* 171 (1971): 972.

Broecker, Wallace. "Enough Air." *Environment* 12 (September 1970): 26.

Brooks, Harvey, and Bowers, Raymond. "The Assessment of Technology." *Scientific American* 222 (February 1970): 13.

Brooks, John. "Mr. White and Mr. Blue: Notes on the New Middle Class." *Harper's* 232 (June 1966): 88.

Buchanan, Scott. "So Reason Can Rule: The Constitution Revisited." *Center Occasional Papers* 1 (February 1967): 1.

Burhoe, Ralph Wendall. "Five Steps in the Evolution of Man's Knowledge of Good and Evil." *Zygon* 2 (1967): 77.

Burnham, Walter Dean. "Crisis of American Political Legitimacy," *Society* 10 (November–December 1972), 24.

Burstein, Samuel M. "Science, Abraham, and Ecology." *Intellectual Digest* 3 (October 1972): 88.

Caldwell, Lynton C. "Biopolitics: Science, Ethics, and Public Policy." *Yale Review* 54 (October 1964): 1.

———. "Environment: A New Focus for Public Policy." *Public Administration Review* 23 (1963): 152.

Calhoun, John B. "Plight of the Ik and Kaiadilt Is Seen as a Chilling Possible End for Man." *Smithsonian* 3 (November 1972): 27.

———. "Population Density and Social Pathology." *Scientific American* 206 (1962): 139.

Callahan, Daniel. "Ethics and Population Limitation." *Science* 175 (1972): 487.

———. "What Obligations Do We Have to Future Generations?" *American Ecclesiastical Review* 164 (1971): 265

Cannon, Walter B. "The Body Physiologic and the Body Politic." *Science* 93 (1941): 1.

Caplow Theodore. "Are the Poor Countries Getting Poorer?" *Foreign Policy*, no. 3 (Summer 1971): 90.

Carroll, James D. "Participatory Technology." *Science* 171 (1971): 647.

Cassinelli, C. W. "The Concept of the Public Interest." *Ethics* 69 (1959): 48.

———. "The National Community." *Polity* 2 (1969): 14.

Cauthen, Kenneth. "Process and Purpose: Toward a Philosophy of Life." *Zygon* 3 (1968): 183.

Cellarius, Richard A., and Platt, John. "Councils of Urgent Studies." *Science* 177 (1972): 670.

Champlin, John R. "Behavior (-ism & -alism) & Theory (Political): A Prolegomenon to Any Future Review of Skinner." *Polity* 5 (1972): 243.

Chomsky, Noam. "The Case Against B. F. Skinner." *New York Review of Books* 17 (30 December 1971): 18.

Clark, Kenneth B. "Leadership and Psychotechnology." *New York Times*, 9 November 1971.

Clarke, Arthur C., and Watts, Alan. "Technology and Mysticism: A Dialog," *Playboy* 19 (January 1972): 94.

Coale, Ansley J. "Man and His Environment." *Science* 170 (1970): 132.

Coates, Joseph F. "Technology Assessment: The Benefits . . . The Costs . . . The Consequences." *The Futurist* 5 (1971): 225.

Cochran, Clarke E. "The Politics of Interest: Metaphysics and the Limitations of the Science of Politics." Unpublished paper, Annual Meeting of American Political Science Association, Chicago, 1971.

Comfort, Alexander. "Sexuality in a Zero Growth Society." *Center Report* 5 (December 1972): 12.

Conrad, Thomas. "Rationality and Political Science." *Polity* 2 (1970): 479.

Collier, Bernard Law. "Brain Power: The Case for Bio-Feedback Training." *Saturday Review* 54 (10 April 1971): 10.

Colton, Timothy. "The 'New Biology' and the Causes of War." *Canadian Journal of Political Science* 2 (1969): 434.

Commoner, Barry; Corr, Michael; and Stamler, Paul J. "The Causes of Pollution." *Environment* 13 (April 1970): 2.

"The Confrontation of the Two Americas." *Time* (2 October 1972): 15.

Cook, Earl. "Energy Sources for the Future." *The Futurist* 6 (1972): 142.

Corning, Peter A. "The Biological Bases of Behavior and Some Implications For Political Science." *World Politics* 23 (1971): 321.

Corr, Michael, and Macleod, Dan. "Getting It Together." *Environment* 14 (November 1972): 2.

Count, Earl W. "The Biological Basis of Human Sociality." *American Anthropologist* 60 (1958): 1049.

Cowhey, Peter F.; Hart, Jeffrey A.; and Schmidt, Janet K. "The Theory of Collective Goods and the Future Regime of Ocean Space." Unpublished paper, International Studies Association meeting, New York City, 1973.

Cowan, Ian McTaggert. "Conservation and Man's Environment." *Nature* 208 (1965): 1145.

Cowen, Robert C. "Life On Earth: The Beginning of the Beginning." *Technology Review* 74 (March–April 1972): 5.

Crenson, Matthew A. "Pollution Isn't THAT Serious." *New York Times*, 21 April 1971.

Cropsey, Joseph. "A Reply to Rothman." *American Political Science Review* 56 (1962): 353.

Crowe, Beryl F. "The Tragedy of the Commons Revisited." *Science* 166 (1969): 1103.

Daly, Herman E. "On Economics as a Life Science." *Journal of Political Economy* 76 (1968): 392.

Davies, James C. "Biology, Darwinism, and Political Science: Some Old and New Frontiers." Unpublished paper, Annual Meeting of American Political Science Association, Chicago, 1971.

———. "Toward a Theory of Revolution." *American Sociological Review* 27 (1962): 5.

Deevey, Edward S., Jr. "The Hare and the Haruspex: a Cautionary Tale." *American Scientist* 48 (1960): 415.

De Towarnicki, Frederic. "Rats, Apes, Naked Apes, Kipling, Instincts, Guilt, the Generations and Instant Copulation—A Talk With Konrad Lorenz." The *New York Times Magazine* (8 July 1970): 4.

Dobzhansky, Theodosius. "An Essay on Religion, Death, and Evolutionary Adaptation." *Zygon* 1 (1966): 317.

Downs, Anthony. "The Public Interest: Its Meaning in a Democracy." *Social Research* 29 (1962): 1.

———. "Up and Down With Ecology—The 'Issue-Attention Cycle.'" *The Public Interest*, no. 28 (Summer 1972): 38.

Drucker, Peter F. "A Key to American Politics: Calhoun's Pluralism." *The Review of Politics* 10 (1948): 412.
———. "Saving the Crusade." *Harper's* 244 (January 1972): 66.
Dubos, Rene J. "Humanizing the Earth." *Science* 179 (1973): 769.
———. "The Human Landscape." *Bulletin of the Atomic Scientists* 26 (1970): 31.
———. "Man Overadapting to the Environment." *Psychology Today* 4 (February 1971): 50.
———. "Man's Creative Touch Often Improves the Land Even Though He Is Beset by Ecological Devils of His Own Making." *Smithsonian* 3 (December 1972): 18.
———. "St. Francis vs. St. Benedict." *Psychology Today* 6 (May 1973): 54.
Duncan, Otis Dudley. "Social Forecasting: The State of the Art." *The Public Interest*, no. 17 (Fall 1969): 88.
Dunn, Halbert L. "Evolution of the Mind and Human Potential." *Man's Emergent Evolution* 3, no. 1 (1970): 53.
Durbin, Paul T. "Technology and Values: A Philosopher's Perspective." *Technology and Culture* 13 (1972): 556.
Dwyer, Archbishop Robert J. "Ecology: Our New Religion." *Twin Circle* (24 August 1973).
Ehricke, Krafft A. "Extraterrestrial Imperative." *Bulletin of the Atomic Scientists* 27 (November 1971): 18.
Ehrlich, Paul R., and Holdren, John P. "Impact of Population Growth." *Science* 171 (1971): 1212.
Eisenberg, L. "The *Human* Nature of Human Nature." *Science* 176 (1972): 123.
Emerson, Alfred E. "Dynamic Homeostasis: A Unifying Principle in Organic, Social, and Ethical Evolution." *Zygon* 3 (1968): 129.
Engel, David E. "Elements in a Theology of Environment." *Zygon* 5 (1970): 216.
"The Environment: A National Mission for the Seventies." Special issue of *Fortune*, 81 (February 1970).
Etzioni, Amitai. "Agency for Technological Development for Domestic Problems." *Science* 164 (1969): 43.
———. "Human Beings Are Not Very Easy to Change After All." *Saturday Review* 55 (3 June 1972): 45.
———. "Sex Control, Science, and Society." *Science* 161 (1968): 1107.
———. "Shortcuts to Social Change." *The Public Interest*, no. 12 (Summer 1968): 40.
Eulau, Heinz. "Skill Revolution and Consultative Commonwealth." *American Political Science Review* 67 (1973): 169.
Evans, Wayne O. "Mind-Altering Drugs and the Future." *The Futurist* 5 (1971): 101.
Farvar, M. Taghi; Thomas, Margaret L.; Boksenbaum, Howard; and Soule, Theodore N. "The Pollution of Asia." *Environment* 13 (October 1971): 10.
Fay, Charles. "Ethical Naturalism and Biocultural Evolution." *Zygon* 4 (1969): 24.
Feagin, Joe R. "Poverty: We Still Believe That God Helps Those Who Help Themselves." *Psychology Today* 6 (November 1972): 101.
Federov, E. K. "Interactions of Man and His Environment." *Bulletin of the Atomic Scientists* 28 (February 1972): 5.
Feinberg, Gerald. "Long Range Goals and the Environment." *The Futurist* 5 (1971): 241.
Ferkiss, Barbara, and Ferkiss, Victor. "Race and Politics in Trinidad and Guyana." *World Affairs* 134 (1971): 5.

Ferkiss, Victor C. "American Youth and the Process of Social Change." *Proceedings* of the Catholic Theological Society 26 (1971): 331.

——. "Christian Political Philosophy: Its Nature and Limitations." *Continuum* 5 (1966): 418.

——. "Man's Tools and Man's Choices: The Confrontation Between Political Science and Technology." *American Political Science Review* 67 (1973): 973.

——. "Technology and the Future of Man." *The Review and Expositor* 69 (1972): 49.

——; Lee, Alfred McClung; and Wooldridge, Dean E. "Mechanical Man: Review Symposium." *Philosophy Forum* 12 (1972): 342.

Ferry, W. H. "Must the Constitution Be Amended to Control Technology?" *Saturday Review* 51 (2 March 1968): 50.

Findley, Tim. "Tom Hayden: Rolling Stone Interview Part I." *Rolling Stone* (26 October 1972): 36. Part II, ibid. (9 November 1972): 28.

Fisher, Anthony C.; Krutilla, John V.; and Cicchetti, Charles J. "The Economics of Environmental Preservation: A Theoretical and Empirical Analysis." *American Economic Review* 62 (1972): 605.

Fisher, John. "Unwritten Rules of American Politics." *Harper's* 197 (November 1948): 27.

Flacks, Richard. "Strategies For Radical Social Change." *Social Policy* 1 (March–April 1971): 7.

——. "Young Intelligentsia in Revolt. "*Trans-Action* 7 (June 1970): 46.

Ford, Amasa B. "Casualties of Our Times." *Science* 167 (1970): 256.

Fox, Elliot M. "Mary Parker Follett: The Enduring Contribution." *Public Administration Review* 28 (1968): 520.

Frank, Lawrence K. "Man's Changing Image of Himself." *Zygon* 1 (1966): 11.

——. "The Need For a New Political Theory." *Daedalus* 96 (Summer 1967): 177.

Frank, Lewis C., Jr. "People vs. Population." *The Center Magazine* 1 (January 1968): 18.

Franks, C. E. S. "Further Thoughts on 'The New Biology' and the Causes of War." *Canadian Journal of Political Science* 3 (1970): 318.

Freedman, Jonathan L. "The Crowd—Maybe Not So Madding After All." *Psychology Today* 5 (September 1971): 58.

Friedmann, John. "Planning, Progress, and Social Values." *Diogenes* (Spring 1957): 98.

Friedrich, Carl J. "The Deification of the State." *The Review of Politics* 1 (1939): 18.

——. "Greek Political Heritage and Totalitarianism." *The Review of Politics* 2 (1940): 218.

Fromm, Erich. "The Erich Fromm Theory of Aggression." The *New York Times Magazine* (27 February 1972): 14.

Fuller, Buckminster. "Planetary Planning." *American Scholar* 40 (Winter 1970–1971): 29.

Fyodorov, Academician Yevgeny. "Against the Limits to Growth." *New Scientist* 57 (1973): 431.

Galle, Omar R., and McPherson, John. "Population Density and Pathology: What Are the Relations for Man?" *Science* 176 (1972): 23.

Gastil, Raymond D. "Social Indicators and the Quality of Life." *Public Administration Review* 30 (1970): 596.

Gaylin, Willard. "The Frankenstein Myth Becomes a Reality—We Have the

Awful Knowledge To Make Exact Copies of Human Beings." The *New York Times Magazine* (5 March 1972): 12.

Gellen, Murray. "The Making of a Pollution-Industrial Complex." *Ramparts* 8 (May 1970): 22.

"Genetic Science and Man." Special issue of *Theological Studies* 33 (September 1972).

Gerard, R. W. "A Biological Basis for Ethics." *Philosophy of Science* 9 (1942): 92.

Gilula, Marshall F., and Daniels, David N. "Violence and Man's Struggle to Adapt." *Science* 164 (1969): 396.

Glass, Bentley. "The Ethical Basis of Science." *Science* 150 (1965): 1254.

———. "Science: Endless Horizons or Golden Age?" *Science* 171 (1971): 23.

"God, Man, and Apollo." *Time* (1 January 1973).

Gofman, John W., and Tamplin, Arthur R. "Radiation: The Invisible Casualties." *Environment* 12 (April 1970): 12.

Golding, Martin P. "Obligations to Future Generations." *The Monist* 56 (1972): 85.

———, and Golding, Naomi H. "Ethical and Value Issues in Population Limitation and Distribution in the United States." *Vanderbilt Law Review* 24 (1971): 495.

Goldman, Marshall I. "The Convergence of Environmental Disruption." *Science* 170 (1970): 37.

———. "Externalities and the Race for Economic Growth in the U.S.S.R.: Will the Environment Ever Win?" *Journal of Political Economy* 80 (1972): 314.

Gomez-Pompa, A.; Vasquez-Yanes, C.; and Guevara, S. "The Tropical Rain Forest: A Nonrenewable Resource." *Science* 177 (1972): 762.

Goodman, Daniel. "Ideology and Ecological Irrationality." *Bioscience* 20 (1970): 1247.

Greeley, Andrew. "The End of American Catholicism?" *America* 127 (1972): 334.

Greene, Wade. "The Militant Malthusians." *Saturday Review* 55 (11 March 1972): 40.

"The Greening of the Astronauts." *Time* (11 December 1972).

Gregory, Dick. "My Answer to Genocide." *Ebony* 26 (October 1971): 66.

Gross, Bertram. "Friendly Fascism: A Model for America." *Social Policy* 1 (December 1970): 44.

———. "Planning in an Era of Social Revolution." *Public Administration Review* 31 (1971): 254.

Grossman, Edward. "The Obsolescent Mother: A Scenario." *The Atlantic* 227 (May 1971): 39.

Guttmacher, Alan F. "Population and Pollution." *The Review and Expositor* 69 (1972): 55.

Hacker, Andrew. "Crime in the Streets." *New York Review of Books* 20 (19 April 1973): 9.

———. "On Original Sin and Conservatives." The *New York Times Magazine* (25 February 1973): 65.

Hale, Myron Q. "The Cosmology of Arthur F. Bentley." *American Political Science Review* 54 (1960): 955.

Hall, Elisabeth. "Will Success Spoil B. F. Skinner?" *Psychology Today* 6 (November 1972): 65.

Hallow, Ralph Z. "The Blacks Cry Genocide." *The Nation* 208 (1969): 635.

Hardin, Garrett. "Population Skeletons in the Environmental Closet." *Bulletin of the Atomic Scientists* 28 (June 1972): 37.

——. "To Trouble a Star: The Cost of Intervention in Nature." *Bulletin of the Atomic Scientists* 26 (January 1970): 17.

——. "The Tragedy of the Commons." *Science* 162 (1968): 1243.

——, and Berry, R. Stephen. "Limits to Growth—Two Views." *Bulletin of the Atomic Scientists* 28 (November 1972): 23.

Hare, Nathan. "Black Ecology." *The Black Scholar* 1 (April 1970): 2.

Harman, Willis W. "Alternative Futures and Habitability." *Man's Emergent Evolution* 3, no. 1 (1970): 19.

Hart, Henry C. "The Natural Environment in Indian Tradition." *Public Policy* 12 (1969): 41.

Harwood, Michael. "We Are Killing the Sea around Us." The *New York Times Magazine* (24 October 1971): 34.

Haberlein, Thomas A. "The Land Ethic Realized: Some Social Psychological Explanations for Changing Environmental Attitudes." *Journal of Social Issues* 28 (1972): 79.

Hefner, Philip W. "Toward a New Doctrine of Man: The Relationship of Man and Nature." *Zygon* 2 (1967): 127.

Heilbroner, Robert L. "Growth and Survival." *Foreign Affairs* 51 (1972): 139.

——. "On the Limited 'Relevance' of Economics." *The Public Interest*, no. 21 (Fall 1970): 80.

——. "Phase II of the Capitalist System." The *New York Times Magazine* (28 November 1971).

Hirst, Eric. "Energy vs Environment: the Coming Struggle." *The Living Wilderness* 36 (Winter 1972–1973): 43.

Hoagland, Hudson. "Biological Aspects of Aggression and Violence." *Zygon* 4 (1969): 206.

——. "The Brain and the Crisis in Human Values." *Zygon* 1 (1966): 140.

——. "Ethology and Ethics—The Biology of Right and Wrong." *Zygon* 2 (1967): 43.

——, and Burhoe, Ralph W., eds. "Evolution and Man's Progress." *Daedalus* 90 (Summer 1961): 411.

Hockett, Charles F., and Asher, Robert. "The Human Revolution." *Current Anthropology* 5 (1964): 135.

Hoelsher, Harold E. "Technology and Social Change." *Science* 166 (1969): 68.

Holden, Constance. "Altered States of Consciousness: Mind Researchers Meet to Discuss Exploration and Mapping of 'Inner Space.'" *Science* 179 (1973): 982.

——. "Environmental Action Organizations Are Suffering from Money Shortages, Slump in Public Commitment." *Science* 175 (1972): 394.

Holdren, John P., and Ehrlich, Paul R. "One-Dimensional Ecology Revisited: A Rejoinder." *Bulletin of the Atomic Scientists* 28 (June 1972): 42.

Horowitz, Irving Louis. "The Environmental Cleavage: Social Ecology vs. Political Economy." *Social Theory and Practice* 2 (1972): 125.

Hubbard, Barbara. "From Meaninglessness to New Worlds." *The Futurist* 5 (1971): 72.

Hubbard, Earl. "The Need for New Worlds." *The Futurist* 3 (1969): 117.

Huddle, Franklin P. "The Social Management of Technological Consequences." *The Futurist* 6 (1972): 16.

Humpstone, Charles Cheney. "Pollution: Precedent and Prospect." *Foreign Affairs* 50 (1972): 325.

Iltis, Hugh H.; Loucks, Orie L.; and Andrews, Peter. "Criteria for an Optimum Human Environment." *Bulletin of the Atomic Scientists* 26 (January 1970): 2.

Inglehart, Robert. "The Silent Revolution in Europe: Intergenerational Change in Post-Industrial Societies." *American Political Science Review* 65 (1971): 991.

Inglis, David Rittenhouse. "Nuclear Energy and the Malthusian Dilemma." *Bulletin of the Atomic Scientists* 27 (February 1971): 14.

Jacobsoen, Sally. "Gunnar Myrdal Comments On: America's Image, Black Rebellion, Limits to Growth and Population Control." *Bulletin of the Atomic Scientists* 28 (November 1972): 5.

Jones, Martin V. "The Methodology of Technology Assessment." *The Futurist* 6 (1972): 19.

Kahn, Herman. "The Squaring of America." *Intellectual Digest* 3 (September 1972): 16.

Kass, Leon B. "Making Babies—The New Biology and the Old Morality." *The Public Interest*, no. 26 (Winter 1972): 18.

———. "The New Biology: What Price Relieving Man's Estate?" *Science* 174 (1971): 779.

Kaysen, Carl. "The Computer That Printed Out W.O.L.F." *Foreign Affairs* 50 (1972): 660.

Keen, Sam, and Raser, John. "Conversation with Herbert Marcuse." *Psychology Today* 4 (February 1971): 35.

Kelley, Ken. "Blissed Out With the Perfect Master." *Ramparts* 12 (July 1973): 32.

Kendall, Willmoore, and Carey, George. "The Intensity Problem and Democratic Theory." *American Political Science Review* 62 (1968): 5.

Kennan, George W. "To Prevent a World Wasteland: A Proposal." *Foreign Affairs* 48 (1970): 401.

Keyfritz, Nathan. "Population Density and the Style of Social Life." *Bioscience* 16 (1966): 868.

Kiefer, David M. "Assessing Technology Assessment." *The Futurist* 5 (1971): 234.

Kirkpatrick, Evron M. "Toward a More Responsible Two-Party System: Political Science, Policy Science, or Pseudo-Science." *American Political Science Review* 65 (1971): 965.

Kluckhohn, Clyde. "Ethical Relativity: Sic et Non." *Journal of Philosophy* 52 (1955): 663.

Kohn, Melvin C. "Social Class and Parental Values." *American Journal of Sociology* 64 (1959): 337.

Kopkind, Andrew. "Mystic Politics: Refugees from the New Left." *Ramparts* 12 (July 1973): 26.

Kress, Paul F. "The Web and the Tree: Metaphors of Reason and Value." *Midwest Journal of Political Science*. 13 (1969): 389.

Krieger, Martin H. "What's Wrong with Plastic Trees?" *Science* 179 (1973): 446.

Krippner, Stanley, and Davidson, Richard. "Parapsychology in the U.S.S.R." *Saturday Review* 55 (18 March 1972): 56.

Kristol, Irving. "About Equality." *Commentary* 54 (November 1972): 41.

———. "A Foolish American Ism—Utopianism," The *New York Times Magazine* (17 November 1971): 31.

Krutch, Joseph Wood. "Conservation is not Enough." *American Scholar* 23 (1954): 295.

Krutilla, John V. "Conservation Reconsidered." *American Economic Review* 57 (1967): 777.

Landsberg, Helmut E. "Man-Made Climatic Changes." *Science* 170 (1970): 1265.

Lane, Robert E. "The Fear of Equality." *American Political Science Review* 53 (1959): 35.

Lasch, Christopher. "Birth, Death, and Technology: The Limits of Cultural Laissez-Faire." *Hastings Center Report*, 2 (June 1972): 1.

Lasswell, Harold D. "The Political Science of Science." *American Political Science Review* 50 (1956): 961.

Laszlo, Erwin. "Reverence for Natural Systems." *Man's Emergent Evolution* 3, no. 1 (1970): 34.

Latham, Earl. "The Group Basis of Politics: Notes for a Theory." *American Political Science Review* 46 (1952): 376.

Leake, C. D. "Ethicogenesis." *Scientific Monthly* 60 (1945): 245.

Leakey, Louis S. B., and Ardrey, Robert. "Man, The Killer: A Dialogue." *Psychology Today* 6 (September 1972): 73.

Leavenworth, May. "On Integrating Fact and Value." *Zygon* 4 (1969): 33.

Lederberg, Joshua. "Government Is Most Dangerous of Genetic Engineers." Washington *Post*, 19 July, 1970.

Lekachman, Robert. "Humanizing GNP." *Social Policy* 2 (September–October 1971): 34.

Levine, Amory S. "The Case Against the Fast Breeder Reactor: An Anti-Nuclear Establishment View." *Bulletin of the Atomic Scientists* 29 (March 1973): 29.

Levine, George. "Paul Goodman, Outsider Looking In." *New York Times Book Review* (18 February 1973): 4.

Lewis, Richard S. "End of Apollo: The Ambiguous Epic." *Bulletin of the Atomic Scientists* 28 (December 1972): 39.

Lindblom, Charles A. "Muddling Through . . . 'Science' or Inertia?" *Public Administration Review* 24 (1964): 153.

———. "The Science of Muddling Through." *Public Administration Review* 19 (1959): 79.

Lorenz, Konrad. "On Killing Members of One's Own Species." *Bulletin of the Atomic Scientists* 26 (October 1970): 2.

Lowenthal, Richard. "What Prospects for Socialism?" *Encounter* 40 (February 1973): 8.

Lowi, Theodore. "The Public Philosophy: Interest-Group Liberalism." *American Political Science Review* 61 (1967): 5.

Lowry, Ritchie P. "The New Religecology: Salvation or Sophorific?" *Social Policy* 1 (July–August 1970): 46.

Mackenzie, W. M. "Pressure Groups: The 'Conceptual Framework.'" *Political Studies* 3 (1955): 247.

MacLeod, Scott. "One Earth and the Third World: A Stockholm Review." *International Development Review* 14, no. 2 (1972): 3.

Makhijani, A. B., and Lichtenberg, A. J. "Energy and Well-Being." *Environment* 14 (June 1972): 10.

Mandel, William. "The Soviet Ecology Movement." *Science and Society* 36 (1972): 385.

Mansfield, Harvey C. Jr. "Disguised Liberalism." *Public Policy* 18 (1970): 605.

Martin, Paul S. "The Discovery of America." *Science* 179 (1973): 969.

Marx, Leo. "American Institutions and Ecological Ideals." *Science* 170 (1970): 945.

Maslow, Abraham H. "Toward a Humanistic Biology." *Man's Emergent Evolution* 3, no. 1 (1970): 4.

McCaull, Julian. "Building a Shorter Life." *Environment* 13 (September 1971): 2.

———. "The Politics of Technology." *Environment* 14 (March 1972): 2.

McDermott, John D. "Technology: The Opiate of the Intellectuals." *New York Review of Books* 13 (31 July 1969): 25.

Mesthene, Emmanuel. "Technological Change and Religious Unification." *Cross Currents* 22 (1972): 239.

Metz, William D. "The Decline of the Hubble Constant: A New Age For the Universe." *Science* 178 (1972): 600.

Meynell, Hugh. "Ethology and Ethics." *Philosophy* 45 (1970): 290.

Michael, Donald N. "On Coping With Complexity: Planning and Politics." *Daedalus* 97 (Fall 1968): 1179.

Michener, James. "The Revolution in Middle Class Values." The *New York Times Magazine* (18 August 1968): 20.

Miles, Rufus E. "Three Ways to Solve the Population Crisis." *The Futurist* 5 (1971): 200.

Miller, Abraham H. "People's Park: Dimensions of a Campus Confrontation." *Politics and Society* 2 (1972): 433.

Mishan, E. J. "Making the Future Safe for Mankind." *The Public Interest*, no. 24 (Spring 1971): 33.

———. "To Grow or Not to Grow? A Debate and a Dilemma." *Encounter* 40 (May 1973): 9.

Moles, Abraham. "The Future Oriented Society: Axioms and Methodology." *Futures* 2 (1970): 312.

Moncrief, Lewis W. "The Cultural Basis of Our Environmental Crisis." *Science* 170 (1970): 508.

Moore, Wilbert E. "But Some Are More Equal Than Others." *American Sociological Review* 28 (1963): 508.

Morgan, Neil. "Running Out of Space." *Harper's* 247 (September 1973): 59.

Morison, Robert S. "Darwinism: Foundation for an Ethical System?" *Zygon* 1 (1966): 347.

Morris, Robin. "Is the Corporate Economy a Corporate State?" *American Economic Review* 62 (1972): 103.

Mosher, Charles A. "Needs and Trends in Congressional Decision-Making." *Science* 178 (1972): 134.

Murray, Bertram G., Jr. "What the Ecologists Can Teach the Economists." The *New York Times Magazine* (10 December 1972): 38.

Mutiso, G. C. M. "Tools Are For People: Towards An Africanized Technology." *The Ecumenical Review* 24 (1972): 31.

Nahel, Thomas. "Reason and National Goals." *Science* 177 (1972): 766.

Nelson, Senator Gaylord A. "Politics and Ecology: Perils and Possibilities." *The Review and Expositor* 69 (1972): 37.

North, Robert C., and Choucri, Nazli. "Population, Technology, and Resources in the Future International System." *Journal of International Affairs* 25 (1971): 244.

O'Brien, Larry. "The 'Ecology' of the Slums." *New York Times*, 21 August 1971.

"The Ocean." Special issue of *Scientific American* 221 (September 1961).

Odegard, Peter H. "The Group Basis of Politics: A New Name for an Ancient Myth." *Western Political Quarterly* 11 (1958): 689.

Odum, Eugene P. "The Strategy of Ecosystem Development." *Science* 164 (1969): 262.

Oglesby, Carl. "The Crisis Is Political." *The Center Magazine* 4 (March–April 1971): 65.

Ophuls, William. "The Return of Leviathan." *Bulletin of the Atomic Scientists* 29 (March 1973): 50.

Orleans, Leo A., and Suttmeier, Richard P. "The Mao Ethic and Environmental Quality." *Science* 170 (1970): 1173.

Paradise, Scott. "The Vandal Ideology." *The Nation* 209 (1969): 729.

Park, James L. "Complementarity Without Paradox." *Zygon* 2 (1967): 382.

Parker, Edwin B., and Dunne, Donald A. "Information Technology: Its Social Potential." *Science* 176 (1972): 1392.

Partridge, P. H. "Philosophy, Politics, Ideology." *Political Studies* 9 (1961): 217.

Pearl, Arthur. "An Ecological Rationale for a Human Services Society." *Social Policy* 2 (September–October 1971): 40.

————, and Pearl, Stephanie. "Strategies for Radical Social Change: Toward an Ecological Theory of Value." *Social Policy* 2 (May–June 1971): 30.

Perl, Martin L. "The Scientific Advisory System: Some Observations." *Science* 173 (1971): 1211.

Petersen, James. "Lessons From the Indian Soul." *Psychology Today* 6 (May 1973): 63.

Petersen, William. "Population Explosion—A Conservative Reacts." *National Review* 24 (1972): 520.

Peterson, Eugene K. "The Atmosphere: A Clouded Horizon." *Environment* 12 (April 1970): 32.

Pilbeam, David. "The Fashionable View of Man as a Naked Ape Is: 1. An Insult to Apes 2. Simplistic 3. Male-Oriented 4. Rubbish." The *New York Times Magazine* (3 September 1972): 10.

Pitts, Jesse. "The Counter Culture: Tranquilizer or Revolutionary Ideology?" *Dissent* 18 (1971): 216.

Plamenetz, John. "Interests." *Political Studies* 2 (1954): 1.

Platt, John. "Can Determinism and Freedom Be Reconciled?" Unpublished paper, Teilhard Center for the Future of Man, London, 1972.

————. "Hierarchical Growth." *Bulletin of the Atomic Scientists* 26 (November 1970): 2.

————. "How Men Can Shape Their Future." *Futures* 3 (1972): 32.

————. "Lock-ins and Multiple Lock-ins in Collective Behavior." *American Scientist* 57 (1969) 961.

————. "Properties of Large Molecules That Go Beyond the Properties of Their Chemical Sub-groups." *Journal of Theoretical Biology* 1 (1961): 342.

————. "Social Traps." *American Psychologist* 28 (1973): 641.

————. "What We Must Do." *Science* 169 (1969): 1115.

————; Black, Max; Toynbee, Arnold; and Skinner, B. F. "Beyond Freedom and Dignity." *The Center Magazine* 5 (March–April 1972): 33.

"Playboy Interview: Dr. Paul Ehrlich." *Playboy* 17 (August 1970): 55.

"Playboy Interview: R. Buckminster Fuller." *Playboy* 19 (February 1972): 59.

Podhoretz, Norman. "Laureate of the New Class." *Commentary* 54 (December 1972): 4.

Polanyi, Michael. "The Republic of Science: Its Political and Social Theory." *Minerva* 1 (1962): 54.

Popper, Karl. "Indeterminism Is Not Enough: A Philosophical Essay." *Encounter* 40 (April 1973): 20.

"Population: The U. S. . . . A Problem. The World . . . A Crisis." Supplement to the *New York Times*, 30 April 1972.

Potter, Van Rensselaer. "Biocybernetics and Survival." *Zygon* 5 (1970): 229.

————. "Disorder As a Built-in Component of Biological Systems: The Survival Imperative." *Zygon* 6 (1971): 135.

Pound, Roscoe. "A Survey of Social Interests." *Harvard Law Review* 57 (1943): 1.

Pryde, Philip R. "Soviet Pesticides." *Environment* 13 (November 1971): 16.
——. "Victors Are Not Judged." *Environment* 12 (November 1970): 30.
Putnam, Robert D. "Studying Elite Political Culture: 'The Case of Ideology.' "
    American Political Science Review 65 (1971): 651.
Rabinowitch, Eugene et al. "Can Man Control His Biological Evolution? A
    Symposium on Genetic Engineering." *Bulletin of the Atomic Scientists* 27
    (December 1972): 12.
Reader, Mark. "Watching the Revolution Uncork." *Politics and Society* 2 (1972):
    337.
Ricci, David M. "Democracy Attenuated: Schumpeter, the Process Theory, and
    American Political Thought." *Journal of Politics* 32 (1970): 239.
Ridker, Ronald G. "Population and Pollution in the United States." *Science* 176
    (1972): 1085.
Riga, Peter J. "Overpopulation and the American Catholic Conscience." *World
    Justice* 12 (1970): 199.
Roback, Herbert. "Do We Need a Department of Science and Technology?"
    Science 165 (1969): 36.
Roberts, Marc J. "Is There an Energy Crisis? *The Public Interest*, no. 31 (Spring
    1973): 17.
Rock, William Pennell. "Alienation, Yes; Patriotism: Yes." *The Center Magazine*
    4 (November–December 1971): 2.
Rodman, John. "The Ecological Perspective and Political Theory." Unpublished
    paper, Annual Meeting of American Political Science Association, Los
    Angeles, 1970.
Romanell, Patrick. "Does Biology Afford a Sufficient Basis for Ethics?"
    Scientific Monthly 81 (September 1955): 138.
Rosenberg, Nathan. "Technology and the Environment: An Economic Explora-
    tion." *Technology and Culture* 12 (1971): 534.
Roszak, Theodore. "Can Jack Get Out of His Box? Or The Counter Cultural Image
    of the Future." *Anticipation*, no. 9 (October 1971): 5.
Rothman, Stanley. "The Revival of Classical Political Philosophy: A Critique."
    American Political Science Review 56 (1962): 341.
Rowan, Carl T. " 'Genocide' Fear Deserves Study." Washington *Star*, 24 June
    1970.
Ruggie, John Gerold. "Collective Goods and Future International Collaboration."
    American Political Science Review 66 (1972): 874.
Russett, Bruce M., and Sullivan, John D. "Collective Goods and International Or-
    ganization." *International Organization* 25 (1971): 845.
——. "Licensing: For Cars and Babies." *Bulletin of the Atomic Scientists* 26
    (November 1970): 15.
Rust, Eric C. "Man and Nature in Theological Perspective." *The Review and
    Expositor* 69 (1972): 11.
Sale, Kirkpatrick. "The New Left—What's Left." Los Angeles *Free Press*, 29 June
    1973, 18.
——. "The World Behind Watergate." *New York Review of Books* 20 (3 May
    1973).
Salomon, Jean-Jacques. "Science Policy and Its Myths: The Allocation of Re-
    sources." *Public Policy* 20 (1972): 1.
Sartori, Giovanni. "Technological Forecasting and Politics." *Survey* 16 (1971):
    60.
Schelling, Thomas. "The Ecology of Micromotives." *The Public Interest*, no. 25
    (Fall 1971): 61.

Schiller, Herbert J. "Social Control and Individual Freedom." *Bulletin of the Atomic Scientists* 24 (May 1968): 16.

Schultz, Gustav H. "People's Park: The Rise and Fall (?) of a Religious Symbol." *Christian Century* 90 (1973): 204.

Science Policy Research Unit, University of Sussex. "The Ideological Background to the Limits to Growth Controversy." *Futures* 5 (1973): 157.

"Science and Society Symposium: Prospects for Survival." *Bulletin of the Atomic Scientists* 27 (May 1971): 19.

Searle, John R. "How to Derive 'Ought' From 'Is.'" *Philosophical Review* 73 (1964): 43.

Sears, Paul B. "The Inexorable Problem of Space." *Science* 127 (1958): 9.

Shourie, Arun. "Growth, Poverty, and Inequalities." *Foreign Affairs* 51 (1973): 340.

Silk, Leonard et al. "Does Economics Ignore You?" *Saturday Review* 55 (22 January 1972): 33.

Singh, Devendra. "The Pied Piper vs. the Protestant Ethic." *Psychology Today* (5 January 1972): 53.

Sittler, Joseph. "Ecological Commitment as Theological Responsibility." *Zygon* 5 (1970): 172.

Skolnikoff, Eugene B. "The International Functional Implications of Future Technology." *Journal of International Affairs* 25 (1971): 266.

————. "Science and Technology: The Implications for International Institutions." *International Organization* 25 (1971): 759.

"Society and Ecology." Special issue of the *American Behavioral Scientist*, 11 (July–August 1968).

Solow, Robert M. "The Economist's Approach to Pollution and Its Control." *Science* 173 (1971): 498.

Sorauf, Frank M. "The Public Interest Reconsidered." *Journal of Politics* 19 (1957): 616.

Spilhaus, Athelstan. "Ecolibrium." *Science* 165 (1972): 711.

Spiro, Melford J. "Human Nature in its Psychological Dimensions." *American Anthropologist* 56 (1964): 19.

Stans, Maurice B. "Environment: Wait a Minute." *Wall Street Journal*, 6 August 1971.

Starr, Chauncey. "Social Benefit vs. Technological Risk." *Science* 165 (1969): 1232.

Stearns, J. Burton. "Ecology and the Indefinite Unborn." *The Monist* 56 (1972): 612.

Stewart, T. D. "Fossil Evidence of Human Violence." *Trans-Action* 6 (May 1969): 48.

Strong, Maurice B. "One Year After Stockholm—An Ecological Approach to Management." *Foreign Affairs* 51 (1973): 690.

Stulman, Julius. "Beyond Crises—A 'Creative Ladder' For Oncoming Generations." *Man's Emergent Evolution* 3, no. 1 (1970): 60.

"Symposium: Do Life Processes Transcend Physics and Chemistry?" *Zygon* 3 (1968): 442.

"Symposium: Does the U.S. Economy Require Imperialism?" *Social Policy* 1 (November–December 1970): 52.

"Symposium: On Human Values and Natural Science." *Zygon* 4 (1969): 251.

Tart, Charles T. "States of Consciousness and State-Specific Sciences." *Science* 176 (1972): 1203.

Taviss, Irene. "Futurology and the Problem of Values." *International Social Science Journal* 21 (1969): 574.

———. "A Survey of Popular Attitudes Toward Technology." *Technology and Culture* 13 (1972): 606.
Taylor, Richard W. "Arthur F. Bentley's Political Science." *Western Political Quarterly* 5 (1952): 214.
Thurow, Lester J. "Towards a Definition of Economic Justice." *The Public Interest*, no. 31 (Spring 1973): 56.
Time Essay. "Second Thoughts About Man: The Rediscovery of Human Nature." *Time*, 2 April 1973.
Timlin, Mabel F. "Theories of Welfare Economics." *Canadian Journal of Economics and Political Science* 15 (1949): 551.
Tompkins, Paul, and Tompkins, C. Bird. "Love Among the Cabbages." *Harper's* 245 (November 1972): 90.
Tribe, Larry. "Legal Frameworks for the Assessment and Control of Technology." *Minerva* 9 (1971): 243.
Tugwell, Rexford Guy et al. "Model for a New Constitution." *The Center Magazine* 3 (September 1970): 1.
Vickers, Geoffrey. "Changing Ethics of Distribution." *Futures* 3 (1972): 116.
Von Foerster, H.; Mora, P. M.; and Amiot, L. W. "Doomsday: Friday 13 November A.D. 2026." *Science* 132 (1960): 1291.
Wade, Nicholas. "Theodore Roszak: Visionary Critic of Science." *Science* 178 (1972): 960. 1
Wagar, J. Alan. "Growth versus the Quality of Life." *Science* 168 (1970): 1179.
Wagner, Walter C. "Futurity Morality." *The Futurist* 5 (1972): 197.
Waldo, Dwight. "Theories of Democratic Planning." *American Political Science Review* 46 (1952): 81.
Wallace, Anthony F. C. "Perceptions of Order and Richness in Human Cultures." *Zygon* 6 (1971): 151.
Watson, James D. "Moving Toward the Clonal Man: Is This What We Want?" *The Atlantic* 227 (May 1971): 50.
Weinberg, Alvin M. "Social Institutions and Nuclear Energy." *Science* 177 (1972): 27.
———. "Science and Trans-Science." *Minerva* 10 (1972): 209.
Weisskopf, Walter A. "The Dialectics of Abundance." *Diogenes*, no. 57 (1967): 1.
Weissman, Steve. "Why the Population Bomb Is a Rockefeller Baby." *Ramparts* 8 (May 1970): 42.
Weizenbaum, Joseph. "On the Impact of the Computer on Society." *Science* 176 (1972): 609.
Westhoff, Charles F., and Bumpass, Larry. "The Revolution in Birth Control Practices of U.S. Roman Catholics." *Science* 179 (1973): 41.
Wharton, Clyde. "Green Revolution: Cornucopia or Pandora's Box?" *Foreign Affairs* 47 (1969): 464.
Wheeler, Harvey. "Bringing Science Under Law." *The Center Magazine* 2 (March 1969): 59.
———. "Making the World One." *The Center Magazine* 1 (November 1968): 34.
———. "The Politics of Ecology." *Saturday Review* 53 (7 March 1970): 51.
———. "Technology: Foundation of Cultural Change." *The Center Magazine* 5 (July–August 1972): 48.
Wieck, David T. "The Violence of Men: Remarks on Konrad Lorenz' *On Aggression*." *Diogenes*, no. 62 (1968): 103.
Willhoite, Fred W., Jr. "Ethology and the Tradition of Political Thought." *Journal of Politics* 33 (1971): 615.
Williams, Gardener. "The Natural Causation of Free Will." *Zygon* 3 (1968): 72.
Wilson, Ian. "How Our Values Are Changing." *The Futurist* 4 (1970): 5.

Winner, Langdon. "On Criticizing Technology." *Public Policy* 20 (1972): 35.
Winthrop, Henry. "The Future of Sexual Revolution." *Diogenes*, no. 70 (Summer 1970): 57.
———. "Total Environment Management: An Approach to the Dilemmas of the Affluent Society." *Futures* 2 (1970): 332.
———. "Environmental Dilemma: Possible Steps toward Its Dissolution." *American Journal of Economics and Sociology* 31 (1972): 387.
Wolman, Abel. "Pollution As an International Issue." *Foreign Affairs* 47 (1968): 164.
Woodcock, Leonard. "Labor and the Politics of the Environment." *Sierra Club Bulletin* 56 (December 1971): 11.
Woodhouse, Edward J. "Re-Visioning the Future of the Third World: An Ecological Perspective on Development." *World Politics* 25 (1972): 1.
Wrong, Dennis H. "The Population Problem: An Unorthodox View." *Dissent* 18 (1971): 357.
Yankelovitch, Daniel. "The New Naturalism." *Saturday Review* 55 (1 April 1972): 32.
———. "Will the Election Change Anything?" *Parade*, 19 November 1972.
Zwerdling, Daniel. "Food Pollution." *Ramparts* 9 (June 1971): 30.

Index

Activity, 171, 181–185
Acton, Lord John, 141
Adams, John, 35, 49
Adler, Mortimer J., 116, 142
Advertising, 251
Aggression, 142–144
Alexander, Samuel, 111
Alien invasion, 3–7
Alienation, 7, 88
Altamont, 78
American Farm Bureau Federation, 45
American Medical Association (A.M.A.), 45
American Revolution, 33–34
Anarchism, 75–80
Antimatter, 20
Aquinas, Saint Thomas, 28, 65, 100, 104, 153, 187
Arab oil. See Oil
Ardrey, Robert, 140
Aristotle: problem of justice, 23, 178–179; ethics, 63, 91–92, 103–104, 121–122, 144, 153, 177–178; physics, 109, 128; metaphysics, 115; on the common good, 187
Atomic Energy Commission, 46
Atomic testing, 81
Atomism, 12, 110
Auden, W. H., 112
Augustine, Saint, 82–83, 127, 140, 153
Automation, 68
Automobiles, 21, 50, 161, 263

Bakunin, Mikhail, 264
Baran, Paul A., 70
Barbados, 216
Barth, Karl, 97, 125

Becker, Ernest, 95, 168
Bellamy, Edward, 42
Benedict, Ruth, 124
Bentham, Jeremy, 35, 187
Berdayev, Nicholas, 164
Berger, Mary, 80
Berger, Peter, 80
Bertalanffy, Ludwig von, 114
Biddle, Nicholas, 36–37
Bidney, David, 122, 124
Big-bang theory, 107–108
Birth control. See Population growth
Black, John, 241
Bohr, Niels, 114, 118
Bookchin, Murray, 75–76, 78, 80
Boulding, Kenneth, 157, 233
Bourne, Randolph, 19
Boyle, Robert, 12
—The Skeptical Chymist, 12
Bradbury, Ray, 277
Brazil, 215
Brecht, Bertolt, 169
Brinton, Crane, 267
Buckley, James, 65–66
Buckley, William, 64, 66
Bureau of the Budget, 229
Bureaucracy, 73
Burke, Edmund, 33, 64

Calhoun, John C., 39–40, 141
—Disquisition on Government, 39
Calley, William, 49
Calvert, Gregory, 72
Calvin, John, 76
Canada, 215, 253
Carey, Henry, 43
Carnegie, Andrew, 42
Carrell, Alexis, 116
Chesterton, G. K., 184

China: detente with Soviet Union, 15, 51; environmental crisis, 69; Mao Tse Tung, 79; atomic testing, 81; great power, 215; imperialism, 279

Christian political thought, 82–84, 97, 103–104, 127, 140, 153–154, 187

Churchill, Winston, 227

Cicero, 23, 98

Clay, Henry, 36

Collingwood, R. G., 110

Common good, 186–192; subsistence, 193–197; social order, 197–200; social purpose, 200–204; world community, 204–205

Communication and public policy, 250–252

Compromise, 59–60

Computer, 17

Congress (U.S.), 219–222

Conservation, 11, 42–43

Conservatism, 6–7, 39–40, 62, 176–177

Constitution (U.S.), 34–35, 91, 140–141, 218, 231

Coolidge, Calvin, 43

Corporate liberalism. See Liberalism

Counterculture, 20, 126, 271

Crime, 88, 173–174

Croly, Herbert, 43

Crusoe, Robinson, 147

Cuba, 216

Cultural pluralism, 248–250. See also Planetary culture

Cultural relativism, 122–124

Cultural revolution, 81–82

Cybernation, 68

Darwin, Charles, 113

Darwinism, 42, 113, 116, 129, 140, 142

Declaration of Independence, 25–26

Davies, James C., 179

De Maistre, Joseph, 140

Democritus, 12, 110

Depression. See Great Depression

Descartes, René, 12, 19, 115

—Discourse on Method, 12

Determinism, 116–118

DeTocqueville, Alexis, 37–38

—Democracy in America, 37

Dewey, John, 190

Didion, Joan, 32

Dirke, Robert, 107

Disarmament, 283

Dobzhansky, Theodosius, 114, 116, 143

Dollo's law, 131

Dropping out, 81–82, 84

Drug addiction, 88

Dubos, René, 75, 168, 181

Dutsche, Rudi, 269

Easton, David, 153

Echeverria, Luis, 133

Ecoanarchists, 75–76

Ecological humanism, 87, 101; philosophy, 206–208; politics, 213–229; economics, 230–247; and cultural pluralism, 249; freedom of speech, 250; future of, 255–260, 267–275; and planetary society, 276–289

Ecophysics, 233, 236

Eichmann, Adolf, 124

Einstein, Albert, 108, 262

Ellul, Jacques, 5, 16, 56, 82–83, 272

—The Technological Society, 82

Energy, 193–195, 291–292. See also Oil

Engels, Friedrich, 67–68, 142

Enlightenment, 103–104, 106, 128, 147

Entropy, 107–108

Environmental degradation, 5, 7, 11, 88

Environmentalism, 50

Ephemeralization, 8

Equality, 177–180

Evolutionary process, 114, 120–121, 129–131, 137–138, 142–143, 286. See also Darwinism

Fitzhugh, George, 39

Follet, Mary Parker, 190

Fourier, Charles, 247
France, 78–79, 81, 200–201, 217, 238
Francis of Assisi, Saint, 103, 128
Francoeur, Robert, 115–116
Free enterprise, 45
Free will, 116–118
Freedom, 156–158; and power, 158; and interdependence, 159–160; and coercion, 160–163; and hope, 163–164
Freud, Sigmund, 121, 140, 143, 167
Fromm, Erich, 70, 72, 169
Fuller, Buckminster, 28, 88, 196, 236
Future shock, 10

Galbraith, John K., 44
—*The New Industrial State*, 44
Gasoline shortage. *See* Oil
Geertz, Clifford, 124
General Accounting Office, 219, 229
Genetic manipulation, 5
Genghis Khan, 5
George, Henry, 42
Ghana, 215–216
Ginsburg, Allen, 271
Ginsburg, Morris, 122
Glass, Bentley, 120
Global village, 16–17
Goldman, Lucian, 71
Goldwater, Barry, 22
Goldwin, Robert, 29
Goodman, Paul, 66, 76, 78, 161, 254
Grant, George, 66, 175
Great Britain, 16, 22, 69, 200, 217, 251
Great Depression, 43, 45
Green, T. H., 187
Greening. *See* Reich, Charles
Gross national product, 27, 232–233, 236

Haldane, J. B. S., 287
Halley, Edmund, 12
Hamilton, Alexander, 34–35, 38, 58, 264
Hardin, Garrett, 132, 235

Harrington, Michael, 70–71, 76
Hartz, Louis, 34
Harvey, William, 12
Health, 92–93
Hegel, Georg Wilhelm, 100, 187
Heisenberg, Werner, 117
Hitler, Adolf, 5, 67, 142, 184
Hobbes, Thomas: political philosophy, 23–27, 29, 55, 90, 96, 100, 103, 122, 140, 143, 154, 172; inspiration, 110; pluralism, 188
Hobhouse, L. T., 187
Holism, 89
Hoyle, Fred, 107
Hughes, Howard, 32
Human nature, 137–138; goodness of man, 138–142; roots of aggression, 142–144; knowledge, reason and virtue, 144–146; social basis, 147–151; society and self, 151–152; the primacy of politics, 153–155
Humanism, 5–6. *See also*, Ecological humanism
Humphrey, Hubert, 22
Hutchins, Robert M., 191, 281
Huxley, Aldous, 213
—*Brave New World*
Huxley, Julian, 117, 137

Identity, 170–171; physical, 174; sexual, 175; community, 175–177; privacy, 180; contact with nature, 181
Immanentism, 89
Income distribution, 245–247
India, 200, 215
Information overload, 219
Intelligence, 199
International Telephone and Telegraph Corp. (I.T.T.), 45
Interstate Commerce Commission, 45–46

Jackson, Andrew, 36–38, 61
Japan, 15, 46–47, 215, 279, 283
Jeffers, Robinson, 104

Jefferson, Thomas, 25–26, 34–36, 42, 61, 91, 128
Jesus Christ, 131
Johnson, Lyndon, 32, 87
Johnson, Samuel, 108–109
Jones, Howard Mumford, 31
—*O Strange New World*, 31
Joyce, James, 286
—*Ulysses*, 286

Kahn, Herman, 14, 267
Kamenka, Eugene, 71
Kant, Immanuel, 91
Kendall, Willmore, 65
Kennedy, John F., 87
Keynes, John M., 43
Kissinger, Henry, 16
Knowledge, 144–147
Koestler, Arthur, 113, 116
Korea, 32
Kristol, Irving, 267
Kropotkin, Peter, 142
Krutch, Joseph W., 109, 130
Kuwait, 216

Leadership, 226–228
Leary, Timothy, 80
Legislative Reference Service, 219
Lenin, V. I., 264, 268
Leopold, Aldo, 134
Lewis, C. S., 292
Liberalism, 6–8, 18, 31–62, 119; root failure, 7, 21, 136, 253–255; corporate, 45, 87; historical meaning, 22–23; philosophical foundations, 23–30; worldview, 147–148; and freedom, 156–158; and the common good, 187
Life. *See* Nature
Lifton, Robert J., 265
Lincoln, Abraham, 40
Lippmann, Walter, 83, 190
Locke, John: 17th century world of, 9–13, 19–20; political philosophy, 23–30, 34–35, 55, 90, 100, 103, 106, 154, 173–174, 179; on property,

27–30; on nature, 27–30, 33, 38, 111; on technology, 27–30; legacy, 31–33, 76, 264
Lockheed Aircraft Corp., 45
Lombardi, Vince, 55
Lucretius, 12
Luther, Martin, 82–83
Lycurgus, 96

McGovern, George, 22, 61, 71, 269
MacLeish, Archibald, 31
McLuhan, Marshall, 16

Machiavelli, Niccolo, 100
Madison, James, 34, 36, 140
Magic, 20
Mailer, Norman, 249
Malthus, Thomas, 36
Manicheanism, 82
Marcuse, Herbert, 70, 72, 74, 81, 252
Margenau, Henry, 118
Marin, Luis Muñoz, 227
Marsh, George Perkins, 42
Marshall, John, 35
—*McCullough vs. Maryland*, 35
Marx, Karl, 44, 63–64, 67–69, 76–77, 96, 99–100, 264, 268
—*The Communist Manifesto*, 63
Marxism, 70–73, 99, 233, 260
Maslow, Abraham, 169–170
Mayans, 101, 197
Mead, George Herbert, 151
Meany, George, 67
Mental health, 167–169
Merton, Thomas, 134, 284
Meyer, Frank S., 65
Military-technological complex, 19
Mill, John Stuart, 57, 161, 236, 252
—*Principles of Political Economy*, 57
Mills, C. Wright, 87
Mind, 115–116, 118–119
Miniaturization, 8
Mishan, E. J., 236
Monopoly capitalism. *See* Neoliberalism

Montesquieu, Charles de Secondat, Baron de la Brède et de, 100
More, Thomas, 99
—*Utopia*, 99
Muir, John, 42
Mumford, Lewis, 79–80
—*The Myth of the Machine*, 79–80
Myrdal, Gunnar, 37
—*American Dilemma*, 37

National Industrial Recovery Act (N.I.R.A.), 45
National Review, 65
Naturalism, 20, 89
Nature: definition, 105–106; origin, 106–108; constitution, 108–112; life, 112–115; value in, 127–132; value of, 132–135; as a model of society, 135–136
Needs, in social perspective, 165–185
Neoliberalism, 44–47
Neoteny, 131
Nepal, 216
New Deal, 45–47
New Left, 72–74, 79, 89, 176, 269, 274
Newton, Isaac, 12, 20, 109; worldview of Newtonian physics, 147–148, 218, 222, 268
—*Principia Mathematica*, 12
Niebuhr, Reinhold, 83
—*Moral Man and Immoral Society*, 83
Nietzsche, Friedrich, 104
Nisbet, Robert, 262–264
Nixon, Richard, 23, 48, 55, 80, 87, 172, 219–220, 237, 292
Northrop, F. S. C., 25, 126
Nuclear holocaust, 7, 14, 51, 53, 140, 183, 204, 215–216, 220, 265, 279–280; Hiroshima, 46
Nye Committee, 46

Oakeshott, Michael, 65
Office of Technology Assessment, 219
Oil: Arab, 15, 279; trans-Alaskan pipeline, 183, 254; gasoline shortage, 5, 291–292

Ombudsman, 229
Oppenheimer, J. Robert, 118
Origin of the universe, 106–108
Ortega y Gasset, José, 104
Orwell, George, 213
—*1984*, 213, 223
Overpopulation. *See* Population growth

Paraguay, 216
Parapsychological phenomena, 119, 288
Patriotism, 253
Peoples Park, 291–292
Pesticides, 50
Petersen, William, 66
Planetary culture, 282–289
Planning, 199, 222–226
Plato: political thought, 23, 96, 98–100, 103, 121–122, 166, 286; cosmology, 106–107; the soul, 115; on virtue, 144; common good, 186
—*Republic*, 153
Platt, John, 235
Play. *See* Activity
Political intelligence, 218–220
Political power, 228–229
Political will, 199, 220–228
Pollution, 5, 7, 11, 21, 68, 292
Pope, Alexander, 217
Popper, Karl, 118
Population control, 241–245
Population growth, 5, 7, 8, 14, 16, 21, 77, 292
Postliberal society, 21, 97, 136
Postmodern era, 8–9
Poverty, 7, 88
Powell, John Wesley, 42
Presidency (U.S.), 220–222
Privacy. *See* Identity
Propaganda, 252
Proudhon, Pierre Joseph, 264
Public interest. *See* Common good
Purpose. *See* Common good

Racism, 7, 37–38, 88
Rand, Ayn, 65

Rationalism, 8, 12
Reagan, Ronald, 22
Reason, 144–147
Recycling, 8
Reductionism, in science, 20
Reich, Charles, 79–80
—*The Greening of America*, 20, 79, 81, 89, 268
Relativity theory, 108
Resource depletion, 5
Reston, James, 220
Revel, Jean François, 80
Ricardo, David, 68
Robespierre, Maximilien, 253
Roosevelt, Franklin, 221
Roosevelt, Theodore, 42
Ropke, Wilhelm, 66
Roszak, Theodore, 81
Rousseau, Jean J., 99–100, 103, 177, 187–188
Russell, Bertrand, 158
Ryle, Martin, 107

Sandage, Allan, 107
Schroedinger, Erwin, 127, 147
Schweitzer, Albert, 124, 133
Security, 170–177
Shepard, Paul, 134
Sherrington, Charles, 117, 128
Sibley, Mulford, 76
Simpson, George Gaylord, 116
Skinner, B. F., 93, 118, 153, 159, 167, 169
Slavery, in U.S., 37–38
Smith, Adam, 264
Smog, 161
Snyder, Gary, 134
Social Darwinism. *See* Darwinism
Social democrats, 22
Social order. *See* Common good
Socialism, 6–7, 67–71, 264–265
Soviet Union: detente with China, 15, 51; environmental pollution, 68–69, 283; great power, 215; space program, 277; new trade relationship with U.S., 279; cold war, 280

Space programs, 183, 276–277, 285–286
Spencer, Herbert, 42
Sperry, R. W., 116
Spinoza, Baruch, 115
Stalin, Joseph, 142, 223
Standard Oil Co., 45
State capitalism. *See* Neoliberalism
Stevenson, Adlai, 253
Strauss, Leo, 27, 64
Students for a Democratic Society (S.D.S.), 72
Subsistence. *See* Common good
Supersonic transport, 16, 65, 183, 254
Surveillance, 18
Sweden, 69
Sweezy, Paul M., 70
Syndicalism. *See* Neoliberalism

Taney, Roger B., 37
Tanzania, 215
Taylor, John, 35
Technological efficiency, 5
Technological man, 88–89, 262
Technology: purposes, 4; threat of, 5; essential characteristic of the modern age, 8–9; in the 21st century, 16–19; scepticism about, 20–21; as ideology, 45; and the end of liberalism, 52–60; and cultural revolution, 81; compared with nature, 105; growth of, 154–155; and subsistence, 195–197; communication, 250–251, 263; and political intelligence, 219–220; control of, 238–241
Technostructure. *See* Neoliberalism
Teilhard de Chardin, Pierre, 112
Television, 17
Theology, and human values, 125–127
Thomas, W. I., 262
Tolkien, J. R. R., 262
Trans-Alaskan pipeline. *See* Oil
Trotsky, Leon, 68
Tugwell, Rexford Guy, 43
Twenty-first century, 13–21

United Nations, 244, 280
Utopianism, 84, 99, 213–214

Value judgments, 120–121
Values, 120–136
Vandal ideology, 30, 40–42, 206, 286
Vickers, Geoffrey, 182, 237, 254
Vietnam war, 32, 36, 72, 74, 83, 87, 182–183, 200, 266
Violence, 273–275
Virtue, 144–147

Waddington, C. W., 121
Wagner Act, 46
Wallace, Anthony, 263
Wallace, George, 80
Wallace, Henry A., 43
Wallach, Henry, 285
War, as a social institution, 18–19
Ward, Lester F., 43
Watergate, 48–49, 221, 249, 266, 292

Watson, James W., 167
Watt, James, 262
Watts, Alan, 5, 126, 134, 151
Weber, Max, 153
Weisberg, Barry, 254
Weiss, Paul, 112
Welles, Orson, 3
Wells, H. G., 3
—War of the Worlds, 3
Westward expansion (U.S.), 40–42
Whigs, 24, 34, 36, 40
White, Lynn, 19
Whitehead, Alfred N., 109–112, 239
Whitney, Eli, 37
Williams, William Appleton, 70
Wills, Garry, 65
Wilson, Woodrow, 42–43
Woodstock, 78
World community. See Common good
World War I, 42–43
World War II, 44–46
Work. See Activity